Emerging Security
Algorithms and Techniques

Emerging Security Algorithms and Techniques

Edited by
Khaleel Ahmad, M. N. Doja,
Nur Izura Udzir, and Manu Pratap Singh

CRC Press
Taylor & Francis Group
Boca Raton London New York

CRC Press is an imprint of the
Taylor & Francis Group, an **informa** business

A CHAPMAN & HALL BOOK

CRC Press
Taylor & Francis Group
52 Vanderbilt Avenue,
New York, NY 10017

© 2019 by Taylor & Francis Group, LLC
CRC Press is an imprint of Taylor & Francis Group, an Informa business

---

Library of Congress Cataloging-in-Publication Data

Names: Ahmad, Khaleel, editor.
Title: Emerging security algorithms and techniques / editors, Khaleel Ahmad, M.N. Doja, Nur Izura Udzir, Manu Pratap Singh.
Description: Boca Raton : Taylor & Francis, a CRC title, part of the Taylor & Francis imprint, a member of the Taylor & Francis Group, the academic division of T&F Informa, plc, 2019. | Includes bibliographical references and index.
Identifiers: LCCN 2019010556 | ISBN 9780815361459 (hardback : acid-free paper) | ISBN 9781351021708 (ebook)
Subjects: LCSH: Computer security. | Data encryption (Computer science) | Cryptography.
Classification: LCC QA76.9.A25 E466 2019 | DDC 005.8—dc23
LC record available at https://lccn.loc.gov/2019010556

Visit the Taylor & Francis Web site at
http://www.taylorandfrancis.com

and the CRC Press Web site at
http://www.crcpress.com

Contents

Preface vii

Editors ix

Contributors xi

1. **Modular Arithmetic** 1
   Afroz

2. **Finite Fields** 11
   Afroz

3. **Prime Number** 27
   Khaleel Ahmad, Afsar Kamal, and Khairol Amali Bin Ahmad

4. **Discrete Logarithm Problem** 47
   Khaleel Ahmad, Afsar Kamal, and Khairol Amali Bin Ahmad

5. **Integer Factorization Problem** 59
   Pinkimani Goswami and Madan Mohan Singh

6. **Symmetric Algorithms I** 79
   Faheem Syeed Masoodi and Mohammad Ubaidullah Bokhari

7. **Symmetric Algorithms II** 97
   Vivek Kapoor and Shubhamoy Dey

8. **Asymmetric Cryptography** 119
   Rajiv Ranjan, Abir Mukherjee, Pankaj Rai, and Khaleel Ahmad

9. **Post-Quantum Cryptography** 139
   Amandeep Singh Bhatia and Ajay Kumar

10. **Identity-Based Encryption** 159
    Tanvi Gautam, Aditya Thakkar, and Nitish Pathak

11. **Attribute-Based Encryption** 183
    Tanvi Gautam, Aditya Thakkar, and Nitish Pathak

12. **Key Management** 197
    Jyotsna Verma

13. **Entity Authentication** 213
    Hamza Mutaher and Pradeep Kumar

14. Message Authentication..225
  Ajay Prasad and Jatin Sethi

15. Digital Signatures 249
  Ajay Prasad and Keshav Kaushik

16. Applications..273
  M. A. Rizvi and Ifra Iqbal Khan

17. Hands-On "SageMath" 293
  Uma N. Dulhare and Khaleel Ahmad

Index 309

Preface

Security is pervasive. It is estimated that 3.3 billion mobile phone users are registered worldwide, which is more than half of the global population. Most of the mobile phones are built-in with Bluetooth, Wi-Fi, cameras, sensors, and many other components. Hence, digital security is no longer limited to the scholarly community but is now the concern of all users of computers worldwide. As e-commerce and electronic transactions have grown exponentially, the development of secure systems has also increased rapidly to protect the digital world against adversaries, giving rise to a vast number of applications of emerging security algorithms, and more applications are developing in proportion to the increasing development and usage of smartphone technology, online banking, m-commerce, share market, cloud computing, etc.

This book covers symmetric, asymmetric and post-quantum cryptography algorithms, and their modern applications in developing secure systems, viz. digital signature, entity authentication, message authentication, cyber security, web security, cloud security, smart card technology, and CAPTCHA. The book has been structured well with simple explanations and illustrations. In other words, this book is designed to support students who are studying computer science and engineering, information technology, electronics and telecommunication engineering at undergraduate and postgraduate levels and teachers, researchers, scientists and engineers who are working in the field of digital security. This book is written with aims to bring out a new paradigm for learning emerging security algorithms and techniques.

This book is organized into 17 chapters. Each chapter covers the following important topics in brief: modular arithmetic, finite fields, prime numbers, discrete logarithm problems, integer factorization problems, symmetric algorithms, asymmetric algorithms, post-quantum cryptography, identity-based encryption, attribute-based encryption, key management, entity authentication, message authentication, digital signatures, applications, and hands-on "SageMath."

Editors

Dr. Khaleel Ahmad is currently an assistant professor in the Department of Computer Science & Information Technology at Maulana Azad National Urdu University, Hyderabad, with over 6 years of experience in Academic and Research. He holds a PhD in computer science & engineering and M.Tech. in information security. His research interests include opportunistic network, information security, and cryptography. He has successfully completed a project of INR 1.05 lakh. He has supervised five M.Tech. dissertations and two PhDs under progress. He has published a total of 35 papers in refereed journals and conferences (viz. Elsevier, ACM, IEEE, and Springer), including 14 book chapters (Taylor & Francis Group, IGI Global). He has edited a book published by CRC Press/Taylor & Francis Group and Proceeding of International Conference published by McGraw Hill. He has authored a book titled *Secure Communication in Opponents Using ECC* published by Lap Lambert Academic Publishing. He has delivered guest lectures at the Central University of Haryana, India, and Telangana University, India. He has also chaired the technical sessions at an international conference in Malaysia and India. He is also a life member of various international/national research societies such as ISTE, CRSI, ISCA, IACSIT (Singapore), IAENG (Hong Kong), IARCS, and ISOC (USA). Besides this, he also serves as an editorial and review board member for several international research organizations.

Dr. M. N. Doja is currently a professor in the Department of Computer Engineering at Jamia Millia Islamia (A Central University), New Delhi, India. In addition, he is also the Honorary Director of FTK-Centre for Information Technology Jamia Millia Islamia (A Central University). He is the founder and first head of the Department of Computer Engineering. He has a total of 33 years of experience in teaching and 3 years in industry. He has been nominated as Chief Information Security by Ministry of Electronics & Information Technology, Government of India, and Nodal Officer of Digital India. He was chairman and member of a number of committees constituted by UGC, MHRD, AICTE, other universities, and other external agencies. He has completed a successful project of INR 65 lakhs and other projects of INR 8.74 crore and INR 40 lakhs are ongoing. His research interests include mobile ad hoc network, information security, computer network, and soft computing. He has supervised 16 PhDs and 8 PhDs under progress. He has filed a patent titled "Method to Select Sext Best Mobility Anchor Point (MAP) before Band Off Process Reducing Handoff Latency" vide application no. 581/DEL/2009 in 2009. He has published more than 150 papers in journal and conference proceedings of international and national repute. He has published two books. He is a reviewer and editor for a number of journals of international and national repute. He has also delivered keynote speeches and expert talks, and also as invited speaker at various international conferences in India and abroad.

Assoc. Prof. Dr. Nur Izura Udzir is an academic staff member at the Faculty of Computer Science and Information Technology, Universiti Putra Malaysia (UPM) since 1998. She received her Bachelor of Computer Science (1995) and Master of Science (1998) from UPM, and her PhD in computer science from the University of York, UK (2006). Her research interests include access control, intrusion detection systems, secure operating systems, coordination models and languages, and distributed systems. She has served as the Head of the

Department of Computer Science and is currently a member of the Information Security Group (which she led in 2008–2013) at the faculty. She is a member of IEEE Computer Society, Malaysian Society for Cryptology Research (MSCR), and the Society of Digital Information and Wireless Communications (SDIWC). She currently serves as a committee member of the Information Security Professionals Association of Malaysia (ISPA.my). She has supervised and co-supervised over 30 PhD students and over 15 Master (by research) students. She has written a book titled *Pengenalan kepada Pengaturcaraan C++* (*Introduction to C++ Programming*) (2001). She has published over 60 articles in journals and book chapters, and over 80 international conference proceedings, thus earning an H-index of 9 in Scopus (H-index 15 in Google Scholar) as of March 2016. In addition, she has won various awards for her contributions in academics and research, including six Best Paper Awards at international conferences and the MIMOS Prestigious Award 2015 for the supervision of her student's doctoral thesis. She has also delivered keynote speeches and expert talks, and also as invited speaker at various international conferences.

Assoc. Prof. Dr. Manu Pratap Singh received his PhD in computer science from Kumaun University Nainital, Uttarakhand, India, in 2001. He has completed his Master of Science in computer science from Allahabad University, Allahabad, in 1995. Furthermore, he obtained his M.Tech. in information technology from Mysore. He currently serves as an associate professor in the Department of Computer Science, Institute of Engineering and Technology, Dr. B. R. Ambedkar University, Agra, UP, India, since 2008. He has been engaged in teaching and research for the last 17 years. He has more than 90 research papers in journals of international and national repute. His work has been recognized widely around the world in the form of citations of many research papers. He has also received the Young Scientist Award in computer science by the International Academy of Physical Sciences, Allahabad, in 2005. He has guided 18 students for their doctorate in computer science. He is also referee of various international and national journals such as *International Journal of Uncertainty, Fuzziness and Knowledge Based Systems* by World Scientific Publishing Cooperation Ltd, *International Journal of Engineering*, Iran, *IEEE Transaction of Fuzzy Systems*, and *European Journal of Operation Research* by Elsevier. His research interests are focused on neural networks, pattern recognition, and machine intelligence. He is a member of the technical committee of IASTED, Canada, since 2004. He is also the regular member of Machine Intelligence Research Labs (MIR Labs), Scientific Network for Innovation and Research Excellence (SNIRE), Auburn, Washington, USA (www.mirlabs.org), since 2012. His Google citation indices are 9, i10-index is 8 and he has 257 citations. He has been invited as keynote speaker and invited guest speaker in various institutions in India and abroad.

Contributors

Afroz
Department of Mathematics
Maulana Azad National Urdu University
Hyderabad, Telangana, India

Khaleel Ahmad
Department of Computer Science &
    Information Technology
Maulana Azad National Urdu University
Hyderabad, Telangana, India

Khairol Amali Bin Ahmad
Department of Electrical & Electronics
    Engineering
National Defence University of Malaysia
Kuala Lumpur, Malaysia

Amandeep Singh Bhatia
Department of Computer Science
    Engineering
Thapar Institute of Engineering &
    Technology
Patiala, Punjab, India

Mohammad Ubaidullah Bokhari
Department of Computer Science
Aligarh Muslim University
Aligarh, Uttar Pradesh, India

Shubhamoy Dey
Department of Information System
Indian Institute of Management Indore
Indore, Madhya Pradesh, India

Uma N. Dulhare
Department of Computer Science
    Engineering
M. J. College of Engineering and
    Technology, Osmania University
Hyderabad, Telangana, India

Tanvi Gautam
Bhai Parmanand Institute of Business
    Studies
Guru Gobind Singh Indraprastha
    University
Delhi, India

Pinkimani Goswami
Department of Mathematics
University of Science and Technology
Meghalaya, India

Afsar Kamal
Department of Computer Science &
    Information Technology
Maulana Azad National Urdu
    University
Hyderabad, Telangana, India

Vivek Kapoor
Information Technology
    Department
Institute of Engineering & Technology
Devi Ahilya University
Indore, Madhya Pradesh, India

Keshav Kaushik
Department of Computer Science &
    Engineering
University of Petroleum and Energy
    Studies
Dehradun, Uttarakhand, India

Ifra Iqbal Khan
Department of Computer Engineering &
    Applications
National Institute of Technical Teachers'
    Training and Research
Bhopal, Madhya Pradesh, India

Ajay Kumar
Department of Computer Science
  Engineering
Thapar Institute of Engineering &
  Technology
Patiala, Punjab, India

Pradeep Kumar
Department of Computer Science &
  Information Technology
Maulana Azad National Urdu University
Hyderabad, Telangana, India

Faheem Syeed Masoodi
Department of Computer Science
University of Kashmir
Srinagar, Jammu and Kashmir, India

Abir Mukherjee
Department of Information Technology
Birsa Institute of Technology Sindri
Dhanbad, Jharkhand, India

Hamza Mutaher
Department of Computer Science &
  Information Technology
Maulana Azad National Urdu University
Hyderabad, Telangana, India

Nitish Pathak
Bharati Vidyapeeth's Institute of
  Computers Applications and
  Management (BVICAM)
Guru Gobind Singh Indraprastha
  University
Delhi, India

Ajay Prasad
Department of Computer Science &
  Engineering
University of Petroleum and Energy
  Studies
Dehradun, Uttarakhand, India

Pankaj Rai
Department of Electrical Engineering
Birsa Institute of Technology Sindri
Dhanbad, Jharkhand, India

Rajiv Ranjan
Department of Information Technology
Birsa Institute of Technology Sindri
Dhanbad, Jharkhand, India

M. A. Rizvi
Department of Computer Engineering &
  Applications
National Institute of Technical Teachers'
  Training and Research
Bhopal, Madhya Pradesh, India

Jatin Sethi
Department of Computer Science &
  Engineering
University of Petroleum and Energy
  Studies
Dehradun, Uttarakhand, India

Madan Mohan Singh
Department of Basic Sciences and Social
  Sciences
North-Eastern Hill University
Shillong, Meghalaya, India

Aditya Thakkar
Bhai Parmanand Institute of Business
  Studies
Guru Gobind Singh Indraprastha
  University
Delhi, India

Jyotsna Verma
Department of Computer Science &
  Engineering
Central University of Rajasthan
Ajmer, Rajasthan, India

1

Modular Arithmetic

Afroz

Maulana Azad National Urdu University

CONTENTS

1.1 Introduction to Number Theory 2
    1.1.1 Integers 2
    1.1.2 Integer Arithmetic 2
    1.1.3 Arithmetic Operations 2
1.2 Modular Arithmetic 3
    1.2.1 Examples 4
1.3 Additive Inverse 4
    1.3.1 Examples 4
1.4 Multiplicative Inverse 4
    1.4.1 Examples 4
1.5 Matrix 5
    1.5.1 Examples 5
1.6 Linear Congruence 6
    1.6.1 Solution of Linear Congruence 6
1.7 Prime and Relative Prime Numbers 6
    1.7.1 Prime Numbers 6
    1.7.2 Relatively Prime Numbers 6
    1.7.3 Examples 7
1.8 Greatest Common Divisor (Euclid's Algorithm, Bézout's Algorithm, and Extended Euclid's Algorithm) 7
    1.8.1 Greatest Common Divisor 7
        1.8.1.1 Examples 7
    1.8.2 Euclid's Algorithm 7
        1.8.2.1 Example 7
    1.8.3 Bézout's Theorem 8
    1.8.4 Extended Euclid's Algorithm 8
        1.8.4.1 Examples 9
1.9 Conclusion 9
References 10

1.1 Introduction to Number Theory

Number theory is a branch of pure mathematics which helps to analyze and study the integers. It is partly experimental and partly theoretical. The experimental part leads to questions and suggests the ways to answer the questions. The theoretical part tries to device an argument that gives a conclusive answer to the questions. It examines and accumulates numerical data, and then finds the relationships between different types of numbers (Anglin, 1995). Number theory is also called "The Queen of Mathematics."

1.1.1 Integers

The positive, negative, and whole numbers with zero together are known as an integer, i.e., ... –3, –2, –1, 0, 1, 2, 3 ..., etc. An integer can be represented along a number line by an arrow (Gay, 2007). A positive number is represented by an arrow pointing to the right. A negative number is represented by an arrow pointing to the left. The total value of these numbers is represented by the length of the arrow.

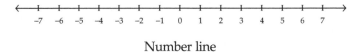

Number line

1.1.2 Integer Arithmetic

The integer arithmetic is indispensable at most every area such as cryptology, computer graphics, mathematics, and several other engineering areas.

1.1.3 Arithmetic Operations

The most basic integer operations are as follows:

1. Addition
2. Subtraction
3. Multiplication
4. Division.

Thus, the mathematical operations of the above integers can be performed as follows.

1. **Addition**: It states that the addition of two integers is always an integer.
   a. $2 + 5 = 7$

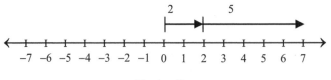

Number line

b. $-7 + 4 = -3$

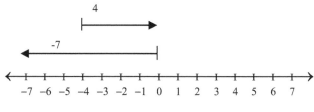

Number line

2. **Subtraction**: It states that the subtraction of two integers is again an integer.
   a. $6 - 4 = 2$

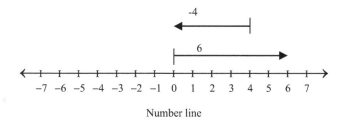

Number line

3. **Multiplication**: It states that multiplying two integers always results in an integer.

Examples	Notations
a. $2 \times 4 = 8$	$+ \times + = +$
b. $3 \times -3 = -9$	$+ \times - = -$
c. $-5 \times 6 = -30$	$- \times + = -$
d. $-8 \times -5 = 40$	$- \times - = +$
	$0 \times a = 0$
	$a \times 0 = 0$
	(where a is any integer)

4. **Division**: It states that dividing ($/$, $\div$) integers with other integers always results in an integer.

Examples	Notations
a. $6 \div 4 = 2$	$0 \div a = 0$
b. $-12 \div 3 = -4$	$a \div 0 =$ meaningless (it is not possible to divide 0 to any
c. $-4 \div -2 = 2$, etc.	number)

1.2 Modular Arithmetic

Let a, b, and m be integers, where $m \neq 0$, such that "a is congruent to b modulom" (Barua, 2017; Mcnamra, 2016).

Thus, we can write:

$$a \equiv b \ (\text{mod } m) \ \text{if}$$

$m/a - b$ i.e., m divides $a - b$

Or

Sometimes we can say, $a \equiv b \pmod{m}$ if

Integers a and b have same reminder, if we divide m or there exists an integer say k such that

$a - b = km$.

1.2.1 Examples

    a. $6 \equiv 11 \pmod{5}$

    b. $23 \equiv 1 \pmod{2}$

    c. $-7 \equiv 5 \pmod{4}$, etc.

1.3 Additive Inverse

In mathematics, when we add two numbers and their sum is zero (0), the number that is added to a given number is called its additive inverse, e.g., $a + (-a) = 0 = -a + a$, where $-a$ is additive inverse of a (Gloag, Gloag, & Kramer, 2015).

1.3.1 Examples

    a. $7 + (-7) = 0$

       Here, -7 is additive inverse of 7.

    b. $-2 + 2 = 0$

       Here, 2 is additive inverse of -2.

    c. $4 + (-4) = 0$

       Here, -4 is additive inverse of 4.

    d. $0 + 0 = 0$

       $-0 + 0 = 0$

       $0 + (-0) = 0$, etc.

       Here, 0 is additive inverse of 0 itself.

Note: Zero is the only number which is additive inverse of itself.

1.4 Multiplicative Inverse

In mathematics, when we multiply two numbers and their sum is 1, the number we multiply to a given number is called its multiplicative inverse, e.g., $a \times 1/a = 1$ or $a \cdot 1/a$, where **1/a** is the multiplicative inverse of **a** (Barua, 2017).

1.4.1 Examples

    a. $7 \times (1/7) = 1$

       Here, 1/7 is the multiplicative inverse of 7.

b. $-2 \times -1/2 = 1$

Here, $-1/2$ is the multiplicative inverse of -2.

c. $1/4 \times (4) = 0$

Here, 4 is the multiplicative inverse of $1/4$.

d. $-8/5 \times -5/8 = 1$

Here, $-5/8$ is the multiplicative inverse of $-8/5$, etc.

1.5 Matrix

In mathematics, a matrix is defined as a rectangular array of numbers arranged in rows and columns:

$$\begin{pmatrix} a_{11}, & \cdots & a_{1n} \\ \vdots & & \vdots \\ a_{1m}, & \cdots & a_{mn} \end{pmatrix}$$

where we count rows from the top and column from the left. For example, see below:

$$\begin{pmatrix} a_{i1}, & \cdots & a_{in} \end{pmatrix} \text{ and } \begin{pmatrix} a_{1j} \\ \vdots \\ a_{mj} \end{pmatrix}$$

which represent the ith row and jth column of the matrix, and a_{ij} represents the entry in the matrix on ith row and jth column.

1.5.1 Examples

a. $\begin{pmatrix} 1 & 2 & 3 \\ 0 & 1 & 2 \\ 1 & 2 & 1 \end{pmatrix}$

is a 3×3 matrix, i.e., three rows and three columns matrix.

b. $\begin{pmatrix} 1 & 2 & 3 & 4 \\ 4 & 3 & 2 & 1 \end{pmatrix}$

is a 2×4 matrix, i.e., two rows and four columns matrix, etc.

Remark

We cannot say that a matrix is a collection of elements, but every element has assigned at a fixed position in a particular row and column (Chandra, Lal, Raghavendra, & Santhanam, 2018).

1.6 Linear Congruence

The standard form of the linear congruence is as follows:

$$ax \equiv b \pmod{m}$$

where a, b, and m are positive integers and x is a variable (or integer) which is called **linear congruence** (Hall & Keynes, 2015).

Examples

    a. $56x \equiv 1 \pmod{93}$
    b. $15x \equiv 12 \pmod{57}$, etc.

1.6.1 Solution of Linear Congruence

1. A linear congruence $ax \equiv b \pmod{m}$ has a solution if and only if $(a, m) | b$.
2. We can solve linear congruence by finding modular inverses, by using the Euclidean algorithm, and by turning the congruence into a linear Diophantine equation.

Remarks

If a linear congruence $ax \equiv b \pmod{n}$ has a solution, then there are always infinitely many solutions, because if $x0$ is a solution, then $x0 + kn$ is also a solution for any integer k. By mod n, we can always assume that $0 \leq x0 \leq n - 1$ (Hall & Keynes, 2015).

1.7 Prime and Relative Prime Numbers

1.7.1 Prime Numbers

A prime number is a positive integer with exactly two positive numbers. If n is prime, then its only two divisors are 1 and n itself. A number is called composite if it is greater than 1 and is the product of two numbers greater than 1.

    Or

An integer n is called prime if $n > 1$, and if the only positive divisors of n are 1 and n, if $n > 1$ and if n is not prime, then n is called composite (Apostol, 1976).

Example

The following are some of the prime numbers, as there are infinite prime numbers: 2, 3, 5, 7, 11, 13, 17, 19, …

1.7.2 Relatively Prime Numbers

Any two numbers are relatively prime if their gcd is 1(one), i.e., $\gcd(m, n) = 1$. But it does not imply that m and n are prime numbers. Neither 10 nor 45 is a prime number, since they both have two prime factors: $10 = 2 \times 5$ and $65 = 5 \times 13$.

1.7.3 Examples

a. 2, 3, 5, 7, 11, 13, 17 are some of the prime numbers, as there are infinite many prime numbers.

b. 7 and 11 are relatively prime.

c. 13 and 14 are relatively prime.

d. 15 and 18 are not relatively prime.

e. 2 and 6 are not relatively prime, etc.

1.8 Greatest Common Divisor (Euclid's Algorithm, Bézout's Algorithm, and Extended Euclid's Algorithm)

1.8.1 Greatest Common Divisor

In mathematics, any two numbers m and n, gcd(m, n) are known as the greatest common divisor (GCD) shared by m and n (Davenport, 2009; Bogomolny, 1996).

1.8.1.1 Examples

a. gcd(25, 45) = 5

b. gcd(6, 54) = 6

c. gcd(11, 5) = 1

d. gcd(9, 12) = 3, etc.

1.8.2 Euclid's Algorithm

In mathematics, the Euclidean algorithm is defined as the GCD of two positive integers a and b. The gcd g is the largest positive integer, which divides both a and b without leaving a remainder. This method also tells us that the GCD of two numbers does not vary if the larger number is substituted by its difference with the smaller number (Bogomolny, 1996; Rosen, 1986).

1.8.2.1 Example

The GCDs by Euclid's algorithm is given as follows:

1. gcd(165, 75)

   $165 - 75 \times 2 = 15$

   $75 - 15 \times 5 = 0$

   Hence, gcd(165, 75) = gcd(75, 15) = 15.

2. The gcd of (266, 85) by Euclid's algorithm is given as follows:

   $266 \times 85 \times 3 = 11$

   $58 - 11 \times 5 = 3$

$11 - 3 \times 3 = 1$

$3 - 1 \times 3 = 0$

Hence, gcd(266, 58) = gcd(58, 11) = gcd(11, 3) = gcd(3, 1) = 1, etc.

1.8.3 Bézout's Theorem

Bézout's theorem is defined as the description of algebraic geometry that states the number of common points, or intersection points, of two plane algebraic curves that do not share a common component (i.e., which do not have infinitely many common points) (Hefferon & Clark, 2003).

Bézout's identity: Bézout's identity (or Bézout's lemma) is the following theorem in elementary number theory: For nonzero integers a and b, let d be the GCD $d = \text{gcd}(a, b)$, then there exist integers x and y such that

$$ax + by = d$$

For $a = 120$ and $b = 168$, the gcd is 24. Thus, $120x + 168y = 24$ for some x and y. Let's find the x and y.

Start with the next to last line of the Euclidean algorithm, $120 = 2(48) + 24$ and write

$$24 = 120 - 2(48)$$

In the line above this one, $168 = 1(120) + 48$. Thus, we have

$$48 = 168 - 1(120)$$

Substitute $168 - 1(120)$ for 48 in $24 = 120 - 2(48)$, and simplify:

$$24 = 120 - 2(48)$$

$$= 120 - 2\left[168 - 1(120)\right]$$

$$= 120 + 2(120) - 2(168)$$

$$= 3(120) - 2(168)$$

Compare this to $120x + 168y = 24$ and we see $x = 3$ and $y = -2$.

Checking these values, we obtain:

$$120x + 168y = 120(3) + 168(-2)$$

$$= 360 - 336$$

$$= 24$$

1.8.4 Extended Euclid's Algorithm

The Extended Euclidean algorithm is an extension to the algorithm. This is a process of finding the GCD of two positive integers a and b and also GCD of an integer linear combination of a and b (Ireland, 2010; Rosen, 1986).

The following are main steps of the algorithm:

1. Set the value of the variable c to the larger of the two values a and b, and set d to the smaller of a and b.
2. Find the quotient and the remainder when c is divided by d. Call the quotient q and the remainder r. Use the division algorithm and expressions for previous remainders to write an expression for r in terms of a and b.
3. If $r = 0$, then $\gcd(a, b) = d$. The expression for the previous value of r gives an expression for $\gcd(a, b)$ in terms of a and b. Stop.
4. Otherwise, use the current values of d and r as the new values of c and d, respectively, and go back to step 2.

1.8.4.1 Examples

1. Find the gcd of 81 and 57 by the Euclidean algorithm:

$$81 = 1(57) + 24$$
$$57 = 2(24) + 9$$
$$24 = 2(9) + 6$$
$$9 = 1(6) + 3$$
$$6 = 2(3) + 0$$

Now if the $\gcd(a, b) = r$, then there exist integers p and s such that

$$p(a) + s(b) = r$$

We find values of p and s by reversing the steps in the Euclidean algorithm by starting with the next to last line of the above example:

$$3 = 9 - 1(6)$$

$$3 = 9 - 1(24 - 2(9)) = 3(9) - 1(24) \quad \text{As } 6 = 24 - 2(9)$$

$$3 = 3(57 - 2(24)) - 1(24) = 3(57) - 7(24) \quad \text{As } 9 = 57 - 2(24)$$

$$3 = 3(57) - 7(81 - 1(57)) = 10(57) - 7(81) \quad \text{As } 24 = 81 - 1(57)$$

Thus, we have $p = -7$ and $s = 10$.

By using the above procedure, we found that 3 is a gcd of 81 and 57 and 3 is a linear combination of 57 and 81.

1.9 Conclusion

In this chapter, we have thrown a little light on basic concept and definition of number theory or numerical theory, and we conclude that number theory is really the "Queen of Mathematics". We conclude that a set of integers is the main part of number theory. We study integers and its arithmetic operations on a number line, modules arithmetic with examples, additive inverse, matrix, linear congruence, a solution of linear congruence, prime and relatively prime number, GCDs with examples to cover all content tasks. We also conclude that Euclid's algorithm and Extended Euclid's algorithm are two main approaches to find out GCD of two positive integers.

References

Anglin, W. S. (1995). *An Introduction to Number Theory*. Montreal: Department of Mathematics and Statistics, McGili University.

Apostol, M. T. (1976). *Introduction to Analytic Number Theory*. New York: Springer. ISBN 0-387-90163-9. http://www.mathsisfun.com/numbers/prime-numbers-advanced.html.

Barua, R. (2017). *Number Theory and Algebra: A Brief Introduction*. Kolkata: Indian Statistical Institute.

Bogomolny, A. (1996). Euclid's Algorithm. Retrieved May, 2018, from http://www.cut-the-knot.org/blue/Euclid.shtml.

Chandra, P., Lal, A. K., Raghavendra, V., & Santhanam, G. (2018). Notes on Mathematics. Retrieved April, 2018, from http://home.iitk.ac.in/~peeyush/102A/Lecture-notes.

Davenport, H. (2009). *The Higher Arithmetic: An Introduction to the Theory of Numbers* (8th ed.). Cambridge: Cambridge University Press. ISBN: 9780521722360.

Gay, E. M. (2007). *Basic College Mathematics with Early Integers* (2nd ed.). New Orleans, LA: University of New Orleans.

Gloag, A., Gloag, A., Kramer, M. (2015). Additive Inverses and Absolute Values. Retrieved from https://www.ck12.org/book/CK-12-Basic-Algebra-Concepts/section/2.2/.

Hefferon, J., & Clark, W. E. (2003). Elementary Number Theory - Saint Michaels College. Retrieved April, 2018, from http://joshua.smcvt.edu/numbertheory/book.pdf.

Ireland, D. (2010). Retrieved May, 2018, from https://www.di-mgt.com.au/euclidean.html.

Keynes, M. (2015) *Number Theory by Walton Hall*. The Open University, MK7 6AA. Copyright 2015.

Mcnamra, P. (2016). Modular Arithmetic. Retrieved May, 2018, from http://www.ucd.ie/t4cms/modular.pdf.

Rosen, K. H. (1986). *Elementary Number Theory and its Applications*. Reading, MA: Addison-Wesley.

2

Finite Fields

Afroz
Maulana Azad National Urdu University

CONTENTS

2.1 Introduction 12
    2.1.1 Background 12
    2.1.2 Binary Operation 12
    2.1.3 Examples 12
2.2 Group 13
    2.2.1 Examples 13
    2.2.2 Cyclic Group 13
    2.2.3 Examples 14
    2.2.4 Results 14
    2.2.5 Abelian Group 14
    2.2.6 Coset of a Group 14
    2.2.7 Lagrange's Theorem 15
2.3 Ring 15
    2.3.1 Examples 16
2.4 Field 16
    2.4.1 Examples 17
    2.4.2 Finite Field 17
    2.4.3 Finite Field of the Type $GF(p)$ 18
    2.4.4 Finite Fields of Order p 19
    2.4.5 The Field of the Form $GF(P^n)$ 20
2.5 Schoof–Elkies–Atkin (SEA) Algorithm 20
    2.5.1 Steps of SEA 20
    2.5.2 Steps of Algorithm 20
2.6 Algebraic Polynomial 21
2.7 Coppersmith's Method 22
    2.7.1 Some Cases of the Coppersmith Method 22
2.8 Wiener's Theorem 23
2.9 Conclusion 25
References 26

2.1 Introduction

2.1.1 Background

The word "algebra" is used to mean many different things. This word is used in the title of al-Khwarizmi's book 1,200 years ago. But as a subject it is studied 4,000 years ago by ancient Babylonia and Egypt. Modern algebra, also called as Abstract algebra, is a branch of mathematics which deals with the algebraic structure of various sets such as real numbers, complex numbers, set of integers under modulo n (Z_n), matrices, modules, vector spaces, lattices, and algebras (Hazewinkel, Gubareni, & Kiričenko, 2005). Knonecker (1888) claimed that the study of modern algebra began with his first paper of Vandermonde. Cauhy states quite clearly that Vandermonde had priority over Lagrange for this remarkable idea which eventually led to the study of group theory (Connor & Robertson, 2001).

Abstract algebra emerged around the start of the 20th century under the name modern algebra. Initially, the assumption in classical algebra on which whole mathematics depends took the form of the axiomatic system. The formal definition of certain algebraic structures began to emerge in the 19th century, such as results on permutation group, algebra of structure, and classification of various mathematical objects. These processes were occurring throughout all of mathematics, but became especially pronounced in algebra. Formal definitions through primitive operations and axioms were proposed for many basic algebraic structures such as group, rings, and fields. Therefore, such things as group theory and ring theory took their place in pure mathematics.

2.1.2 Binary Operation

A mapping $*: A \times A \rightarrow A$ is called a binary composition on the set A, where $A \times A = \{(a,b): a,b \in A\}$.

In simple words, binary operation on a set A is a calculation that combines two elements of set A such that the result is also the element of the same set A. Image of an element (a, b) $\in A \times A$ under $*$ is usually represented by $a * b$ (Kumar, 2009).

2.1.3 Examples

1. Addition and multiplication are binary operations on the set of Natural numbers, because $a + b \in N$ and $a \cdot b \in N$.

   However, the operation subtraction is not a binary operation in N, since $3 \in N$ and $5 \in N$ but $3 - 5 = -2 \notin N$

2. Let $a * b = a + b + 1$, where $a, b \in N$

   $*$ is a binary composition in the set of all odd integers, but $*$ is not a binary composition in the set of all even integers, since $2 * 4 = 2 + 4 + 1 = 7$ which is not an even integer.

3. Similarly, $a * b = a + b - ab$ for all $a, b \in Z$ is a binary composition in Z.

Remark

Binary operation is also called binary composition.

2.2 Group

A nonempty set G with binary composition $*$ is called a **group** if the following conditions are satisfied (Bhattacharya, Jain, & Nagpaul, 1999):

1. $a*b \in G \forall a,b \in G$ (Cloure law)
2. $a*(b*c) = (a*b)*c \forall a,b,c \in G$ (Associative law)
3. There exists an element $e \in G$ such that

$$a*e = e*a = a \forall a \in G \text{ (Existence of identity)}$$

e is called an identity element in G

4. For each

$$a \in G, \text{ there exixts an element } b \in G \text{ such that}$$

$$a*b = b*a = e \text{ (Existence of inverse)}$$

2.2.1 Examples

1. Set $A = \{1,-1\}$ under the binary operation multiplication with identity 1 and with −1 as inverse of −1, 1 as inverse of 1 known as self-inverse.
2. Set $A = \{1,-1,i,-i\}$ under the binary operation multiplication is a group. Multiplicative inverse 1 and inverse if 1.
3. $(I, +)$ is a group, with 0 as its additive identity and the inverse of $a \in I$ is $-a \in I$.
4. $(R,+),(Q,+)$ are groups.

Remark

$(N,+)$ is not a group as the two conditions, namely existence of identity and existence of inverse, are not satisfied in N under +.

2.2.2 Cyclic Group

A group is known as the cyclic group (Hazewinkel, 2001), if there exists an element $a \in G$ that generates all its elements and each element of G is expressed as

$$x = a^n = a \cdot a \cdot a \cdot a \cdots a \, (n \text{ times})$$

The element of $a \in G$ is called a generator of G and G is written as

$$G = \langle a \rangle \text{ or } G = (a).$$

Remark

A group $\{G,+\}$ is cyclic, if there exists some element a of G such that each element x of G is expressed as follows:

$$x = na = a + a + a + \cdots + a \, (n \text{ times}) \text{ for some integer } n$$

2.2.3 Examples

1. The set $A = \{1, -1\}$ under binary operation multiplication is a cyclic group with -1 as its generator.
2. The set $A = \{1, -1, i, -i\}$ with binary operation multiplication is a cyclic group. Where $G = \langle i \rangle$ since $i^1 = 1, i^2 = -1, i^3 = -i, i^4 = 1 \le$.
3. $(z, +)$ is a cyclic group, where $Z = \langle 1 \rangle$.
4. The nth root of unity is a cyclic group.

2.2.4 Results

1. A subgroup of cyclic group is a cyclic.
2. The number of generators of an infinite cyclic group is 2, namely a generator and its inverse.
3. If $G = \langle a \rangle$ be a finite cyclic group of order n, then a^m is a generator of G iff $0 < m < n$ and $(m, n) = 1$.

Illustrated Example

Find the generators of a cyclic group of order 8.

Solution: We have $G = \langle a \rangle$, $a^n = e$
Then by the above theorem, all the generators of a, a^3, a^5, a^7 since gcd of 1, 3, 5, 7 with 8 is 1. Thus, $G = \langle a \rangle = \langle a^3 \rangle = \langle a^5 \rangle = \langle a^7 \rangle$.

2.2.5 Abelian Group

An abelian group is a set A with a binary operation $*$ satisfying the following conditions:

1. For all $a, b, c \in A$, we have $a * (b * c) = (a * b) * c$ (the Associative law).
2. There is an element $e \in A$ such that $a * e = a$ for all $a \in A$.
3. For any $a \in A$, there exists $b \in A$ such that $a * b = e$.
4. For all $a, b \in A$, we have $a * b = b * a$ (the commutative law).

Thus, an Abelian group is a group satisfying the commutative law (Kumar, 2009).

Examples

1. $(Z, +), (R, \times), (Q, +), (C, +)$ are all infinite abelian group.
2. All the above cyclic groups are an abelian group.

2.2.6 Coset of a Group

The coset of a group is of two types: one is called left coset, and the other is called right coset. It can be defined as follows: if H is a subgroup of group G and $a \in G$, then $aH = \{ah \mid h \in H\}$ is called left coset and $Ha = \{ha \mid h \in H\}$ is called right coset (Bhattacharya, Jain, & Nagpaul, 1999).

Note both cosets are subsets of G, and these cosets partition the group into equivalence classes.

2.2.7 Lagrange's Theorem

Statement

The order of a subgroup of a finite group divides the order of a group.

In other words, if G is a finite group and H is a subgroup of G, then $o(H)$ is a divisor of $o(G)$ (Gallian, 2012).

Proof

Let G be a group such that $|G|$ = number of elements of $G = n$

Also suppose that H is any subgroup of G such that $|G| = m$, we have to prove that m divides G.

Take $G' = \{g_1, g_2, g_3, \ldots, g_m\}$ be m distinct elements of G.

Let $a \in G$ be any element of G, then clearly aG' is called left coset of H in G.

Define $\varphi : G' \to aG'$ defined as $\varphi(g) = aG'$, here we can easily prove that φ is a bijective function from G' to aG', such that $|G'| = |aG'|$, i.e., every left coset of G' has the same number of elements as that of G'.

Since we know that cosets of G' in G partition G, i.e., $G'a = \{aG' \mid h \in G'\}$ $G = a_1G' \cup a_2G' \cup \ldots \cup a_rG'$

$\therefore |G| = r|G'|$, where r is the number of left cosets of G.

Thus, the order of G' divides the order of G.

This proves the theorem.

Remark 1

The converse of Lagrange's theorem is not true, i.e., if a number m divides the order of G, then it is not necessary that G may have a subgroup of order m.

Remark 2

The converse of Lagrange's theorem holds in the case of finite cyclic groups.

2.3 Ring

A ring is a nonempty set R with two binary compositions denoted by $+$ and $\cdot$, satisfying the following properties:

1. $a + b \in R$ for all $a, b \in R$
2. $a + b = b + a$ for all $a, b \in R$
3. $a + (b + c) = (a + b) + c$ for all $a, b, c \in R$
4. There exists an element denoted by $0 \in R$ such that $a + 0 = 0 + a = a$ for all $a \in R$ (0 is called additive identity or zero elements in R)
5. For each $a \in R$, there exists an element $b \in R$ such that $a + b = 0$

6. $a \cdot b \in R$ for all $a,b \in R$

7. $a \cdot (b \cdot c) = (a \cdot b) \cdot c$ for all $a,b,c \in R$

8. $a \cdot (b + c) = a \cdot b + a \cdot c$ (Left distributive law), $\forall a,b,c \in R$

9. $(a + b) \cdot c = a \cdot c + b \cdot c$ (Right distributive law), $\forall a,b,c \in R$ (Kumar, 2009).

Remark

Axioms 1–5 means that $(R, +)$ is an abelian group under addition and 6–7 means that $(R, \cdot)$ is a semi-group.

2.3.1 Examples

1. The set Z of integers is a commutative ring under the usual addition and multiplication of integers.
2. The ring of integers modulo n, i.e., Z_n.
3. The ring of Gaussian Integers $Z[i]$

$$Z[i] = \left\{ m + ni : m,n \in Z, i = \sqrt{-1} \right\}$$

4. The set of all 2×2 matrices

$$M = \left\{ \begin{bmatrix} a & b \\ c & d \end{bmatrix} : a,b,c,d \in R \text{ (all reals)} \right\}$$ is a noncommutative ring with unity,

under matrix addition and matrix multiplication.

2.4 Field

A field is one of the fundamental algebraic structures. In formal language, we can say a field is a set F together with two operations, namely addition and multiplication, which satisfy the following properties (Beachy & Blair, 2006):

1. **Associative property of addition and multiplication**:

$$x + (y + z) = (x + y) + z \text{ and } x \cdot (y \cdot z) = (x \cdot y) \cdot z.$$

2. **Commutative property of addition and multiplication**:

$$x + y = y + x \text{ and } x \cdot y = y \cdot x.$$

3. **Additive and multiplicative identity**: there exist two different elements 0 and 1 in F such that $x + 0 = x$ and $x \cdot 1 = x$ $\forall x$.

4. **Additive inverses**: for all 'x' in F, there exists an element '$-x$' in F, called additive inverse of x, such that $x + (-x) = 0$.

5. **Multiplicative inverses**: for every $x \neq 0$ in F, there exists an element in F, denoted by x^{-1}, $1/x$, called the multiplicative inverse of x, such that

$$x \cdot x^{-1} = 1.$$

6. **Distributive property of multiplication over addition**:

$$x \cdot (y + z) = (x \cdot y) + (x \cdot z).$$

2.4.1 Examples

1. The set of real numbers R, with the usual addition and multiplication, forms a field.
2. The set of complex numbers C also forms a field under usual addition and multiplication.
3. The set of four elements, namely {0, I, x, y}, where O stands additive identity, I stands for multiplicative identity, and A, B are the two elements in the set with the two binary operations defined in the composition tables (Tables 2.1 and 2.2), forms a field denoted by F_4.
4. The subset consisting of 0 and I in the above example is also a field, which is known as the binary field F_2 or $GF(2)$.

2.4.2 Finite Field

Finite fields (or Galois fields) are the field with finite elements; such cardinality is referred to as the order of the field. The above example F_4 is a field with four elements. Its subfield F_2 is the smallest field, because by definition a field has at least two distinct elements $1 \neq 0$.

Finite fields with prime order are the most directly accessible fields using modular arithmetic. For a fixed positive integer n, the arithmetic "modulo n" means to work with the numbers $Z/nZ = \{0, 1, \ldots, n - 1\}$.

TABLE 2.1

Addition

+	0	I	x	y
0	0	I	x	y
I	I	0	y	x
x	x	y	0	I
y	y	x	I	0

TABLE 2.2

Multiplication

×	0	I	x	y
0	0	0	0	0
I	0	I	x	y
x	0	x	y	I
y	0	y	I	x

The mathematical operations, addition and multiplication, on this set are done by performing the operation in question in the set Z of integers, dividing by n and taking the remainder as result. This construction yields a field precisely if n is a prime number. For example, taking $n = 2$ in the above-mentioned field, i.e., F_2.

For $n = 4$ and more generally, for any composite number $n = r \cdot t$, Z/nZ is not a field: the product of two non-zero elements is zero since $r \cdot t = 0$ in Z/nZ, which was explained above, prevents Z/nZ from being a field. The field Z/pZ with p elements (p being prime) constructed in this way is usually denoted by Fp.

In modular arithmetic modulo 12, $9+4 = 1$ since $9+4 = 13$ in Z, which divided by 12 leaves remainder 1. Since 12 is not a prime, $Z/12Z$ is not a field.

Remark

- The number of elements in any finite field F has p^n elements, where p is prime and $n \geq 1$.
- The group of positive integers modulo a prime p given by

$$Z_p \equiv \{1, 2, 3, \ldots, p-1\} \text{ is a field.}$$

Satisfaction of Required Properties

- Closure. Yes.
- Associatively. Yes.
- Identity. 1.
- Inverse. Yes.

Example

$$Z_7 = \{1, 2, 3, 4, 5, 6, 7\}$$

$$1-1 = 1, 2^{-1} = 4, 3^{-1} = 5, 6^{-1} = 6, 7^{-1} = 7$$

2.4.3 Finite Field of the Type $GF(p)$

The field with p elements can be constructed as the splitting field of the polynomial of the form $f(x) = x^p - x$, where p is a prime number.

This splitting field is an extension of F_p in which the polynomial f has p zeros, i.e., f has as many zeros as possible because the degree of f is p. For $p = 2^2 = 4$, it can be checked step by step by using the above multiplication table that all four elements of F_4 satisfy the equation $x^4 = x$, so they are zeros of f. However in F_2, f has only two zeros (namely 0 and 1), so f does not split into linear factors in this smaller field. Elaborating further on basic field-theoretic notions, it can be shown that two finite fields with the same order are isomorphic. It is thus customary to speak of the finite field with p elements, denoted by F_p or $GF(p)$ (Bussey, 1910).

The simplest finite field is $GF(2)$ by (Tables 2.3–2.5).

TABLE 2.3

Addition

+	0	1
0	0	1
1	1	0

TABLE 2.4

Multiplication

×	0	1
0	0	0
1	0	1

TABLE 2.5

Inverses

w	$-w$	w^{-1}
0	0	-
1	1	1

2.4.4 Finite Fields of Order *p*

$$GF(7): \{0,1,2,3,4,5,6\}$$

Arithmetic in *GF*(7) by Tables 2.6–2.8.

TABLE 2.6

Addition Modulo 7

+	0	1	2	3	4	5	6
0	0	1	2	3	4	5	6
1	1	2	3	4	5	6	0
2	2	3	4	5	6	0	1
3	3	4	5	6	0	1	2
4	4	5	6	0	1	2	3
5	5	6	0	1	2	3	4
6	6	0	1	2	3	4	5

TABLE 2.7

Additive and Multiplicative Inverse Modulo 7

w	$-w$	w^{-1}
0	0	-
1	6	1
2	5	4
3	4	5
4	3	2
5	2	3
6	1	6

TABLE 2.8

Multiplication Modulo 7

×	0	1	2	3	4	5	6
0	0	0	0	0	0	0	0
1	0	1	2	3	4	5	6
2	0	2	4	6	1	3	5
3	0	3	6	2	5	1	4
4	0	4	1	5	2	6	3
5	0	5	3	1	6	4	2
6	0	6	5	4	3	2	1

2.4.5 The Field of the Form $GF(P^n)$

The polynomials $Z_p[x] \bmod p(x)$, where $p(x) \in Z_p[x]$, $p(x)$ is irreducible, and $\deg(p(x)) = n$ (i.e., $n+1$ coefficients), form a finite field. Such a field has p^n elements.

These fields are called Galois fields or $GF(p^n)$ (Benvenuto, 2012).

The special case $n = 1$ reduces to the fields Z_p

$$GF(2^n)$$

The coefficients are bits {0, 1}.

For example, the elements of $GF(2^n)$ can be represented as a byte, one bit for each term, and $GF(2^{64})$ as a 64-bit word:

$x^6 + x^4 + x + 1 = 01010011.$

2.5 Schoof–Elkies–Atkin (SEA) Algorithm

In finding the order or calculating the number of points on an elliptical curve over a finite field, SEA algorithm is used, i.e., an algorithm which determines the order of an elliptic curve E over the finite field F_p. This algorithm is an extension of Schoof's algorithm by Noam Elkies and A. O. L. Atkin to improve its efficiency (under heuristic assumptions) (Schoof, 1995).

Let $L = \{l_1, l_2, \ldots, l_s\}$ be a set of primes of certain kinds known as Elkies primes and Atkin primes. Note a prime l is called an Elkies prime if the characteristic equation $\phi^2 - t\phi + q = 0$ splits over F_l while as Atkin prime is a prime which is not Elkin.

2.5.1 Steps of SEA

Let a prime order of a finite field, p, and an Elliptic curve E: $f = y^2 = x^3 + Ax + B$ over that field. We want to find $\#E(F_p) = p + 1 - a_p$. For each Elkies prime, we will keep a residue $E_l \equiv a_p$ modulo l. For each Atkin prime, we will keep a set A_l of possible residues of a_p modulo l.

2.5.2 Steps of Algorithm (Gaski, 2010)

1. Compute the j-invariant $j = j(E)$.

2. Loop over primes l while a_p is not fully determined. For each prime l:
   a. Compute $\gcd\left(\Phi_l(x,j), x^p - x\right)$.
   b. If the degree of the gcd is 0, this is an Atkin prime.
      I. Find degree r of p-power Frobenius τ acting on $E[l]$.
      II. Choose a non-square element d of F_l.
      III. Find a generator g of $F_{l^2}^*$.
      IV. Create $T = \left\{g^n : \gcd(n,r) = 1, n \in F_l\right\}$.
      V. For each $\gamma \in T$:
         A. Express γ as $g_1 + g_2\sqrt{d}$.
         B. Check if $p(g_1 + 1)/2$ is a square in F_l. If not, move to next element of T. If so, calculate a_1 such that $a_1^2 = p(g_1 + 1)/2$ and add $\{\pm 2a_1\}$ to the set A_l, possible residues a_p mod l.
   c. Otherwise, this is an Elkies prime.
      I. Find the polynomial F_l factor of f_l.
      II. Find $\lambda \in F_l$ such that $\gcd\left(\psi_2^\lambda(x-x) + \psi_{\lambda-1}\psi_{\lambda+1}, F_l\right) \neq 1$.
      III. Save $[a_p] = \lambda + p/\lambda$ as E_l.
3. Recover a_p from the A_l and E_l residues, as the unique integer satisfying those congruence in the range $-2\sqrt{q} \leq a_p \leq 2\sqrt{q}$.

2.6 Algebraic Polynomial

Polynomials are one-variable algebraic expressions that consist of terms in the form ax^n, where n is a nonnegative (i.e., positive or zero) integer and a is a real number and is called the coefficient of the term. The nth degree polynomial is as follows:

$$P(x) = a_n x^n + a_{n-1} x^{n-1} + \cdots + a_2 x^2 + a_1 x^1 + a_0, \quad a_n \neq 0, a_i \in R$$

The degree of a polynomial in one variable is the largest exponent in the polynomial. The following are some examples of polynomials and their degrees:

$$12x^8 - 7x^5 + 8x^2 - 4 \qquad \text{degree} = 8$$
$$x^5 + x^4 - x^3 + x^2 - x + 1 \qquad \text{degree} = 5$$
$$4x^6 \qquad \text{degree} = 6$$
$$7x^{42} - 4 \qquad \text{degree} = 42$$
$$3x - 2 \qquad \text{degree} = 1$$
$$9 \qquad \text{degree} = 0$$

A polynomial is not required to have all powers of x as we see in the first example. It may contain any finite number of terms depending on the demand of the context.

Note that a polynomial is any algebraic expression that consists of terms in the form ax^n. Therefore, another way to write the last example is $9x^0$. Thus, it is clear that the exponent on the x is zero (this also explains the degree equal to 0) and so it is really a polynomial in one variable.

The following are some examples that are not polynomials:

$$5x^3 + 3x^{-4} + 4$$

$$4\sqrt{x} + x - x^2$$

$$\frac{2}{3x} + 4x^5 + 1$$

Because they cannot be put in the expression like in equation 2.1.

Polynomials in two variables are algebraic expressions consisting of terms in the form $ax^m y^n$.

The degree of each term in a polynomial in two variables is the sum of the exponents in each term, and the degree of the polynomial is the largest such sum. The following are some examples of polynomials in two variables and their degrees:

$$5x^6 y^7 - 7x^2 y^4 + 8x^2 - 12y^3 + \frac{1}{2} \qquad \text{degree} = 13$$

$$10x^5 + 23y^7 - xy \qquad \text{degree} = 7$$

$$13x^9 + 7y^8 - 4x + 6y^2 + 2 \qquad \text{degree} = 9$$

A monomial is a polynomial that consists of exactly one term. A binomial is a polynomial that consists of exactly two terms. Finally, a trinomial is a polynomial that consists of exactly three terms.

2.7 Coppersmith's Method

The Coppersmith method, proposed by Don Coppersmith, is a method to find small integer zeros of univariate or bivariate polynomials modulo a given integer.

Statement

Let $F(x) = x^n + a_{n-1}x^{n-1} + \cdots + a_1 x + a_0$ and assume that $F(x_0) \equiv 0 \pmod{M}$ for some. Coppersmith's algorithm can be used to find this integer solution x_0. Lattice basis reduction techniques in time polynomial in $\log M$ and n (Coppersmith, 1996).

2.7.1 Some Cases of the Coppersmith Method

1. The Problem Univariate Modular (Case I):

    Input:

    a. A polynomial $f(x) = x^\delta + a_{d-1}x^{\delta-1} + \cdots + a_1 x + a_0$

      b. *M* an integer of unknown factorization

Find:

      a. All integers x_0 such that $f(x_0) \equiv 0 \bmod M$

2. Coppersmith's theorem for the Univariate Modular (Case II)
   The solutions x_0 can be found in polynomial time in $\log(M)$ if

$$|x_0| < N^{1/\delta}$$

3. The Problem (Bivariate Integer Case):

Input:

      a. A polynomial $f(x, y)$

      b. *M* an integer of unknown factorization

Find:

      All integers x_0, y_0 such that $f(x_0, y_0) = 0$ over Z

4. The Problem (Multivariate Modular Case):

Input:

      a. A polynomial $f(x_1, x_2, \ldots, x_m)$

      b. *M* an integer of unknown factorization

Find:

      a. All integers $x_1, x_2, \ldots, x_n$ such that

$$f(x, x, \ldots, x) \equiv 0 \bmod M$$

Applications of Coppersmith's method: Coppersmith's method is usually used to find small solutions to polynomial equations:

a. Univariate modular

b. Multivariate modular (heuristic)

c. Bivariate over the integers

d. Multivariate over the integers (heuristic).

Example

$$\text{Consider } P(x) = \left(2^t k + x\right)^e - C \equiv 0 \bmod M$$

$$\text{Coppersmith's condition: } |x| < M^{1/e}$$

If $e = 3$, recover *m* provided that 2/3 of the bits are known.

2.8 Wiener's Theorem

Let $N = pq$ with $q < p < 2q$. Let $d < 1/3N^{1/4}$. Given (N, e) with $ed = 1 \bmod \varphi(N)$, an attacker can efficiently recover *d*.

Proof

The proof is based on approximations using continued fractions. Since $ed = 1$ mod $\varphi(N)$, there exists a k where $ed - k \cdot \varphi(N) = 1$. Therefore,

$$\left| \frac{e}{\varphi(N)} - \frac{k}{d} \right| = \frac{1}{d\varphi(N)}$$

Since $\varphi(N) = N - p - q + 1$ and $p + q - 1 < 3\sqrt{N}$ an attacker can use N to approximate $\varphi(N)$.

In order to avoid this attack, and since N is 1,024 bits, d must be at least 256 bits long. This is unfortunate for smart cards or low-powered devices.

Now we have $|p + q - 1| < 3\sqrt{N}$

$$\left| N + 1 - \varphi(N) - 1 \right| < 3\sqrt{N}$$

Also on replacing $\varphi(N)$ by N we have

$$\left| \frac{e}{N} - \frac{k}{d} \right| = \left| \frac{ed - kN}{Nd} \right|$$

$$= \left| \frac{ed - k\varphi(N) - kN + k\varphi(N)}{Nd} \right|$$

$$= \left| \frac{1 - k(N - \varphi(N))}{Nd} \right|$$

$$\leq \left| \frac{3k\sqrt{N}}{Nd} \right| = \frac{3k\sqrt{N}}{\sqrt{N}\sqrt{Nd}} = \frac{3k}{d\sqrt{N}}$$

Here we have $k\varphi(N) = ed - 1 < ed$; therefore, $k\varphi(N) < ed$. Since $e < \varphi(N)$, so $k\varphi(N) < ed < \varphi(N)$ which gives

$$k\varphi(N) < \varphi(N)d$$

$$k < d$$

we have $k < d$ also $d < \frac{1}{2}N^{1/4}$

Hence, we get

$$\left| \frac{e}{N} - \frac{k}{d} \right| \leq \frac{1}{dN^{1/4}} \tag{2.1}$$

Since $d < \frac{1}{3}N^{1/4}$, $2d < 3d$, then $2d < 3d < N^{1/4}$.

Hence, $2d < N^{1/4}$,

$$\frac{1}{2d} > \frac{1}{N^{1/4}} \tag{2.2}$$

From (2.1) and (2.2), we get

$$\left| \frac{e}{N} - \frac{k}{d} \right| \le \frac{3k}{d\sqrt{N}} < \frac{1}{d \cdot 2d} = \frac{1}{2d^2}$$

By the theorem, if it satisfies the above condition, then $\frac{k}{d}$ is among the convergent of $\frac{e}{N}$. Hence, theorem finds $\frac{k}{d}$.

Example

Suppose pick keys are $\langle N, e \rangle = \langle 90581, 17993 \rangle$

Here attack determine **d**

Now on using winter's theorem and continued fraction to approximation d.

Let's first to find continued fraction expansion of $\frac{e}{N}$. Here $\frac{e}{N} = \frac{17{,}993}{90{,}581} =$

$$\cfrac{1}{5 + \cfrac{1}{29 + \cdots + \cfrac{1}{3}}} = [0, 5, 29, 4, 1, 3, 2, 3]$$

As per continued fraction expansion $\frac{e}{N}$ all convergent $\frac{k}{d}$ are as follows:

$$\frac{k}{d} = 0, \frac{1}{5}, \frac{29}{146}, \frac{117}{589}, \frac{146}{735}, \frac{555}{2{,}794}, \frac{1{,}256}{6{,}323}, \frac{5{,}579}{28{,}086}, \frac{17{,}993}{90{,}581}$$

We get first convergent does not produce factorization of N.

$$\varphi(N) = \frac{ed - 1}{k} = \frac{17{,}993 \times 5 - 1}{1} = 89{,}964$$

Now if we solve these equations

$$x^2 - (N - \varphi(N)) + N = 0$$
$$x^2 - ((90{,}581 - 89{,}964) + 1)x + 90{,}581 = 0$$
$$x^2 - (618) + 90{,}581 = 0$$

Now we find roots which are

$$x = 379; 239$$

Hence, factorization are, $N = 90{,}581 = 379 \times 239 = \mathrm{p} \times \mathrm{q}$.

2.9 Conclusion

In this chapter, one will come to know about the brief history of Abstract algebra and binary operation defined on a set with counter example. The chapter sheds light on the concepts of modern algebra like Group, rings, fields, and algebraic polynomials in terms of the structure of a set along with the postulates. This plays an important role in the development of abstract algebra. These concepts not only develop the interest in modern algebra but also act as the gateways of understanding higher development of Abstract algebra. The chapter also briefly discusses order, finiteness, generators, and properties of an important group helping students in understanding the associated concepts of finite fields, polynomials, and their execution in computer science, including important theorems, methods, and algorithms.

References

Beachy, J. A., & Blair, W. D. (2006). *Abstract Algebra*. Prospect Heights, IL: Waveland Press.

Benvenuto, C. J. (2012). Galois Field in Cryptography - sites.math.washington.edu. Retrieved from https://sites.math.washington.edu/~morrow/336_12/papers/juan.pdf.

Bhattacharya, P. B., Jain, S. K., & Nagpaul, S. R. (1999). *Basic Abstract Algebra*. Cambridge: Cambridge University Press.

Bussey, W. H. (1910). Tables of Galois fields of order less than $1,000$. *Bulletin of the American Mathematical Society, 16*(4), 188–207. doi:10.1090/s0002-9904-1910-01888-7.

Connor, J. J., & Robertson, E. F. (2001). Alexandre-Théophile Vandermonde. Retrieved April, 2018, from www-groups.dcs.st-and.ac.uk/history/Biographies/Vandermonde.html.

Coppersmith, D. (1996). Finding a small root of a univariate modular equation. *Advances in Cryptology—EUROCRYPT '96 Lecture Notes in Computer Science*, 155–165. doi:10.1007/3-540-68339-9_14.

Gallian, J. A. (2012). *Contemporary Abstract Algebra* (8th ed.). Thomson Brooks. ISBN 13: 9781133599708.

Gaski, J. (2010). Retrieved May, 2018, from https://wstein.org/edu/2010/581d/projects/joanna_gaski/schoofs.pdf.

Hazewinkel, M. (2001). *Encyclopaedia of Mathematics*. Dordrecht: Kluwer Academic Publishers.

Hazewinkel, M., Gubareni, N., & Kiričenko, V. V. (2005). *Algebras, Rings and Modules*. Dordrecht: Kluwer Academic Publishers.

Kumar, R. (2009). *Abstract Algebra Group Theory*. New Delhi: Vardhman's Publication.

Schoof, R. (1995). Counting points on elliptic curves over finite fields. *Journal De Théorie Des Nombres De Bordeaux, 7*(1), 219–254. doi:10.5802/jtnb.142.

3

Prime Number

Khaleel Ahmad and Afsar Kamal
Maulana Azad National Urdu University

Khairol Amali Bin Ahmad
National Defence University of Malaysia

CONTENTS

3.1 Introduction 27
3.2 Prime Number 28
3.3 Co-Prime Number 29
3.4 Composite Number 30
3.5 Fermat's Theorem 30
    3.5.1 Multiplicative Property 30
3.6 Euler's Totient Function 31
3.7 Euler's Theorem 32
3.8 Divisibility Rules 33
3.9 Square Root Primality Test 35
3.10 Miller–Rabin for Primality Testing 36
3.11 AKS Primality Test 37
3.12 Mersenne Number 39
3.13 Chinese Remainder Theorem 40
3.14 Solovay–Strassen Test 43
3.15 Conclusion 43
References 43

3.1 Introduction

The prime number has a fundamental position in mathematics and number theory. Hence, it behooves us to know as much as possible about the different properties of this number. There are several efficient methods with different characteristics to determine whether a number is prime or not. For example, the AKS algorithm is deterministic as opposed to the Miller–Rabin and Fermat approaches that are based on the polynomials as well as having slow speed. However, the Miller–Rabin test is more accurate almost for all numbers.

Among the set of integers, it is very difficult to identify which number is reducible or not through the multiplicative process. Suppose **a** and **b** are integers where $\mathbf{a} \neq 0$. If $\mathbf{b} = \mathbf{a} \cdot k$ for some value k, then we can distinguish that **a** is the devisor of **b**, and we simply write **b**|**a**.

A simple deterministic algorithm to test an odd integer **g** to determine whether it is a prime or not is given as follows:

$$\text{for } d = 2, 3, \ldots, \lfloor \sqrt{I} \rfloor \text{ do}$$
$$\text{if } d | g$$
$$\text{then output "composite number" and stop.}$$
$$\text{Otherwise, output "prime number"}$$

The loop has $O(\sqrt{n})$ iteration without polynomial running time in a number of input bits.

3.2 Prime Number

A prime number is an integer that can only be divided without remainder by positive and negative values of itself and 1 (Stein, 2009; Aaronson, 2014; Agrawal, Kayal, & Saxena, 2004; Crandall & Pomerance, 2005).

- Prime numbers only have divisors of 1 and itself.
- They cannot be written as a product of other numbers.
- A prime number is always greater than 1.
- Rules to find out prime numbers between 1 and 100
    1. 1 is not prime, not composite.
    2. 2 is prime but all numbers divisible by 2 are not prime.
    3. 3 is prime but all numbers multiples of 3 are not prime.
    4. 5 is prime but diagonally all numbers multiples of 5 are not prime.
    5. 7 is prime but diagonally all numbers multiples of 7 are not prime, as depicted in Table 3.1.
- Every even number greater than 2 can be written as the sum of two prime numbers.
  For example: $8 = 7 + 1$, $20 = 19 + 1$, $36 = 31 + 5$

Method to check whether a given number is prime or not:

- A prime number that has exactly two factors.
- 1 is a unit not specified as a prime or composite number; it has also only one factor itself.
- Every prime number satisfies the rule: $(6n + 1) = P$ or $(6n - 1) = P$ except 2 and 3, where n is the natural number, as shown in Table 3.2.
- It is not the final condition to check the primality of a given number, but all the prime numbers are in the form of $(6n + 1)$ or $(6n - 1)$.
  For example, 35 is not a prime number but satisfies the rule $6n - 1 = 6 * 6 - 1 = 35$.
- For 3 or 4 digits number: To check whether the given number is prime or not just finds the perfect square closest to the given number.
  For example, checking 197 is prime or not
  The perfect square closest to 197 is $14 * 14 = 14^2 = 196$

TABLE 3.1

Prime Numbers between 1 and 100

Rules	R-1	R-2	R-3	R-2		R-2
R-5	1	2	3	4	5	6
	7	8	9	10	11	12
	13	14	15	16	17	18
	19	20	21	22	23	24
	25	26	27	28	29	30
R-4	31	32	33	34	35	36
	37	38	39	40	41	42
R-5	43	44	45	46	47	48
	49	50	51	52	53	54
	55	56	57	58	59	60
R-4	61	62	63	64	65	66
	67	68	69	70	71	72
	73	74	75	76	77	78
	79	80	81	82	83	84
R-5	85	86	87	88	89	90
R-4	91	92	93	94	95	96
	97	98	99	100		

TABLE 3.2

Every Prime Number is in the Form of $(6n + 1)$ or $(6n - 1)$

Prime No.	5	7	11	13	59	61	89	157
Rule	$6n - 1$	$6n + 1$	$6n - 1$	$6n + 1$	$6n - 1$	$6n + 1$	$6n - 1$	$6n + 1$
	$6 * 1 - 1$	$6 * 1 + 1$	$6 * 2 - 1$	$6 * 2 + 1$	$6 * 10 - 1$	$6 * 10 + 1$	$6 * 15 - 1$	$6 * 26 + 1$

Now divide 197 with all the prime numbers smaller than 14, i.e., (2, 3, 5, 7, 11, 13) < 14

Finally, if 197 is not divisible by them, then it is a prime number.

3.3 Co-Prime Number

A set of two natural numbers having only 1 as a common factor is called a co-prime number. It is also known as relatively prime numbers (Erdős, Luca, & Pomerance, 2008; Berliner, 2016; Raggi, Montes, & Wisbauer, 2005; Camina & Camina, 2009), as depicted in Table 3.3.

- Co-prime numbers always exist in a set or group
- The highest common factor (HCF) of the co-prime numbers is 1. In other words, the greatest common divisor gcd(**a**, **b**) = 1 where **a** and **b** are natural numbers and **a** > **b**.

    Suppose **a** and **b** are co-prime and **d** is a common factor, then *d* has to be 1.
- The sum of two consecutive odd numbers is always co-prime numbers, as shown in Table 3.4.
- Co-prime numbers can be of any odd or even number.

TABLE 3.3

Test Common Factor to Check Co-Prime Numbers

Numbers	Factors	Result
8	1, 2, 4, 8	There is only one common factor;
9	1, 3, 9	so, they are co-prime numbers

TABLE 3.4

Two Consecutive Odd Numbers Must Have HCF 1

Consecutive Odd Numbers	Factors	gcd/HCF	Result
15	1, 3, 5, 15	1	They are co-prime numbers
17	1, 7		
61	1, 61	1	They are co-prime numbers
63	1, 3, 7, 9, 21, 63		

3.4 Composite Number

A whole number that has more factors than itself and 1 can be defined as a composite number (Andrews & Berndt, 2012).

- Every even number greater than 2 is always a composite number.
- Every composite number can be written as the product of two or more primes, as shown in Table 3.5.

3.5 Fermat's Theorem

It states that if **a** and **p** are co-prime numbers where **a** belongs to a natural number and **p** is prime, then the remainder of a^{p-1}/p will be always 1. It is also known as Fermat's little theorem (Faltings, 1995; Robinson, 2011; Korukov, 2014).

In other words, $a^{p-1} \bmod p = 1$ or $a^p \bmod p = a$ where p is prime and $\gcd(a, p) = 1$

For example: $5^{16} \bmod 17 = 1$, $25^{96} \bmod 97 = 1$, $42^{130} \bmod 131 = 1$.

3.5.1 Multiplicative Property

The multiplicative property states that $(a * b) \bmod c = ((a \bmod c) * (b \bmod c)) \bmod c$. It breaks down the numerator into factors and finds out the remainder of each factor then after calculating their products gets the final remainder. It means when the modulo operation is

TABLE 3.5

A Number Having More Than Two Factors Is a Composite Number

Numbers > 1	Factors	Result
20	1, 2, 4, 5, 10, 20	Composite number having more than two factors
33	1, 3, 11, 33	

performed on any composite number, it is subdivided into some factors and then applied modulo on them. At last modulo is again applied after calculating product of their remainders. It gives the same result as whatever is gained after applying modulo on the composite number.

$$\mathbf{a}^{m+n} \bmod P = \left(\mathbf{a}^m \bmod P\right) * \left(\mathbf{a}^n \bmod P\right) \quad \left[\mathbf{a}^{m+n} = \mathbf{a}^m * \mathbf{a}^n\right]$$

For example:

$8^{48} \bmod 17$

$= \left(8^{16+16+16}\right) \bmod 17$

1. $= \left(8^{16} \bmod 17\right) * \left(8^{16} \bmod 17\right) * \left(8^{16} \bmod 17\right) A^{\varnothing(B)}$

$= 1 * 1 * 1$

$= 1$

$25^{97} \bmod 97$

$= \left(25^{96+1}\right) \bmod 97$

2. $= \left(25^{96} \bmod 97\right) * \left(25 \bmod 97\right) A^{\varnothing(B)}$

$= 1 * 25$

$= 25$

3.6 Euler's Totient Function

If N is a positive integer, then function $\phi(N)$ counts all the positive integers relatively prime to N and that is less than or equal to N. The number N can be represented as follows (Turner, 2008; Gorgui-Naguib & Dlay, 2000; Dineva, 2005):

$$N = \left(P_1\right)^{a1} * \left(P_2\right)^{a2} * \left(P_3\right)^{a3} \ldots * P_k^{ak}$$

This form denotes the prime factors of number N. Hence, the formula for Totient(N) is as follows:

$\Phi(N) = N(1 - 1/P_1)(1 - 1/P_2)(1 - 1/P_3) \ldots ((1 - 1/P_k)$ and this is called Euler's product formula. For example:

- $\Phi(35) = 35(1 - 1/7)(1 - 5) = 35 * 6/7 * 4/5 = 24$
  Co-prime numbers to 25: 1, 2, 3, 4, 6, 8, 9, 11, 12, 13, 16, 17, 18, 19, 22, 23, 24, 26, 27, 29, 31, 32, 33, and 34 (not multiples of 5 and 7)

- $\Phi(N) = N - 1$, if N is a prime number, e.g., $\Phi(7) = 7 - 1 = 6$
  Co-prime numbers to 7: 1, 2, 3, 4, 5, 6
- $\Phi(25) = 25(1 - 1/5) = 25 * 4/5 = 20$
  Co-prime numbers to 25: 1, 2, 3, 4, 6, 7, 8, 9, 11, 12, 13, 14, 16, 17, 18, 19, 21, 22, 23, and 24 (not multiples of 5)
- $\Phi(20) = 20(1 - 1/2)(1 - 1/5) = 20 * 1/2 * 4/5 = 8$
  Co-prime numbers to 20: 1,3,7,9,11,13,17, and 19 (not multiples of 2 and 5)
- $\Phi(N) = N - 1$, if N is a prime number, e.g., $\Phi(7) = 7 - 1 = 6$
  Co-prime numbers to 7: 1, 2, 3, 4, 5, and 6.

This formula only needs the distinct prime factor of the number N to find out the totient. As in the case of 25 and 20, the repeated numbers 5 and 2 are written only one time. This is because the factors of 25 and 20 are as follows:

$$25: 5 * 5 \text{ or } 5^2$$

$$20: 2 * 2 * 5 \text{ or } 2^2 * 5$$

- For two prime numbers A and B: $\Phi(AB) = \Phi(A) * \Phi(B) = (A - 1)(B - 1)$,
  e.g., $\Phi(3 * 5) = \Phi(3) * \Phi(5) = 2 * 4 = (3 - 1)(5 - 1) = 8$
  Co-primes to $\Phi(3 * 5) = \Phi(15)$: 1, 2, 4, 7, 8, 11, 13, and 14 (not multiples of 3 and 5)
- $\Phi(N^k) = N^k - N^{k-1}$, if N is prime and $k > 0$,
  e.g., $\Phi(2^3) = 2^3 - 2^{3-1} = 8 - 4 = 4$
  Co-primes to $\Phi(2^3) = \Phi(8)$: 1, 3, 5, and 7 (not multiples of 2)

3.7 Euler's Theorem

Let A and B are relatively prime where $A < B$, then multiplying A with itself $\Phi(B)$ times and dividing the result by B gives the remainder 1 (Conard, 2009, 2011, 2016; Lewis, 1969; Křížek, Luca, & Somer, 2002; Euler, 1996).

Suppose $\gcd(A,B) = 1$
then $A^{\varnothing(B)} \equiv 1 \ (\text{mod } B)$

Example 1

$A = 9$ and $B = 13$, where $A < B$ and $\gcd(9, 13) = 1$
then $\Phi(B) = \Phi(13) = 12$ [$\because$ 13 is prime number, $\therefore \Phi(13) = 13 - 1 = 12$]
Thus, $A^{\Phi(B)} = 9^{12} = 282{,}429{,}536{,}481 = 282{,}429{,}536{,}480 + 1 = 21{,}725{,}348{,}960 * 13 + 1$
or $282{,}429{,}536{,}481 = 1 \ (\text{mod } 13)$ or $282{,}429{,}536{,}481 \mod 13 = 1$.

Example 2

$A = 3$ and $B = 5$
then $\Phi(B) = \Phi(5) = 4$
$A^{\Phi(B)} = 3^4 = 81 \equiv 1 \ (\text{mod } 5)$.

3.8 Divisibility Rules (Briggs, 1999; Langford, 1974; P & A, 1970)

1. Divisibility by 2: A last digit is an even number (2, 4, 6, 8, or 0), as depicted in Table 3.6.
2. Divisibility by 3: The sum of its digits is divisible by 3, as shown in Table 3.7.
3. Divisibility by 4: The last two digits of the given number are divisible by 4, as portrayed in Table 3.8.
4. Divisibility by 5: The last digit of the number is 5 or 0, as portrayed in Table 3.9.
5. Divisibility by 6: The given number is divisible by 2 and 3 which satisfy the divisibility rule of both 2 and 3 which is portrayed in Table 3.10.
6. Divisibility by 7: Multiplying the last digit of the number by 5 and adding the remaining digits into the result, as depicted in Table 3.11.
7. Divisibility by 8: The last three digits of the given number are divisible by 8, as shown in Table 3.12.
8. Divisibility by 9: The sum of its digits is equal to multiple of 9. In other words, the sum of the digits is divisible by 9, as portrayed in Table 3.13.
9. Divisibility by 10: The last digit (Ones) is 0, as depicted in Table 3.14.

TABLE 3.6

Divisibility by 2

Numbers	Last Digit	Result
25,468	8 (even number)	Divisible by 2
33,892	2 (even number)	Divisible by 2
6,589	9 (odd number)	Not divisible by 2

TABLE 3.7

Divisibility by 3

Numbers	Sum of Digits	Result	Remarks
25,468	$2 + 5 + 4 + 6 + 8 = 25$	25,468 is not divisible by 3	25 is not divisible by 3
33,891	$3 + 3 + 8 + 9 + 1 = 24$	33,891 is divisible by 3	24 is divisible by 3
65,809	$6 + 5 + 8 + 0 + 9 = 28$	65,809 is not divisible by 2	28 is not divisible by 3

TABLE 3.8

Divisibility by 4

Numbers	Last Digits	Result	Remarks
25,468	68	25,468 is divisible by 4	68 is divisible by 4
33,891	91	33,891 is not divisible by 4	91 is not divisible by 4
65,809	09	65,809 is not divisible by 4	9 is not divisible by 4

TABLE 3.9

Divisibility by 5

Numbers	Last Digits	Result
25,460	0	25,460 is divisible by 5
33,895	5	33,895 is divisible by 5
65,809	9	65,809 is not divisible by 5

TABLE 3.10

Divisibility by 6

Numbers	Last Digits	Sum of Digits	Result	Remarks
25,462	2 (even no.)	$2 + 5 + 4 + 6 + 2 = 19$	25,462 is not divisible by 6	The sum of digits 19 is not divisible by 3
3,309	9 (odd no.)	$3 + 3 + 0 + 9 = 15$	3,309 is not divisible by 6	The last digit 9 is not divisible by 2
52,314	4 (even no.)	$5 + 2 + 3 + 1 + 4 = 15$	52,314 is divisible by 6	The last digit 4 and sum of the digits 15 are divisible by 2 and 3, respectively

TABLE 3.11

Divisibility by 7

Numbers	Operations	Result	Remarks
811	$81 + (1 * 5)=$	25,462 is not divisible by 9	The sum of digits 19 is not divisible by 9
26543	$3 + 3 + 0 + 9 + 3 = 18$	33,093 is divisible by 9	The sum of digits 18 is divisible by 9

TABLE 3.12

Divisibility by 8

Numbers	Last Digits	Result	Remarks
25,464	464	25,464 is divisible by 8	464 is divisible by 8
33,891	891	33,891 is not divisible by 8	891 is not divisible by 8
65,800	800	65,800 is divisible by 8	800 is divisible by 8

TABLE 3.13

Divisibility by 9

Numbers	Sum of Digits	Result	Remarks
25,462	$2 + 5 + 4 + 6 + 2 = 19$	254,62 is not divisible by 9	The sum of digits 19 is not divisible by 9
33,093	$3 + 3 + 0 + 9 + 3 = 18$	33,093 is divisible by 9	The sum of digits 18 is divisible by 9
57,825	$5 + 7 + 8 + 2 + 5 = 27$	57,825 is divisible by 9	The sum of digits 27 is multiple of 9

TABLE 3.14

Divisibility by 10

Numbers	Last Digit	Result
254,680	0	Divisible by 10
338,902	2	Not divisible by 10
658,900,000	0	Divisible by 10

3.9 Square Root Primality Test

Every composite number N has at least one factor that is less than or equal to its square root ($\sqrt{N}$).

Proof: Suppose N is a composite number, then it can be written as $N = p \cdot q$, where both p and q are between 1 and N. If both p and q are greater than ($\sqrt{N}$), then the product of both p and q will be greater than N as $p \cdot q > \sqrt{N} \sqrt{N}$. This contradiction proofs that at least one factor will be less than $\sqrt{N}$ (Krauel, 2013; Pomerance & Gagola, 1984).

Square Roots:

Let $m^2 = k \bmod N$.

Here, m is the square root of k, mod N.

Considering the Euler–Fermat theorem in which $\gcd(a, N) = 1$, then $a^{\Phi(N)} \equiv 1 \bmod N$, but if N is the prime number, then $\Phi(N) = N - 1$ and $a^{N-1} \equiv 1 \bmod N$.

Now if N is a prime number then it is an odd number and $N - 1$ will be even.

Let $m^2 = k \bmod N$, giving a power of $(N - 1)/2$ to both sides, it becomes:

$$\left(m^2\right)^{(N-1)/2} = k^{(N-1)/2}$$

$m^{N-1} = k^{(N-1)/2}$ since N is prime number, $m^{N-1} \equiv 1$ hence $k^{(N-1)/2} = 1$.

Note: If the prime number N is greater than 2, then $m^2 = k \bmod N$ gives the solution if and only if $k^{(N-1)/2} \equiv 1$.

Proof:

1. 1: $m^2 \equiv 17 \bmod 23$
   If $m^2 \equiv 17 \bmod 23$ has a solution, then it satisfies $17^{(23-1)/2} \equiv 1$
   Checking it $17^{(23-1)/2} \equiv 17^{11} \equiv 22$
   Since it $17^{(23-1)/2!} \equiv 1$, hence $m^2 \equiv 17 \bmod 23$ has no solution.

2. 1: $m^2 \equiv 31 \bmod 83$
   If $m^2 \equiv 31 \bmod 83$ has a solution, then it satisfies $31^{(83-1)/2} \equiv 1$
   Checking it $31^{(83-1)/2} \equiv 31^{41} \equiv 1$
   Since $31^{41} \equiv 1$, hence $m^2 \equiv 17 \bmod 23$ has a solution.
   Suppose prime number $N \equiv 3$, then this satisfies $N = 4n + 3$
   If $m^2 \equiv k$ has a solution, then by Euler's criterion $k^{(N-1)/2} \equiv 1$

and so $k^{(N+1)/2} \equiv k$. This means:

$$m^2 \equiv k$$

$$m^2 \equiv k^{(N+1)/2}$$

$$m^2 \equiv k^{2n+2}$$

$$m \equiv \pm k^{n+1}$$

$$m \equiv \pm k^{(N+1)/4}$$

for the solution $m^2 \equiv 31 \bmod 83$, finding it by evaluating $31^{(83+1)/4}$

$$m \equiv \pm 31^{(83+1)/4}$$

$$\equiv \pm 31^{21}$$

$$\equiv \pm 23$$

$$\equiv 23 \text{ or } 60$$

3.10 Miller–Rabin for Primality Testing

This test is typically used for a large number to check whether it is prime or composite. It is based on Fermat's theorem and more advance in some sense (Conard, 2009, 2011, 2016; Schoof, 2008).

Algorithm

1. Find out $N - 1 = m * 2^k$ where k and m as an integer with $k > 0$ and m as odd
2. Select any random integer number such that $1 < a < N - 1$
3. Apply the test
   if $a^m \bmod N = 1$ then return ("Indefinite Result")
   for $j = 0$ to $k - 1$ do
       if $(a^{2^{jm}} \bmod N = N - 1)$
               return ("Indefinite Result")
   else
               return ("composite number")

Example 1: Test 63

1. $N = 63$ and $a = 2$

$$N - 1 = m * 2^k$$

$$63 - 1 = m * 2^k$$

$$62 = 31 * 2^1$$

So $k = 1$ and $m = 31$

2. Selecting a random integer, $a = 2$ where $(1 < a < 62)$
3. 2^{31} mod $63 = 2{,}147{,}483{,}648$ mod $63 = 2! = 1$ means indefinite result; it may be prime number or composite. Now continue the test for next calculation:

$$2^{2^{0}31} \bmod 31 = 2^{31} \bmod 63 = 2{,}147{,}483{,}648 \bmod 63 = 2! = 62$$

The test returns an indefinite or inclusive result indicating that 63 may be prime.

Example 2: Test 81 where $a = 2$

$$\text{Here } N = 27$$

$$N - 1 = m * 2^{k}$$

$$81 - 1 = m * 2^{k}$$

$$80 = m * 2^{k}$$

$$5 * 2^{4} = m * 2^{k}$$

$$\text{So } m = 5 \text{ and } k = 2$$

Now find out $a^{m} \bmod N = 2^{5} \bmod 81 = 32! = 1$

$$2^{5*2} \bmod 81 = 2^{10} \bmod 81 = 52! = 81$$

Note: If $k = 1$ or the resultant mode $= +1$, then it is the composite number and the prime number in the case of -1.

3.11 AKS Primality Test

This test is also known as a Cyclomatic test or Agrawal–Kayal–Saxena primality and published by Manindra Agrawal, Neeraj Kayal, and Nitin Saxena at IIT Kanpur on 6 August 2002 in a paper titled "PRIMES is in P." This test is based on $(x + a)^{n} \equiv (x^{n} + a)(\bmod n)$ theorem and perfectly defines whether the given number is prime or not within the polynomial time. A number N is prime if all the coefficients of polynomial expansion $(x - 1)^{n} - (x^{n} - 1)$ are divisible by N (Klappenecker, 2002; Cao, 2013; Rotella, 2005; Turchet et al., 2007; Wu et al., 2016; Dietzfelbinger, 2004).

Example 1: Checking 5 for Primality

$$\text{Polynomial expansion: } (x - 1)^{n} - (x^{n} - 1)$$

$$= (x - 1)^{5} - (x^{5} - 1)$$

$$= (x - 1)^{3}(x - 1)^{2} - (x^{5} - 1)$$

$$= \left(x^3 - 3x^2 + 3x - 1\right)\left(x^2 - 2x + 1\right) - \left(x^5 - 1\right)$$

$$= x^5 - 5x^4 + 10x^3 - 10x^2 + 5x - 1 - x^5 + 1$$

$$= -5x^4 + 10x^3 - 10x^2 + 5x$$

Here all the coefficients –5, 10, –10, and 5 are divisible by 5; hence, 5 is a prime number.

Example 2: Checking 8 for Primality

Polynomial expansion: $(x - 1)^n - \left(x^n - 1\right)$

$$= (x - 1)^8 - \left(x^8 - 1\right)$$

Consider the general form of $(x + y)^n$ and apply the first row of Pascal's triangle, looking at the 9th row of Pascal's triangle for the coefficient:

$$1$$

$$1 \quad 1$$

$$1 \quad 2 \quad 1$$

$$1 \quad 3 \quad 3 \quad 1$$

$$1 \quad 4 \quad 6 \quad 4 \quad 1$$

$$1 \quad 5 \quad 10 \quad 10 \quad 5 \quad 1$$

$$1 \quad 6 \quad 15 \quad 20 \quad 15 \quad 6 \quad 1$$

$$1 \quad 7 \quad 21 \quad 35 \quad 35 \quad 21 \quad 7 \quad 1$$

$$1 \quad 8 \quad 28 \quad 56 \quad 70 \quad 56 \quad 28 \quad 8 \quad 1$$

After expanding:

$$1x^8y^0 + 8\,x^7y^1 + 28x^6y^2 + 56x^5y^3 + 70x^4y^4 + 56x^3y^5 + 28x^2y^6 + 8x^1y^7 + 1x^0y^8$$

To solve $(x - 1)^8$, substitute $y = -1$ and simplify:

$$1x^8 (-1)^0 + 8x^7 (-1)^1 + 28x^6 (-1)^2 + 56x^5 (-1)^3 + 70x^4 (-1)^4$$

$$+ 56x^3 (-1)^5 + 28x^2 (-1)^6 + 8x^1 (-1)^7 + 1x^0 (-1)^8$$

$$= x^8 - 8x^7 + 28x^6 - 56x^5 + 70x^4 - 56x^3 + 28x^2 - 8x + 1$$

Now, $(x - 1)^8 - (x^8 - 1)$

$$= x^8 - 8x^7 + 28x^6 - 56x^5 + 70x^4 - 56x^3 + 28x^2 - 8x + 1 - x^8 + 1$$

$$= -8x^7 + 28x^6 - 56x^5 + 70x^4 - 56x^3 + 28x^2 - 8x + 2$$

Hence, not all coefficients are divisible by 8; therefore, 8 is a composite number.

3.12 Mersenne Number

The Mersenne numbers are those that are generated by the formula: $M_n = 2^n - 1$ where $n > 2$. If the value of M_n is prime, then it is called the Mersenne number (refer to Table 3.15). This classification is named after a religious and French mathematician Marin Mersenne (1588–1648) who produced a list of prime numbers in this form (Murata & Pomerance, 2004; Terzich, 2012; Granger & Moss, 2013).

1. If n is a prime number, then M_n may be prime or composite number.
2. If M_n is prime, then n is prime.
3. If n is composite, then M_n is a composite number.

Proof: Let n is a composite number with two factor r and s, then

$$2^n - 1 = 2^{rs} - 1 = \left(2^r - 1\right)\left(\left(2^r\right)^{s-1} + \left(2^r\right)^{s-2} + \left(2^r\right)^{s-3} + \cdots + \left(2^r\right) + 1\right)$$

Take an example for this formula:

$$65,535 = 2^{16} - 1 = 2^{4*4} - 1 = \left(2^4 - 1\right)\left(\left(2^4\right)^3 + \left(2^4\right)^2 + \left(2^4\right)^1 + 1\right)$$

$$= 15 * \left(4,096 + 256 + 16 + 1\right)$$

$$= (15) * (4,369)$$

TABLE 3.15

Checking the Mersenne Number

n	2^n	$2^n - 1$	State
2	4	3	Prime
3	8	7	Prime
4	16	15	Composite
5	32	31	Prime
6	64	63	Composite
7	128	127	Prime
8	256	255	Composite
9	512	511	Composite
10	1,024	1,023	Composite
11	2,048	2,047	Composite
12	4,096	4,095	Composite
13	8,192	8,191	Prime
14	16,384	16,383	Composite
15	32,768	32,767	Composite
16	65,536	65,535	Composite
17	131,072	131,071	Prime
18	262,144	2,621,143	Composite
19	524,288	524,287	Prime
20	1,048,576	1,048,575	Composite
21	2,097,152	2,097,151	Composite

Here, the factors 15 and 4,369 verify that if n is a composite number, then $2^n - 1$ will be composite.

Primality test for Mersenne Number: This test was given by Lucas (Palimar & Shankar, 2012) and after that modified by Lehmer (Robinson, 1954). In this test, M_n is prime when it divides S_{n-2} where S is in the following form:

$$S_0 = 4$$

$$S_1 = (S_0)^2 - 2$$

$$S_2 = (S_1)^2 - 2$$

$$S_3 = (S_2)^2 - 2$$

$$\vdots \quad \vdots$$

$$S_{n-2} = (S_{n-3})^2 - 2$$

Example: $M_7 = 127$, needed $S_{7-2} = S_5$

Hence,

$S_0 = 4$

$S_1 = 4^2 - 2 = 16 - 2 = 14$

$S_2 = 14^2 - 2 = 196 - 2 = 194$

$S_3 = 194^2 - 2 = 37{,}636 - 2 = 37{,}634$

$S_4 = 37{,}634^2 - 2 = 1{,}416{,}317{,}956 - 2 = 1{,}416{,}317{,}954$

$S_5 = 1{,}416{,}317{,}954^2 - 2 = 2{,}005{,}956{,}546{,}822{,}746{,}116 - 2 = 2{,}005{,}956{,}546{,}822{,}746{,}114$

Now, 2,005,956,546,822,746,114 is perfectly divisible by 127 with quotient 15,794,933,439,549,182; therefore, $M_7 = 127$ must be prime.

3.13 Chinese Remainder Theorem

This theorem is used to find out an integer when it is divided by given divisors, and leaves the given remainders. It is a perfect solution for the linearly simultaneous congruence with co-prime divisors (Childs, 2009; Piazza, 2018; Ru, 2012; Lady, n.d.):

$$N \equiv a_1 \, (\bmod \, m_1)$$

$$N \equiv a_2 \, (\bmod \, m_2)$$

$$N \equiv a_3 \, (\bmod \, m_3)$$

$$\vdots \quad \vdots$$

$$N \equiv a_k \, (\bmod \, m_k)$$

Steps to find out N:

1. $M = m_1 * m_2 * m_3 * \cdots * m_k$

   $M_1 = M/m_1$

   $M_2 = M/m_2$

2. $M_3 = M/m_3$

   $\vdots \quad \vdots$

   $M_k = M/m_k$

3. Find the inverse:

$$Y_1 = M_1^{-1} \bmod m_1$$

$$Y_2 = M_2^{-1} \bmod m_2$$

$$Y_3 = M_3^{-1} \bmod m_3$$

$$\vdots \quad \vdots$$

$$Y_k = M_k^{-1} \bmod m_k$$

4. $N = \left(a_1 * M_1 * Y_1 + a_2 * M_2 * Y_2 + a_3 * M_3 * Y_3 + \cdots + a_k * M_k * Y_k\right)$

Example 1

Find out the number N which is divided by 11, 7, 9, and 5 and leaves the remainders of 4, 2, 3, and 1, respectively.

Solution:

$$x \equiv 4 (\bmod 11)$$

$$x \equiv 2 (\bmod 7)$$

$$x \equiv 3 (\bmod 9)$$

$$x \equiv 1 (\bmod 5)$$

Here, $a_1 = 4, a_2 = 2, a_3 = 3, a_4 = 1$ and $m_1 = 11, m_2 = 7, m_3 = 9, m_4 = 5$

Now,

$$M = m_1 * m_2 * m_3 * m_4 = 11 * 7 * 9 * 5 = 3{,}465 \text{ and}$$

$$M_1 = M/m_1 = 3{,}465/11 = 315$$

$$M_2 = M/m_2 = 3{,}465/7 = 495$$

$$M_3 = M/m_3 = 3{,}465/9 = 385$$

$$M_4 = M/m_4 = 3{,}465/5 = 693$$

$$Y_1 = 315^{-1} \bmod 11 = 8$$

$$Y_2 = 495^{-1} \bmod 7 = 3$$

$$Y_3 = 385^{-1} \bmod 9 = 4$$

$$Y_4 = 693^{-1} \bmod 5 = 2$$

Therefore,

$$N = \left(a_1 * M_1 * Y_1 + a_2 * M_2 * Y_2 + a_3 * M_3 * Y_3 + a_4 * M_4 * Y_4 \right) \bmod M$$

$$= \left(4 * 315 * 8 + 2 * 495 * 3 + 3 * 385 * 4 + 1 * 693 * 2 \right) \bmod 3{,}465$$

$$= \left(10{,}080 + 2{,}970 + 4{,}620 + 1{,}386 \right) \bmod 3{,}465$$

$$= 19{,}056 \bmod 3{,}465 = 1{,}731$$

Example 2

Find the smallest number which, when divided by 8, 5, and 7, produces 4, 2, and 3 as remainders.
  Solution:

$$x \equiv 4 \,(\bmod 8)$$

$$x \equiv 2 \,(\bmod 5)$$

$$x \equiv 3 \,(\bmod 7)$$

Here, $a_1 = 4$, $a_2 = 2$, $a_3 = 3$ and $m_1 = 8$, $m_2 = 5$, $m_3 = 7$.
  Now,

$$M = m_1 * m_2 * m_3 = 8 * 5 * 7 = 280 \text{ and}$$

$$M_1 = M/m_1 = 280/8 = 35$$

$$M_2 = M/m_2 = 280/5 = 56$$

$$M_3 = M/m_3 = 280/7 = 40$$

$$Y_1 = 35^{-1} \bmod 8 = 35$$

$$Y_2 = 56^{-1} \bmod 5 = 1$$

$$Y_3 = 40^{-1} \bmod 7 = 3$$

So,

$$N = \left(a_1 * M_1 * Y_1 + a_2 * M_2 * Y_2 + a_3 * M_3 * Y_3 \right) \bmod M$$

$$= \left(4 * 35 * 35 + 2 * 56 * 1 + 3 * 40 * 3 \right) \bmod 280$$

$$= \left(4{,}900 + 112 + 360 \right) \bmod 280$$

$$= 5{,}372 \bmod 280$$

$$= 52$$

3.14 Solovay–Strassen Test

The Solovay–Strassen test is basically an Euler–Jacobi pseudo-primality test introduced by Robert M. Solovay and Volker Strassen (Dietzfelbinger). It is a probability test that gives the result as either the given number is composite or probably prime. The concept of this algorithm comes from the Euler theorem such as $a^{(p-1)/2} \equiv (a/p)$ (mod p), for any value of integer a and prime p, where (a/p) is the Legendre symbol. If we take the value of p to any odd integer n, then it forms the Jacobi symbol as $a^{(p-1)/2} \equiv (a/n)$ (mod p) to observe the congruence. If n is the prime number, then this congruence satisfies all the value of a. The base a is called the Euler witness if n is composite. It is also possible to return an incorrect result (Conrad, 2009, 2011, 2016).

Algorithm

Let an integer n to test the primality and a value t to check the accuracy. If n is composite, then gives the result as composite, otherwise probably prime.

> For t times:
> Choosing a random integer, a form the range $\{2, n - 1\}$
> $x \leftarrow (a/n)$
> if $x = 0$ then return "n is composite".
> $y \leftarrow a^{(n-1)/2}$ (mod n)
> if $x \equiv y$ then return "n is prime".
> otherwise, return "n is composite".

3.15 Conclusion

There are so many ways to determine whether a number is prime or composite. In this chapter, we have covered only a few methods conclusively or practically with the help of examples. The test based on the practical example has not been conclusive and vice versa. The Fibonacci series and Elliptic primality test are not covered here. Furthermore, there are so many references given in this chapter to improve the knowledge and further study the topics in detail.

References

Aaronson, S. (2014, February 16). The Prime Facts: From Euclid to AKS. Retrieved July 01, 2018, from www.scribd.com/document/207338411/The-Prime-Facts-From-Euclid-to-AKS.

Agrawal, M., Kayal, N., & Saxena, N. (2004, September). PRIMES is in P. Retrieved March 05, 2018, from www.jstor.org/stable/pdf/3597229.pdf

Andrews, G. E., & Berndt, B. C. (2012). Highly Composite Numbers. *Ramanujans Lost Notebook*, 359–402. doi:10.1007/978-1-4614-3810-6_10.

Berliner, A. (2016). Coprime and Prime Labelings of Graphs.

Briggs, C. C. (1999, December 25). Simple Divisibility Rules for the 1st 1000 Prime Numbers. Retrieved July 24, 2018, from https://arxiv.org/pdf/math/0001012.

Camina, A. R., &Camina, R. D. (2009). Coprime Conjugacy Class Sizes. *Asian-European Journal of Mathematics,* 02(02), 183–190. doi:10.1142/s1793557109000157.

Cao, Z. (2013). A Note on the Storage Requirement for AKS Primality Testing Algorithm. *IACR Cryptology ePrint Archive,* 2013, 449.

Childs, L. N. (2009). The Chinese Remainder Theorem. *A Concrete Introduction to Higher Algebra,* 194–207. doi:https://doi.org/10.1007/978-1-4419-8702-0_12.

Conrad, K. (2009). Euler's Theorem. Retrieved March 01, 2019, from https://www.math.uconn.edu/~kconrad/blurbs/ugradnumthy/eulerthm.pdf

Conrad, K. (2011). The Miller–Rabin Test. Retrieved April 01, 2019, from https://kconrad.math.uconn.edu/blurbs/ugradnumthy/millerrabin.pdf

Conrad, K. (2016). THE Solovay–Strassen Test. Retrieved April 01, 2019, from https://www.math.uconn.edu/~kconrad/blurbs/ugradnumthy/solovaystrassen.pdf

Crandall, R. E., & Pomerance, C. (2005). *Prime Numbers: A Computational Perspective.* New York: Springer.

Dietzfelbinger, M. (2004). Deterministic Primality Testing in Polynomial Time. *Primality Testing in Polynomial Time Lecture Notes in Computer Science,* 115–131. doi:10.1007/978-3-540-25933-6_8.

Dineva, R. (2005, July 29). The Euler Totient, the Möbius and the Divisor Functions. Retrieved July 22, 2018, from www.mtholyoke.edu/~robinson/reu/reu05/rdineva1.pdf.

Erdős, P., Luca, F., & Pomerance, C. (2008). On the Proportion of Numbers Coprime to a Given Integer. *CRM Proceedings and Lecture Notes Anatomy of Integers,* 47–64. doi:10.1090/crmp/046/03.

Euler, R. (1996). Examples Using the Maximum Modulus Theorem. *The Mathematical Gazette,* 80(488), 394–394. doi:10.2307/3619587.

Faltings, G. (1995, July). The Proof of Fermat's Last Theorem by R. Taylor and A. Wiles. Retrieved July 17, 2018, from www.ams.org/notices/199507/faltings.pdf.

Gorgui-Naguib, R. N., & Dlay, S. S. (2000). Properties of the Euler Totient Function Modulo 24 and Some of Its Cryptographic Implications. *Lecture Notes in Computer Science Advances in Cryptology — EUROCRYPT '88,* 267–274. doi:10.1007/3-540-45961-8_24.

Granger, R., & Moss, A. (2013). Generalised Mersenne Numbers Revisited. *Mathematics of Computation,* 82(284), 2389–2420. doi:10.1090/s0025-5718-2013-02704-4.

Klappenecker, A. (2002, September 04). An Introduction to the AKS Primality Test. Retrieved July 26, 2018, from http://faculty.cs.tamu.edu/klappi/629/aks.pdf.

Korukov, V. (2014, May 15). Fermat's Last Theorem - University of Washington. Retrieved July 19, 2018, from https://sites.math.washington.edu/~morrow/336_14/papers/vladimir.pdf.

Krauel, J. (2013). Some Methods of Primality Testing - Lakehead University. Retrieved July 25, 2018, from www.lakeheadu.ca/sites/default/files/uploads/77/docs/KrauelFinal.pdf.

Křížek, M., Luca, F., & Somer, L. (2002). Fermat's Little Theorem, Pseudoprimes, and Superpseudoprimes. *17 Lectures on Fermat Numbers,* 130–146. doi:10.1007/978-0-387-21850-2_12.

Lady, E. L. (n.d.). CHINESE REMAINDER THEOREM - University of Hawaii. Retrieved April 01, 2019, from http://www.math.hawaii.edu/~lee/courses/Chinese.pdf, "unpublished".

Langford, W. J. (1974). Tests for Divisibility. The Mathematical Gazette, 58(405), 186. doi:10.2307/3615958.

Lewis, J. P. (1969). Homogeneous Functions and Euler's Theorem. In: *An Introduction to Mathematics.* London: Palgrave Macmillan.

Murata, L., & Pomerance, C. (2004). On the Largest Prime Factor of a Mersenne Number. *CRM Proceedings and Lecture Notes Number Theory,* 209–218. doi:10.1090/crmp/036/16.

P & A (1970, January 01). Some Notes on Divisibility Rules. Retrieved July 25, 2018, from https://iris.univr.it/handle/11562/913385.

Palimar, S., & Shankar, B. R. (2012). Mersenne Primes in Real Quadratic Fields. *Integer Sequences,* 15(12.5.6), 01–16. Retrieved September 27, 2018, from http://emis.ams.org/journals/JIS/VOL15/Palimar/palimar5.pdf.

Piazza, N. (2018, April 20). The Chinese Remainder Theorem. Retrieved July 29, 2018, from https://digitalcommons.sacredheart.edu/cgi/viewcontent.cgi?article=1217&context=acadfest

Pomerance, C., & Gagola, S. M. (1984). *Lecture Notes on Primality Testing and Factoring: A Short Course at Kent State University.*

Raggi, F., Montes, J. R., & Wisbauer, R. (2005). Coprime Preradicals and Modules. Journal of Pure and Applied Algebra, 200(1–2), 51–69. doi:10.1016/j.jpaa.2004.12.040.

Robinson, J. (2011). *Elaboration on History of Fremat's Theorem and Implications of Euler's Generalization by Means of the Totient Theorem* (Unpublished master's thesis). The University of Arizona.

Robinson, R. M. (1954). Mersenne and Fermat Numbers. Proceedings of the American Mathematical Society, 5(5), 842–842. doi:10.1090/s0002-9939-1954-0064787-4.

Rotella, C. (2005, May 06). An Efficient Implementation of the AKS Polynomial-Time Primality Proving Algorithm. Retrieved September 26, 2018, from www.cs.cmu.edu/afs/cs/user/mjs/ftp/thesis-program/2005/rotella.pdf.

Ru, M. (2012). Chinese Remainder Theorem, pp. 01–08, www.math.uh.edu/~minru/6322-12/webnumber12/chinese.pdf.

Schoof, R. (2008). Four Primality Testing Algorithms. Algorithmic Number Theory, 44, 101–126. Retrieved July 26, 2018, from www.math.leidenuniv.nl/~psh/ANTproc/05rene.pdf.

Stein, W. A. (2009). *Elementary Number Theory: Primes, Congruences, and Secrets.* New York: Springer.

Terzich, P. (2012). On The Form of Mersenne Numbers. Retrieved July 27, 2018, from http://vixra.org/pdf/1202.0024v1.pdf.

Turchet, A., Scholl, T., Hiatt, R., Mícovíc, D., Patīno, B. V., & Quah, B. T. P. (2007, Winter). WXML Final Report: AKS Primality Test. Retrieved September 26, 2018, from http://depts.washington.edu/uwmxl/wordpress/wp-content/uploads/2017/05/AKS_Final_Report.pdf.

Turner, C. S. (2008, November 07). Euler's Totient Function and Public Key Cryptography Introduction. Retrieved July 20, 2018, from https://studylib.net/doc/10323952/euler's-totient-function-and-public-key-cryptography-intr.

Wu, H. W., Li, C. M., Li, H. L., Ding, J., & Yao, X. M. (2016). An RSA Scheme Based on Improved AKS Primality Testing Algorithm. *MATEC Web of Conferences*, 44, 01032. doi:10.1051/matecconf/20164401032.

4

Discrete Logarithm Problem

Khaleel Ahmad and Afsar Kamal

Maulana Azad National Urdu University

Khairol Amali Bin Ahmad

National Defence University of Malaysia

CONTENTS

4.1 Introduction..47
4.2 Discrete Logarithm Problem in Subgroup of Z_p^* 48
4.3 Baby-Step Giant-Step Algorithm 49
4.4 Function Field Sieve..51
4.5 Index Calculus Algorithm 51
4.6 Number Field Sieve 53
4.7 Pohlig–Hellman Algorithm..53
4.8 Pollard's Rho Algorithm for Logarithms 54
4.9 Pollard's Kangaroo Algorithm...55
4.10 Conclusion 55
References...56

4.1 Introduction

The term "logarithm" is nothing but just an exponent in the sense of real number. For example, $3^2 = 9$ (where 2 is called exponent). Now translating into the language of Logarithm it becomes $\text{Log}_3 9 = 2$ (where $\text{Log}_3 9$ is the exponent equal to 2 and 3 is the base of the logarithm). Basically, the exponent is used to calculate the product as $x = 3^2 = 9$, while Logarithm is to compute the exponent as $x = \log_3 9 = 2$. This prompts to the definition of Logarithm as $\text{Log}_x y = z$, or $xz = y$, assuming x and y as positive besides $x \neq 1$ in algebraic form. Indeed, $\text{Log}_x y$ is generally not an integer number. For example, the solution of logarithm $\text{Log}_2 3 = ? \leftrightarrow 2^? = 3$ gives the approximation result? ≈ 1.58496 as an exponent because of $2^{1.58496} \approx 2.99999$ that is exactly 3. The word "discrete" in discrete logarithm refers to the aspect in discrete group $\{1, \ldots, P - 1\}$ only integer numbers rather than fractions (real numbers) (Pomerance, 2008). Now coming to the point of DLP.

In discrete logarithm, we apply the modulo operation on logarithm such as $x = \text{Log}_3 9 \pmod 7$. It can be also noted as $9 = 3^x \pmod 7$ or $x = \text{dLog}_{3,7} (9)$, where exponent $= x$, base $= 2$, modulus $= 7$, and remainder $= 9$. Suppose x and y are two integer numbers greater than 0 and a prime number P. Then, the methodology to find out the solution of z such that $x^z \equiv y \pmod P$ is called the discrete logarithm problem or DLP. Suppose n is the smallest integer number that satisfies $x^n \equiv 1 \pmod P$ assuming $0 \leq x \leq n$, then it can be signified as

TABLE 4.1

Mapping Values for Distinct Remainders

x	$x^1 \bmod 5$	$x^2 \bmod 5$	$x^3 \bmod 5$	$x^4 \bmod 5$
1	1	1	1	1
2	2	4	3	1
3	3	4	2	1
4	4	1	4	1

$z = \text{Log}_x y$ and called the discrete log of y with respect to $x \bmod P$. For example, if $P = 7$, $x = 3$, $y = 6$, then $z = \text{Log}_3 6 = 3$. In the field of cryptography, the reconciliation of the discrete logarithm is to provide the exponent having a distinct remainder (Odlyzko, 1984).

Example: Let $4^z \pmod 5 = 4$, it gives the result $z = 1$ or 3 where $z = \{1, \ldots, 4\}$. Because $4^1 \bmod 5 = 4$ and $4^3 \bmod 5 = 4$.

If P is a large prime number with certain base values (x), then it is able to generate distinct remainders for the different exponents $(x = 1, \ldots, P - 1)$ and $\text{Log}_x y$ is called the one-way function.

Let's calculate $x^z \pmod 5 = \text{remainder}$, where $z = \{1, \ldots, 4\}$ and modulus $P = 5$, depicted in Table 4.1.

As evident in the table, the discrete logarithm for modulus 5 generates the distinct remainders in the range $\{1, \ldots, 4\}$ using the base value 2 or 3. To solve the problem of $x^z \equiv y \bmod P$, x is recommended to be the primitive root mod P. The base values 2 and 3 are called the primitive roots or generators of modulus 5 usually symbolized by α. It performs the multiplication operation on a single element α^z and generates all elements in the discrete group $\{1, \ldots, P - 1\}$. That's why it is referred to as a generator (Stern, 2001).

4.2 Discrete Logarithm Problem in Subgroup of Z_p^*

Discrete logarithm has a tendency to be associated with the multiplicative cyclic group. If g is the generator of multiplicative group G, then every element e in G can be written as g^y for some y. The DLP has basically three most important parts: a group G, generator g, and an element e to find out the discrete logarithm to the base g of element e in the group G. For example, if a group is Z_7^* with generator 4, then the discrete logarithm of 1 is 3. $\because 4^3 \equiv 1 \bmod 7$. Calculation of the discrete logarithm is not always a tedious task. It depends on the group structure. Z_p^* gives a better selection of groups for discrete logarithm where p is a prime number. Calculation of $4^3 \bmod 7 = ?$ is very simple and easy, but the calculation of its inverse or opposite for finding the logarithm $3 = 4^? \bmod 7$ is very difficult (Chiou, 2016). A system tries on each exponent? $= 0, 1, 2, 3 \ldots$ until the equation matches. For example, generator $\alpha = 3$, modulus $P = 7$, and the discrete group $= \{1, \ldots, P - 1\}$.

$$3^1 \bmod 7 = 3 \qquad 3^7 \bmod 7 = 3$$
$$3^2 \bmod 7 = 2 \qquad 3^8 \bmod 7 = 2$$
$$3^3 \bmod 7 = 6 \qquad 3^9 \bmod 7 = 6$$
$$3^4 \bmod 7 = 4 \qquad 3^{10} \bmod 7 = 4$$
$$3^5 \bmod 7 = 5 \qquad 3^{11} \bmod 7 = 5$$
$$3^6 \bmod 7 = 1 \qquad 3^{12} \bmod 7 = 1$$

All the elements in the discrete group {1, 2, 3, 4, 5, 6} are covered. After reaching the exponent $P - 1 = 6$, it continues to repeat the same results in the same order. This process is called the cyclic group of generator α. After a finite number of exponentiations, the generated elements repeat themselves such that it becomes a loop. When it gets the remainder value as 1, the cycle starts from the origin in the same order. In this example $P = 7$, the cyclic group is represented in the form of notation: Z_p^* or Z_7^*, where * has the meaning of no zero and the total number of elements e is $P - 1 = 6$ (Meshram & Meshram, 2011).

Properties of Z_p^*:

1. If α, β are the elements of Z_p^*, then their product also belongs to Z_p^*.
   Suppose α, $\beta \in Z_p^*$, then $\alpha \cdot \beta \in Z_p^*$ (closed under)

2. If α, β, λ are the elements of Z_p^*, then
   $\alpha \cdot (\beta \cdot \lambda) = (\alpha \cdot \beta) \cdot \lambda$ (associative under)

3. If α is the elements of Z_p^*, then $1 \cdot \alpha = \alpha$ (Z_p^* has identity element)

4. If α is the elements of Z_p^*, then there will be an extra element as $\beta \in Z_p^*$ such that $\alpha \cdot \beta = 1$ (every element in Z_p^* has an inverse)

Therefore, any collection G that satisfies any of the properties above with product ($\cdot$) operation is called the group.

Example: If $3 \in Z_{13}^*$, then there is $\beta \in Z_{13}^*$ (Property 4)

Such that $3 \cdot \beta = 1 \Leftrightarrow 3\beta \bmod 13 = 1 \Leftrightarrow 3\beta \equiv 1 \bmod 13$, Hence $\beta = 9$.

$(Z_p^*, \cdot)$ is a special kind of group called cyclic group. This cyclic group is the backbone of discrete logarithm in a cryptography system (Ilić, 2010).

4.3　Baby-Step Giant-Step Algorithm

As mentioned above, finding the remainder through modulus is easy but doing its inverse to get logarithm is very difficult. To solve this problem, the baby-step giant-step algorithm introduced by Shanks is one of the most popular algorithms that deal with discrete logarithm in any finite abelian group. This algorithm is widely used in many applications of computational theory and invariants of the elliptic curve factorization techniques. It is a time-memory trade-off method with complexity $O(\sqrt{n})$ group multiplication. This algorithm works on any cyclic group G with n order to reduce the time at the cost of extra storage (Kushwaha & Mahalanobis, 2017).

Basic strategy of the baby-step giant-step algorithm. We know that $\mathrm{Log}_\alpha{}^\beta = x \Leftrightarrow \alpha^x = \beta$, where exponent x can be represented in a two-digit numbers X_g and X_b as follows:

$$x = X_g \cdot m + X_b, \tag{4.1}$$

where $m = \lceil \sqrt{n} \rceil$ and $0 \leq X_g \leq X_b < m$ (n is the number of elements in the group and $\lceil \ \rceil$ is the ceiling symbol to round off the fraction number). Now $\alpha^x = \beta$ can be expressed through equation (4.1) as $\alpha^{X_g \cdot m + X_b} = \beta$

$$\Leftrightarrow \alpha^{X_g \cdot m} \cdot \alpha^{X_b} = \beta$$

$$\Leftrightarrow \alpha^{X_b} = \beta \cdot \alpha^{-X_g \cdot m}$$

$$\Leftrightarrow \alpha^{X_b} = \beta \cdot \left(\alpha^{-m}\right)^{X_g} \left(\text{Base of the Baby-Step}\right)$$

To find out the values of the given numbers X_g and X_b, use the divide-and-conquer method (Stein & Teske, 2005; Kameswari, Surendra, & Ravitheja, 2016).

Baby step. Computing and storing all the values α^{Xb} for $0 \leq X_b \leq m$.

Giant step. Checking X_g for $0 \leq X_g \leq n$ if it matches with α^{Xb} as $\beta \cdot (\alpha^{-m})^{Xg} = \alpha^{Xb}$

Algorithm

Let a generator $\alpha \in Z_p^*$ thus $\beta \in Z_p^*$.

Step 1. Calculate $m := \lceil \sqrt{n} \rceil$

Step 2. Construct a hash table to store the pair (X_b, α^{Xb}) for $0 \leq X_b \leq m$.

Step 3. Calculate alpha inverse: $\alpha^{-m} = (\alpha^{-1})^m$ using the extended Euclidean algorithm also store the Beta value into lambda as $\lambda := \beta$

Step 4. Perform loop operation such that:

> For all i, where $0 \leq i \leq m - 1$ and check
> If $\lambda == \alpha^{Xb}$ in the table, then return $X_g \cdot m + X_b$ and stop.
> > Else return $\lambda := \lambda \cdot \alpha^{-m}$
> End for

Example

To find $\text{Log}_5 7$ in the cyclic group Z_{23}^*.

The number 23 is a prime number, and the exponents in this group are $Z_{23}^* = \{1, 2, \ldots, 22\}$.

Step 1. Calculate $m := \lceil \sqrt{n} \rceil = \lceil \sqrt{23} \rceil = \lceil 4.7958 \rceil = 5$

Step 2. Constructing a table for entries that are shown in Table 4.2:

$$\left(X_b, 5^{Xb} \right) \text{ for } 0 \leq X_b < 5$$

where $X_b = 0, 1, \ldots, 4$

Step 3. Compute $\alpha^{-m} = (\alpha^{-1})^m = (5^{-1})^5$

gcd$(\alpha, n) = $ gcd$(5, 23) = 1$ (applying extended Euclidean algorithm) and $5 * 14 + 23 * (-3) = 1$. Here, $\alpha^{-1} = 5^{-1} = 14$, then $\alpha^{-m} = 14^5 \bmod 23 = 15$.

Step 4. Compute $\beta \cdot \left(\alpha^{-m} \right)^{Xg}$ for $X_g = \{0, 1, \ldots, 4\}$ until a number matches in the second row of the hash table.

$$\beta \cdot (\alpha^{-m})^{Xg} = 7 \cdot 15^{Xg} \bmod 23$$

$$7 \cdot 15^0 \bmod 23 = 7$$

$$7 \cdot 15^1 \bmod 23 = 13$$

$$7 \cdot 15^2 \bmod 23 = 11$$

$$7 \cdot 15^3 \bmod 23 = 4$$

$$7 \cdot 15^4 \bmod 23 = 14$$

TABLE 4.2

Comparison on Hash Table

X_b	0	1	2	3	4
5^{X_b} mod 23	1	5	2	10	4

Here, 4 matches in the second row of the table. Hence, $X_g = 3$ and $X_b = 4$. Therefore,
$$x = X_g \cdot m + X_b = 3 \cdot 5 + 4 = 19.$$
Result: $\text{Log}_5 7 = 19$ in Z_{23}^* and it gives the result as 5^{19} mod $23 = 7$.

4.4 Function Field Sieve

The function field sieve is used to compute the discrete logarithm in F_p^{*n} within random time, as introduced by Leonard Adleman in 1994. It generalizes the algebraic number field into the function field sieve. It is the faster method for factoring integers than other older algorithm such as Coppersmith's algorithm in the finite field of small characteristics. It is very similar to number field sieve (NFS) in the prime field, where a ring of integer Z is replaced by a ring of polynomial $K[t]$ with a finite field of size $\kappa = \#K$ (Adleman & Huang, 1999). The probable running time of this algorithm is $L_p^n[1/3, c]$ for a constant $c > 0$, if log $p \leq n^{g(n)}$ where g is a function that satisfies $\lim_{n \to \infty} g(n) = 0$ and $0.98 > g(n) > 0$ replace κ. The calculation starts by taking a finite field F and a non-reducible polynomial $p \in F[x]$. The goal is to find out the absolute non-reducible polynomial $A \in F[x, y]$ with $m \in F[x]$ and degreed in y that satisfies $H(x, m) \equiv 0$ mod p. Continuously working in $F[x]$ and the function field $L = \text{Quotient}(F[x, y])/(H))$ needs double smooth pair $< r$ and $s > \in F[x] * F[x]$, where r and s are relatively prime. The intersection of $ry + s$ with H, $rm + s$ and $r^d \cdot H(x, -s/r)$ are smooth in $F[x]$ (Schirokauer, 2008; Joux & Lercier, 2002).

4.5 Index Calculus Algorithm

The index calculus algorithm is a more efficient and powerful technique for computing the discrete logarithms in groups with a notion of smoothness. It applies to only a chosen group and not to all groups, and gives a sub-exponential time algorithm. The base of this algorithm is Z_p^*. This algorithm needs a factor base S that is selected as a relatively small subset $S \in Z_p^*$, such that the significant fraction of elements of Z_p^* can be capably written as the product of elements of S (Padmavathy & Bhagvati, 2012).

Algorithm

It requires a generator α of Z_p^* and an element β as inputs and gives the discrete logarithm $x = \log_\alpha \beta$ as output.

1. Select a base factor S where subset $S = \{P_1, P_2, P_3, ..., P_t\} \in Z_p^*$.
2. Assemble the linear relationship among logarithms of elements in S.

3. Select a random integer r such that $0 \leq r \leq n - 1$ and calculate α^r.
4. Check α^r as the product of elements in S.

$$\alpha^r \equiv \prod_{i=1}^{t} p_i^{a_i} (\text{mod}\, p), c_i \geq 0, \text{ where } c \text{ is a small positive integer to a unique solution.}$$

If satisfies, obtain logarithms of both sides of the equation as a linear relation $r \equiv$

$$\sum_{i=1}^{t} c_i \log_\alpha p_i \, \text{mod}(p - 1)$$

5. Repeat Steps 3 and 4 until $t + c$ relationship is obtained.
6. Obtain the logarithm of elements in S with modulo $P - 1$ and solve the linear equation $t + c$ gotten in Step 2 to find the values of $\text{Log}_\alpha P_i$, where $1 \leq i \leq t$.
7. Calculate y.
8. Choose an integer r randomly, where $0 \leq r \leq n - 1$ and calculate $\beta \cdot \alpha^r$.
9. Check $\beta \cdot \alpha^r$ as the product of elements in S.

$$\beta \cdot \alpha^r \equiv \prod_{i=1}^{t} p_i^{d_i} (\text{mod}\, p), d_i \geq 0.$$

If the condition satisfies, take the logarithm of both sides of the equation, yields

$$\log_\alpha \beta = \left(\sum_{i=1}^{t} d_i \cdot \log_\alpha p_i - r \right) \text{mod}(p - 1) = x \text{ and return } (x), \text{ otherwise repeat Step}$$

4 (Yan, 2003; Faugère, Perret, Petit, & Renault, 2012).

Example

Considering $\beta = 13$, compute $\log_6 13$, when $P = 229$ and generator $\alpha = 6$.

Solution: Choosing the base factor of first 5 primes as $S = \{2, 3, 5, 7, 11\}$.

$$6^{12} \, \text{mod}\, 229 \equiv 165 \equiv 3 \cdot 5 \cdot 11$$

$$6^{18} \, \text{mod}\, 229 \equiv 176 \equiv 2^4 \cdot 11$$

$$6^{62} \, \text{mod}\, 229 \equiv 154 \equiv 2 \cdot 7 \cdot 11$$

$$6^{100} \, \text{mod}\, 229 \equiv 180 \equiv 2^2 \cdot 3^2 \cdot 5$$

$$6^{143} \, \text{mod}\, 229 \equiv 198 \equiv 3 \cdot 3^2 \cdot 11$$

$$6^{206} \, \text{mod}\, 229 \equiv 210 \equiv 2 \cdot 3 \cdot 5 \cdot 7 \, (\text{Only successful attempts are chosen})$$

Now the logarithms of elements in the factor base are as follows:

$$12 \equiv \log_6 3 + \log_6 5 + \log_6 11 (\text{mod}\, 228)$$

$$18 \equiv 4 \cdot \log_6 2 + \log_6 11 (\text{mod}\, 228)$$

$$62 \equiv \log_6 2 + \log_6 7 + \log_6 11 (\text{mod}\, 228)$$

$$100 \equiv 2 \cdot \log_6 2 + 2 \cdot \log_6 3 + \log_6 5 (\text{mod}\, 228)$$

$$143 \equiv \log_6 2 + 2 \cdot \log_6 3 + \log_6 11 (\text{mod}\, 228)$$

$$206 \equiv \log_6 2 + \log_6 3 + \log_6 5 + \log_6 7 (\text{mod}\, 228)$$

Solving the linear equations using logarithm $x_i = \log_6 P_i$, gives the result as $\log_6 2 = 21$, $\log_6 3 = 208$, $\log_6 5 = 98$, $\log_6 7 = 107$, $\log_6 11 = 162$.

If an integer $r = 77$ is chosen, then $\beta \cdot \alpha^r \equiv 13 \cdot 6^{77}$ (mod 229) $\equiv 147 \equiv 3 \cdot 7^2$ because $\log_6 13 = (\log_6 3 + 2 \cdot \log_6 7 - 77)$ (mod 228) $= 117$.

4.6 Number Field Sieve

NFS is used in public key cryptosystem. It is an extremely fast classical algorithm for factoring a large integer greater than 100 digits to produce a group of prime numbers, developed by Pollard (Stevenhagen, 2011). It is also called general NFS and later version of special number sieve. It is most similar in nature to the quadratic sieve algorithm. It was primarily used to factor RSA-130 number. It has the complexity over $O\{\exp[c(\log n)^{1/3} \cdot (\log \log n)^{2/3}]\}$. This algorithm basically consists of five steps depending on each other (Briggs, 1998; Hoang, 2008).

a. Inputs: Taking composite number N without a small factor.
b. Selection of polynomial: It is the major and open problem for NFS. The purpose of this step is to select the best polynomial $f(x)$ from a large group of usable polynomials such as f_1, f_2 with the common root m modulo N in the homogeneous form: $b^{\deg(f_k)} f_k(a/b) = F_k(a, b)$.
c. Factor bases: It chooses the base factor for the sieving step as

   RFB: $(p_0, p_0 \bmod m), (p_1, p_1 \bmod m), (p_2, p_2 \bmod m), \ldots$ and

   AFB: $(p_0, r_0), (p_1, r_1), (p_2, r_2), \ldots$ such that $f(r_i) = 0 \bmod p_i$
d. Seiving: It finds the relationship (a_i, b_i) where both homogeneous polynomial values $|F_k(a_i, b_i)|$ are smooth for the values $k = 1, 2$. It is normalized and gives out $\gcd(a_i, b_i) = 1$ where $0 < b_i$. The $b^{\deg(f)} f(a/b)$ is smooth over AFB and $f(x)$ is polynomial, $m \rightarrow f(m) = 0 \bmod n$.
e. Linear algebra and matrix creation: It gathers the elements that are fully square product from the relation set.
f. Square root: It finds the algebraic and relational square root from the matrix to obtain the solution.

4.7 Pohlig–Hellman Algorithm

This algorithm is also used to compute prime factorization of N and group order in DLPs discovered by Ronald Silver. It was first published by Pohling and Hellman in 1978. This is applicable only when the $(P - 1)$ has small factors where P is a prime number. It computes the discrete logarithm to find x such that $\beta = \alpha^x$ where $0 \leq x \leq P - 1$. The worst-case time complexity of this algorithm is $[(\sqrt{n})]$ (Mihalcik, 2010; Wong, 2016; Pohlig & Hellman, 1978).

Algorithm

It finds the value of z such that $x^z \equiv y$ (mod P) using the following steps:
1. Compute the prime factors of P in the group as $\varphi(P) = P_1, P_2, P_3, \ldots, P_n$.
2. Assume $\varphi(P) = P_1 * P_2$ and take $z = x_0 + P_1 x_1$.

Hence, $x^{(x_0 + P_1 x_1)P_2} = y^{P_2} \bmod P$ or $x^{P_2 x_0} * x^{P_1 P_2 x_1} = y^{P_2} \bmod P$. This gives the result as $x^{P_2(x_0)} = y^{P_2} \bmod P$. The values of x^{P_2} and y^{P_2} are easy to find through the trial for the value x_0. So, $z = x_0 \bmod P_1$ ...... (a)

3. Similarly, take $z = y_0 + P_2 y_1$ and processing the same as above in Step 2, it gives: $z = y_0 \bmod P_2$ ...... (b)

4. To get the value of z, use the Chinese remainder theorem.

Example

$$7^z \equiv 26 \bmod 41$$

Solution: Here, $\Phi(41) = 40 = 5 * 8$.

Let $z = x_0 + 8x_1$ while given $7^z \equiv 26 \bmod 41$.

Hence, $\left(7^{x_0 + 8x_1}\right)^5 \equiv 26^5 \bmod 41 \Rightarrow 7^{5x_0} * 7^{40x_1} \equiv 26^5 \bmod 41$. Using the Fermat's little theorem $a^{p-1} \equiv 1 \pmod{P}$, turned into $7^{5x_0} \equiv 26^5 \bmod 41$. Now it can be expressed as $38^{x_0} \equiv 27 \bmod 41$ (where $7^5 \bmod 41 = 38$ and $26^5 \bmod 41 = 27$). The solution of x_0 in the equation $38^{x_0} \equiv 27 \bmod 41$ can be obtained by trial and error. Here, $x_0 = 7$ satisfies the equation, $38^7 \equiv 27 \bmod 41$. So, $z = 7 + 8x_1$ and $z \equiv 7 \bmod 8$ ...... (a)

Similarly, let $z = y_0 + 5y_1$, we have $\left(7^{y_0 + 5y_1}\right)^8 \equiv 26^8 \bmod 41$. It gives $7^{8y_0} * 7^{40y_1} \equiv 26^8 \bmod 41 \Rightarrow 7^{8y_0} \equiv 26^8 \bmod 41 \Rightarrow 37^{y_0} \equiv 18 \bmod 41$. Again, applying trial and error on this equation, we find $37^3 \equiv 18 \bmod 41$ where $y_0 = 3$. Then, $z = 3 + 5y_1$. So, $x \equiv 3 \bmod 5$ ...... (b)

Now solving the simultaneous congruence using the Chinese remainder theorem on both equations:

$$z \equiv 7 \bmod 8 \ldots \ldots (a) \text{ and } x \equiv 3 \bmod 5 \ldots \ldots (b)$$

$$175 \equiv 3 \bmod 8$$

$$48 \equiv 3 \bmod 5$$

Here, $175 + 48 = 223$ is the solution for $7^z \equiv 26 \bmod 41$. This solution can also be reduced as $5 * 8 = 40$ to $z = 23$. There is an infinite number of solutions and very difficult to find the unique value of z.

4.8 Pollard's Rho Algorithm for Logarithms

Pollard's rho algorithm is the most efficient and fastest method to solve the DLP. It runs easily on a powerful workstation for solving the DLP in different abelian sets considering large group orders. This algorithm was invented by British mathematician John Pollard in 1975 (Hegde & Deepthi, 2015). It was primarily proposed for integer factorization having a small prime factor but later modified to solve DLP. It has less multiplication calculation than other methods. It is suitable and best applicable on Elliptic curves DLP. The time complexity of this algorithm is approximated as $\sqrt{n}$, where n is the modulo to the size of inputs in bits $O(2^{k/2})$ approximately same as the baby-step giant-step algorithm (Wang & Zhang, 2012). In this algorithm, an iterative function f is used for defining the sequence by $s_{i+1} = f(s_i)$, where $i = 0, 1, 2, 3, \ldots$ with some initial value s_0. The most basic idea of this algorithm is to find k value that satisfies $Q = [k]P$ through partitioning the group points into three disjoint sets S_1, S_2, S_3 having almost equal size. The iterative function is defined on the point T as follows:

$$f(T) = R \oplus P \text{ if } T \in S_1, f(T) = [2]R \text{ if } T \in S_2, f(T) = R \oplus Q, \text{ if } T \in S_3$$

Starting with the point $T_i = [a_i]P \oplus [b_i]Q$ where $i = 0$ and a_i, $b_i \in [1, n-1]$, they are chosen randomly to make the sequence T_i. The sequence of the group is calculated as $a_{i+1} = (a_i + 1)$ mod n if $T_i \in S_1$, $a_{i+1} = 2a_i$ mod n if $T_i \in S_2$, $a_{i+1} = a_i$ mod n if $T_i \in S_3$ and $b_{i+1} = b_i$ if $T_i \in S_1$, $b_{i+1} = 2b_i$ mod n if $R_i \in S_2$, $b_{i+1} = (b_i + 1)$ mod n if $R_i \in S_3$. The total number of points on the curve forming a cyclic group is finite. If $\gcd(b_j - b_i, n) = 1$, then the value is obtained as $k = \log_P Q = \{(a_i - a_j)/(b_j - b_i)\}$ mod n and matching of the points is computed as $T_i = T_j \Rightarrow [a_i]P \oplus [b_i]Q = [a_j]P \oplus [b_j]Q$ (Teske, 1998).

4.9 Pollard's Kangaroo Algorithm

Pollard's Kangaroo algorithm is a variant of the Pollard's rho algorithm discovered by J. M. Pollard in 1978. It is also called Pollard's lambda (λ) algorithm and most suitable to solve the DLP in an arbitrary cyclic group. It has the complexity $O(\sqrt{\omega})$, where $\omega = $ upper $-$ lower interval (Pollard, 2000). It uses a distinct type of r-adding iterative function, in which the multiplier is formed as g^α and the point method is used to detect the collision. Suppose a finite cyclic group G with order n generated by an element α to get the discrete logarithm $x \in Z_n$ such as $\alpha^z = \beta$. Then this algorithm allows us a way to find from the subset $\{a, ..., b\} \subset Z_n$ (Stein & Teske, 2001; Cheon, Hong, & Kim, 2010).

Algorithm

1. Defining a pseudorandom map $f: G \rightarrow S$, where S is a set of integers.
2. Choosing an integer N and computing the sequence of group element as x_0, x_0, x_1, x_2, ..., x_n where $x_0 = \alpha^c$ and $x_{i+1} = x_i \alpha^{f(x_i)}$ for $0 < i < N - 1$.
3. Calculating $d = \sum_{i=0}^{n-1} f(x_i)$ and observing $x_N = x_0 \alpha^d = \alpha^{b+d}$.
4. Computing the other sequence as: $y_0 = \beta$ and $y_{i+1} = y_i \alpha^{f(y_i)}$ for $0 < y < N - 1$ with corresponding sequence of integers $d_0, d_1, ...$ to check $d_i > c - b + d$.
5. Compare x_n with y_i to find the solution.

4.10 Conclusion

In this chapter, we have presented the general DLP in the finite cyclic group G and acquainted the different algorithms with a brief introduction to solving DLP. The different types of algorithm to solve DLP describe the importance of DLP in cryptography. The complexity of DLP is the same as the factorization problem.

References

Adleman, L. M., & Huang, M. (1999). Function field sieve method for discrete logarithms over finite fields. *Applications of Curves over Finite Fields Contemporary Mathematics*, 151(1–2), 5–16. doi:10.1090/conm/245/03728.

Briggs, M. E. (1998, April 17). *An Introduction to the General Number Field Sieve*, 01–84.

Cheon, J. H., Hong, J., & Kim, M. (2010). Accelerating Pollard's Rho algorithm on finite fields. *Journal of Cryptology*, 25(2), 195–242. doi:10.1007/s00145-010-9093-7.

Chiou, S. (2016). Novel digital signature schemes based on factoring and discrete logarithms. *International Journal of Security and Its Applications*, 10(3), 295–310. doi:10.14257/ijsia.2016.10.3.26.

Faugère, J., Perret, L., Petit, C., & Renault, G. (2012). Improving the complexity of index calculus algorithms in elliptic curves over binary fields. *Advances in Cryptology – EUROCRYPT 2012 Lecture Notes in Computer Science*, 27–44. doi:10.1007/978-3-642-29011-4_4.

Hegde, N., & Deepthi, P. (2015). Pollard RHO algorithm for integer factorization and discrete logarithm problem. *International Journal of Computer Applications*, 121(18), 14–17. doi:10.5120/21639-4969.

Hoang, V. P. (2008). *Integer Factorization with the General Number Field Sieve: Thesis*. Rovaniemi: Rovaniemi University of Applied Sciences.

Ilić, I. (2010). *The Discrete Logarithm Problem in Non-abelian Groups*.

Joux, A., & Lercier, R. (2002). The function field sieve is quite special. *Lecture Notes in Computer Science Algorithmic Number Theory*, 2369, 431–445. doi:10.1007/3-540-45455-1_34.

Kameswari, P. A., Surendra, T., & Ravitheja, B. (2016). Shank's Baby-Step Giant-Step Attack Extended to Discrete Log with Lucas Sequences. *Journal of Mathematics*, 12(1), 9–16. doi:10.9790/5728-12110916.

Kushwaha, P., & Mahalanobis, A. (2017). A probabilistic baby-step giant-step algorithm. *Proceedings of the 14th International Joint Conference on E-Business and Telecommunications*. doi:10.5220/0006396304010406.

Meshram, C., & Meshram, S. (2011). A public key cryptosystem based on IFP and DLP. *International Journal of Advanced Research in Computer Science*, 2(5), 616–619. Retrieved August 09, 2018.

Mihalcik, J. P. (2010). An Analysis of Algorithms for Solving Discrete Logarithms in Fixed Groups.

Odlyzko, A. M. (1984). Discrete Logarithms in Finite Fields and Their Cryptographic Significance. *EUROCRYPT*.

Padmavathy, R., & Bhagvati, C. (2012). Discrete logarithm problem using index calculus method. *Mathematical and Computer Modelling*, 55(1–2), 161–169. doi:10.1016/j.mcm.2011.02.022.

Pohlig, S., & Hellman, M. (1978). An improved algorithm for computing logarithms over GF(p) and its cryptographic significance (Corresp.). *IEEE Transactions on Information Theory*, 24(1), 106–110. doi:10.1109/tit.1978.1055817.

Pollard, J. M. (2000). Kangaroos, monopoly and discrete logarithms. *Journal of Cryptology*, 13(4), 437–447. doi:10.1007/s001450010010.

Pomerance, C. (2008). Elementary thoughts on discrete logarithms. *Algorithmic Number Theory*, 44, 385–396. Retrieved August 15, 2018, from www.msri.org/people/staff/levy/files/Book44/11carl.pdf.

Schirokauer, O. (2008). The Impact of the Number Field Sieve on the Discrete... Retrieved August 12, 2018, from www.researchgate.net/publication/228744286_The_impact_of_the_number_field_sieve_on_the_discrete_logarithm_problem_in_finite_fields.

Stein, A., & Teske, E. (2001). The parallelized Pollard kangaroo method in real quadratic function fields. *Mathematics of Computation*, 71(238), 793–815. doi:10.1090/s0025-5718-01-01343-6.

Stein, A., & Teske, E. (2005). *Optimized Baby Step-Giant Step Methods* (1st ed., Vol. 20). Waterloo: Faculty of Mathematics, University of Waterloo.

Stern, J. (2001). Evaluation Report on the Discrete Logarithm Problem over Finite Fields.

Stevenhagen, P. (2011). Number Field Sieve. *Encyclopedia of Cryptography and Security*.

Teske, E. (1998). Speeding up Pollards rho method for computing discrete logarithms. *Lecture Notes in Computer Science Algorithmic Number Theory*, 541–554. doi:10.1007/bfb0054891.

Wang, P., & Zhang, F. (2012). Improved Pollard rho method for computing discrete logarithms over finite extension fields. *Journal of Computational and Applied Mathematics*, 236(17), 4336–4343. doi:10.1016/j.cam.2012.03.019.

Wong, D. (2016). How to Backdoor Diffie-Hellman. *IACR Cryptology ePrint Archive*, 2016, 644.

Yan, S. (2003). Computing prime factorization and discrete logarithms: From index calculus to xedni calculus. *International Journal of Computer Mathematics*, 80(5), 573–590. doi:10.1080/00207160210 00059151.

5

Integer Factorization Problem

Pinkimani Goswami

University of Science and Technology

Madan Mohan Singh

North-Eastern Hill University

CONTENTS

5.1   Introduction 59
5.2   Special-Purpose Algorithms 61
    5.2.1   Trial Division 61
    5.2.2   Fermat's Method 61
    5.2.3   Pollard's $p-1$ Method 61
    5.2.4   Pollard's ρ Method 62
    5.2.5   Elliptic Curve Method 64
5.3   General-Purpose Algorithms 65
    5.3.1   Random Squares Method 65
    5.3.2   Quadratic Sieve Method 67
    5.3.3   Number Field Sieve Method 68
5.4   Factorization Method Based on Square Root Approximation 71
5.5   Prime Factorization Algorithm Without Using Any Approximation 73
5.6   Factoring Records 73
5.7   Conclusion 75
References 75

5.1 Introduction

Integer factorization of a positive integer n is the decomposition of n into a product of two small positive integers a, b (say), i.e., $n = a \cdot b$. Here, a and b are called factors of n. If $a, b \neq 1, n$, then they are the nontrivial factors of n and n is called a composite number; otherwise, n is a prime number. If a, b are primes (or prime power), then the factorization is called prime factorization of n. Note that integer factorization of a number is not unique, but the prime factorization of n is unique. The fundamental theorem of arithmetic states that for any positive integer $n > 1$ can be expressed as the product of prime numbers (not necessarily distinct). Therefore, it is trivial to get n from the given primes, but it is difficult to factor a large n. It is believed that the factoring a large composite number is difficult as there is no efficient algorithm to find the factors. Note that the problem of finding integer factors of a composite number is called integer factorization problem. The concept of integer factorization has been around since Euclid. He defined the idea of unique factorization in around 300 BC. The practical importance of the integer factorization problem arises with the development of public key cryptography. For example, the security of RSA cryptosystem (Rivest, Shamir & Adleman, 1978) is based on integer factorization problem, i.e., RSA cryptosystem will break if integer factorization problem is solved. Consequently, finding an efficient algorithm for integer factorization has practical importance.

Currently, the size of numbers which can be factored using modern technology is around 232 digits. Factoring such a number required an enormous amount of computing power. The records of integer factorization are categorized in two categories (Lenstra & Lenstra, 1993). In the first category, the general composite numbers such as RSA modulus are considered (Lenstra, 2017) and in the second category, the composite numbers with special form such as Fermat numbers (i.e., $F_m = 2^{2^m} + 1$, $m \geq 0$) and Cunningham numbers (i.e., $C^{\pm}(b, m) = b^m \pm 1$, where b and m are integers and b is not a perfect number) (Brillhart et al., 2002; Cunningham & Woodall, 1925; Lenstra, 2017) are considered. Most recently, in the first category, a 232-digit number RSA-768 was factored on 12 December 2009 (Kleinjung et al., 2010) and previously a 200-digit numbers RSA-200 was factored on 9 May 2005. In the second category, numbers of the form $2^m - 1$, where $m \in \{1007, 1009, 1081, 1111, 1129, 1151, 1153, 1159, 1177, 1193, 1199\}$, are factored in the year 2014 (Coppersmith, 1993; Kleinjung, Bos & Lenstra, 2014; Lenstra, 2017).

The integer factorization methods are distinguished into two types: special-purpose method and general-purpose method. The running time of the special-purpose method depends on the size of unknown factors of n, while the running time of the general-purpose method depends on the size of the composite number n whose factors has to be determined. The most important special-purpose methods are trial division method, Fermat's factorization method, Pollard's ρ method, Pollard's $p-1$ method, and the elliptic curve method. Random squares method, Quadratic sieve (QS) method, and Number field sieve (NFS) method are examples of general-purpose methods. The general-purpose methods are basically used for evaluating the security of factoring-based cryptosystems.

This chapter is a survey of various integer factorization methods, which is for the people who want to get an idea of how the modern factoring algorithms work. We will illustrate the basic concept of these methods along with examples. We will provide the algorithms and discuss the time complexity of these methods.

The rest of this chapter is organized as follows. Section 5.2 is devoted to special-purpose methods, which consist of five subsections namely trial division method, Fermat's factorization method, Pollard's ρ method, Pollard's $p-1$ method, and the elliptic curve method. In Section 5.3, we discuss the general-purpose methods. This section consists of three main subsections. In the first subsection, we will discuss the random squares method. The second subsection is devoted to the QS method, which was considered as the most practical general-purpose method from 1982 to 1994. In the year 2001, a 135-digit number was factored with the help of this method. In the third subsection, we consider the NFS method, the fastest algorithm for integer factorization. Note that both RSA-768 and RSA-200 were factored with the help of NFS method. Sections 5.4 and 5.5 are devoted two new factorization methods proposed in 2011 and 2012 (Balasubramaniam, Muthukumar & Othman, 2012; Zalaket & Hajj-Boutros, 2011).

As the integer factorization is the core of many public key cryptosystems, so it is important to know the size of the number which can be factored in current technology. Section 5.6 provides the list of some integers that are factored in between 1990 and 2017. We conclude the chapter in Section 5.7.

5.2 Special-Purpose Algorithms

The special-purpose algorithms are those whose running time depends on the size of the unknown factors of n. There are various integer factorization algorithms of this kind. In this section, we discuss some of them.

5.2.1 Trial Division

This is the simplest method for finding prime factors of a composite integer n. The prime number p will be the smallest factor of n if $p \mid n$ but n is not divisible by any prime number which is less than p. Therefore, to find p one has to divide n by all the prime numbers until p is reached.

5.2.2 Fermat's Method

In 1643, Fermat proposed a factoring algorithm, which is also called as Fermat's difference of squares method. The main idea of this method is that if we have an $n \in \mathbb{N}$, such that $a^2 \equiv b^2 \bmod n$ with $a \not\equiv b \bmod n$ for some $a, b \in \mathbb{Z}$, then n is a composite for $\gcd(a \pm b, n) \neq 1$ and $\gcd(a \pm b, n)$ gives the factors of n.

In Fermat's method, we consider an odd natural number $n = rs$ such that $1 < r < \sqrt{n}$. Then, n can be expressed as $n = a^2 - b^2$, where $a = \dfrac{r+s}{2}$ and $b = \dfrac{r-s}{2}$. Thus, in order to find the factors of n, we need to find the values of $a^2 = b^2 - n$ for $b = \lfloor \sqrt{n} + 1 \rfloor, \lfloor \sqrt{n} + 2 \rfloor, \ldots$ until $b^2 - n$ is a perfect square. For example, we will apply the Fermat method to find the factors of $n = 63{,}787$. The first b equal to $\lfloor \sqrt{63{,}787} \rfloor + 1 = 254$ and $b^2 - n = 729 = 27^2$, which is a perfect square. So, $a = 27$ and $b = 254$, which gives $63{,}787 = (27 + 254)(254 - 27) = 281 \cdot 227$.

This method gives a good result if factors of a number are near the square root of the number. The running time of this method is $O(n^{1/2})$. Lehman had reduced time complexity of this method to $O(n^{1/3})$, by combining this method with the trial division method. There are various methods to speed up Fermat's factorization method (e.g., Brillhart & Selfridge, 1967; Lehman, 1974).

5.2.3 Pollard's $p - 1$ Method

Pollard (1974) proposed a new technique for factorization called Pollard's $p - 1$ method, which is based on Fermat's little theorem. Fermat's little theorem states that for a prime p, $a^{p-1} \equiv 1 \bmod p$. Now, suppose $n = pq$ be a composite integer to find the factors of n. Since $p \mid n$, and p is prime, so $\gcd(a^p - 1, n) = p$ and this is a prime factor of n. But we cannot compute $a^{p-1} \equiv 1 \bmod p$ directly as p is unknown. We can calculate a^m for $m = 1, 2, 3, \ldots$ until $\gcd(a^m - 1, n) > 1$, i.e., until $m = p - 1$. In this case, one has to execute $p - 1$ exponential operations to find the factors of n, which is not more efficient than the trial division method. The algorithm can be improved in the following way.

Note that, for any integer k, $a^{k(p-1)} \equiv 1 \bmod p$. So, instead of finding m for which $\gcd(a^m - 1, n) > 1$, it is sufficient to find m such that $(p-1) \mid m$. We can find such m if for some relatively small bound B, $(p-1)$ is B–smooth, i.e., $(p-1)$ has only the prime factors which are less than or equal to B. Consider $m = \displaystyle\prod_{i=1}^{\pi(B)} p_i^{e_i}$, where $p_i^{e_i} \leq B \leq p_i^{e_i+1}$ for $i = 1, 2, \ldots, \pi(B)$

and p_i are primes $\leq B$, i.e., $e_i = \dfrac{\log n}{\log p_i}$ (also, we can write $m = \displaystyle\prod_{\substack{p \text{ prime} \\ 1 \leq p \leq B}} p^{\left\lfloor \frac{\log n}{\log p} \right\rfloor}$). Clearly, m is multiple of $(p-1)$ and therefore $\gcd(a^m - 1, n) > 1$, which gives a nontrivial factor of n. Note that, if $(p-1)$ is not B-smooth, then $a^m \not\equiv 1 \bmod p$ for at least half of $a \in \mathbb{Z}_n$. Therefore, this algorithm can be used to find the factor p of n, when $(p-1)$ have a small prime factor. The algorithm is as follows:

1. Choose a composite number n and a bound B.

2. Compute $m = \displaystyle\prod_{\substack{p \text{ prime} \\ 1 \leq p \leq B}} p^{\left\lfloor \frac{\log n}{\log p} \right\rfloor}$

3. Choose a random positive integer a relatively prime to n.

4. Compute $d = \gcd(a^m - 1, n)$.
   i. If $d = 1$, then increase B.
   ii. If $d = n$, then change a.
   iii. d will be a nontrivial factor of n if $d \neq 1$ or n.

As an example, if we apply the Pollard's $p-1$ algorithm for $n = 63{,}787$ with $B = 20$ and $a = 2$, then for $m = 7!$, $\gcd(a^m - 1, n) = 281$. Therefore, the prime factors of $n = 63{,}787$ are 227 and 281.

The running time of the algorithm is $O\left(B \log B \left(\log n\right)^2 + \left(\log n\right)^3\right)$ (Stinson, 2013). Since the exponent m depends on the bound B, so the time required to factor n depends on B. The probability of finding the factors of n will increase with increasing the bound B. But it will also increase the time required to factor n. There are various technical improvements to make the algorithm faster, as can be found in Montgomery (1987).

5.2.4 Pollard's ρ Method

Pollard (1975) proposed a factoring algorithm (called Pollard's ρ method), which is more efficient than the trial division method. This method is based on two ideas: the first idea is Birthday Paradox. According to this paradox, in a group of at least 23 randomly chosen people, the probability of sharing the same birthday at least two people is more than $\dfrac{1}{2}$ (Lenstra, 2000, Stinson, 2013). More generally, if we choose some numbers randomly from a set of containing p elements, then after the selection of $1.17\sqrt{p}$ numbers, the probability of choosing the same number becomes more than $\dfrac{1}{2}$. The second idea is that if p is an unknown divisor of a composite number n and $x_1, x_2 \in \mathbb{Z}$ such that $x_1 \neq x_2$ and $x_1 \equiv x_2 \bmod p$, then $p \leq \gcd(x_2 - x_1, n) < n$ and so one can obtain a nontrivial factor of n by computing a gcd.

We consider the following iteration in Pollard's ρ method:

$$x_i = f(x_{i-1}) \bmod n \text{ for } i = 1, 2, \ldots$$

with a random integer x_0 from $(0, n-1)$ and a polynomial function $f(x)$ with integer coefficients. This iteration gives a sequence of integer $\{x_i\}$ modulo n for $i = 1, 2, \ldots$. Now for each $j < i$, we compute $p = \gcd(x_i - x_j, n)$ until $p > 1$. This nontrivial p is a factor of n.

Note that for a large i, it is computationally expensive to compute $\gcd(x_i - x_j, n)$ for every $j < i$ as one has to check $(i-1)$ gcds for each i. This algorithm can be improved by computing only the following gcds:

$$\gcd(x_{2i} - x_i, n) \quad \text{for } i = 1, 2, \ldots$$

This is because if $x_i \equiv x_j \bmod p$, then $x_{i+1} \equiv x_{j+1} \bmod p$, and we can extend $x_{i+k} \equiv x_{j+k} \bmod p$ for all integer $k \geq 0$. Therefore, there exists i such that $x_{2i} \equiv x_i \bmod p$. In this case, for each i, one has to check only one gcd, i.e., $\gcd(x_{2i} - x_i, n)$ instead of checking $(i-1)$ gcds. The algorithm is as follows:

1. Choose a composite number n.
2. Choose a polynomial $f(x)$ with integer coefficient.
3. Choose an integer x_0 modulo n.
4. Compute x_i form $x_i = f(x_{i-1}) \bmod n$ for $i = 1, 2, \ldots$
5. Find i for which $\gcd(x_{2i} - x_i, n) \neq 1$.

Then, the nontrivial $\gcd(x_{2i} - x_i, n)$ will give the factors of n.

This algorithm can also be improved by computing $\gcd(Q_k, n)$ where for every k steps, $Q_k = \prod_{i=1}^{k}(x_{2i} - x_i) \bmod n$. Note that if $\gcd(x_{2i} - x_i, n) > 1$, then $\gcd(Q_k, n) > 1$.

Now if we construct a graph on $\mathbb{Z}_p$ for all $i \geq 1$, then we have a diagram as in Figure 5.1.

That is the figure looks like the letter ρ and hence the algorithm is named as ρ algorithm. The expected running time of this algorithm is proportional to the square root of the size of the smallest prime factor p of n. Hence, the running time of this method is $O\left(n^{\frac{1}{4}} \cdot (\log n)^2\right)$.

Brent (1980) optimized this method. The most remarkable success of Pollard's ρ method is finding the factors of the eight Fermat numbers, $2^{2^8} + 1$ (Brent & Pollard, 1981). For practice, f is considered as a simple quadratic polynomial.

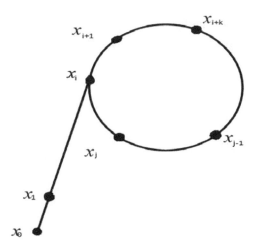

FIGURE 5.1
Pollard's ρ algorithm.

TABLE 5.1

Values of $x_i = f(x_{i-1}) \bmod n$

i	0	1	2	3	4	5	6	7	8	9	10
x_i	2	3	8	63	3,698	53,421	36,647	31,110	55,735	27,111	52,506
i	11	12	13	14	15	16	17	18	19	20	
x_i	5,895	1,132	5,683	20,266	50,049	50,697	16,217	61,074	24,863	8,951	

As an example, if we apply Pollard's ρ algorithm to find factors of $n = 63{,}787$ with $f(x) = x^2 - 1$ and $x_0 = 2$, then we have (Table 5.1):

$\gcd(x_{2i} - x_i, n) = 281$, for $i = 10$, which is a prime factor of 63,787.

5.2.5 Elliptic Curve Method

The major drawback of Pollard's $p-1$ method is that the factors of n cannot find efficiently if it doesn't have a prime factor p where $p-1$ is B-smooth. Lenstra (1987) proposed a new factorization method which avoids the above drawback. He used the group of a random elliptic curve over the field $\mathbb{Z}_p$ instead of the multiplicative group of $\mathbb{Z}_p$.

Note that for a field K of characteristic 0 or characteristic greater than 3, the elliptic curve E over K is a non-singular equation of the form $y^2 = x^3 + ax + b$ where $a, b, x, y \in K$, i.e., $E : y^2 = x^3 + ax + b$ such that $4a^3 + 27b^2 \neq 0$. Also, the set of points (or K-rational points) on E is defined as $E(K) = \left\{ (x, y) : x, y \in K, \ y^2 = x^3 + ax + b \right\} \cup O_E$, where O_E denotes a point at infinity and the number of points on $E(K)$ is denoted by $\#E(K)$. Recall that in Pollard's $p-1$ algorithm, we have to find m such that $(p-1) \mid m$ and then for a nonzero element $a \in \mathbb{Z}_p$, $\gcd(a^m - 1, n)$ gives a nontrivial factor of n. If $(p-1) \nmid m$, then we have to increase m such that $(p-1) \mid m$. The main idea of the elliptic curve method is that the nonzero elements of $\mathbb{Z}_p$ are replaced by F_p-rational points of some elliptic curve E over F_p. Also, we have to find m such that $\#E'(F_p)$ divides m, then for $P \in E(F_p)$, $mP = O_E$ will give a factor of n. If $\#E(F_p)$ does not divide m, instead of increasing m one can define a random elliptic curve E' such that $\#E'(F_p)$ divides m. We can find such m by computing $m = \displaystyle\prod_{1 \leq p \leq B}^{p \text{ prime}} p^{\left\lfloor \frac{\log n}{\log p} \right\rfloor}$, where B is a given bound. If $\#E(F_p)$ is B-smooth, then $\#E(F_p)$ divides m, and therefore, for any point $P \in E(F_p)$, $mP = O_E$. The method is given as follows.

We know that the sum $P + Q$ for the points $P = (x_1, y_1)$ and $Q = (x_2, y_2)$ of $E(F_p)$ is (x_3, y_3), where $x_3 = s^2 - x_1 - x_2$, $y_3 = s(x_1 - x_3) - y_1$ modulo p with $s = (y_2 - y_1)(x_2 - x_1)^{-1}$ mod p if $P = Q$ and $s = (3x_1^2 + a)(2y_1)^{-1}$ mod p if $P \neq Q$. Now, the computation of mP involves the computation of many sum $P + Q$. At some point of time, we would certainly compute $P + Q = O_E$, i.e., $P = -Q$, i.e., $x_2 \equiv x_1 \pmod{p}$ and therefore $\gcd(x_2 - x_1, n)$ gives a nontrivial factor of n. Note that if $\gcd(x_2 - x_1, n) \neq 1$, then $(x_2 - x_1)^{-1}$ mod n does not exist and so the computation $P + Q$ fails and hence we have a nontrivial factor of n. The algorithm is defined as follows:

1. Choose a composite number n and a bound B.

2. Compute $m = \displaystyle\prod_{1 \leq p \leq B}^{p \text{ prime}} p^{\left\lfloor \frac{\log n}{\log p} \right\rfloor}$

3. Choose the random integers $a, x, y \in \mathbb{Z}_n$ and compute $b = y^2 - x^3 - ax$.

   i. If $\gcd\left(4a^3 + 27b^2, n\right) = 1$, then make the new choice for a, x, y.

   ii. If $\gcd\left(4a^3 + 27b^2, n\right) \neq 1$, then $E : y^2 = x^3 + ax + b$ is a random elliptic curve.

   iii. Choose a point $P = (x, y)$ on E.

4. Compute $mP = \underbrace{P + P + \cdots + P}_{m}$. For each computation, we will get either a new point on the curve or a factor of n.

5. If a factor of n was not found in Step (4), then choose a new point on E or go back to Step (3) and choose new a, x, y or go back to Step (1) to increase B.

As an example, if we apply the elliptic curve method to find the factors of $n = 2{,}773$ with the elliptic curve $E : y^2 = x^3 + 4x + 4$ and $P = (1, 3)$. Then, $2P = (1771, 705)$ and the computation of $3P$ fail as $\gcd(1771 - 1, 2775) \neq 1$. Therefore, $\gcd(1770, 2773) = 59$ is a factor of n.

The elliptic curve method is the third fastest method for factoring a large number. Two remarkable factorizations obtained by using the elliptic curve method are Fermat tenth $(2^{2^{10}} + 1)$ (Brent, 2000) and eleventh $(2^{2^{11}} + 1)$ numbers (Stinson, 2013). The largest factor found by the elliptic curve method has 83 digits and was found by R. Propper on 7 September 2013. The running time of this algorithm is $O\left(e^{(1+o(1))\sqrt{2\ln p \ln \ln p}}\right)$, where p is the smallest prime factor of n and the notation $o(1)$ denotes a function of n that approaches 0 as $n \to \infty$ (Stinson, 2013). The various improvement of this method is suggested by Montgomery (1987; 1992) and others (Alkin & Morain, 1993; Brent, 1986). For the general description of ECM, refer to Brent (1996), Montgomery (1994), and Silverman and Wagstaff (1993).

5.3 General-Purpose Algorithms

In this section, we will focus on general-purpose algorithm, whose running time depends on the size on the number n to be factored. General-purpose algorithm is used for the cryptographic application.

5.3.1 Random Squares Method

If n is an odd number, then n can be expressed as the difference of two square, i.e., for $a, b > 0$, $n = a^2 - b^2 = (a - b)(a + b)$. Therefore, the problem of finding factors of n is now transformed to the problem of finding a and b. The above relation can also be written as $a^2 - n = b^2$. For each $a > n$, one can compute $a^2 - n$ and can check whether it is a perfect number or not. If one can find such a and b, then $a + b$ and $a - b$ are factors of n. Note that we cannot find the factors of n if $a \equiv \pm b \bmod n$. Therefore, the problem of finding factors of n is equivalent to finding the integers a and b such that $a^2 \equiv b^2 \bmod n$ with $a \not\equiv b \bmod n$. Then, $\gcd(a \pm b, n)$ will gives the nontrivial factors of n.

Note that finding the perfect number from the sequence of integers $a^2 - n$, where $a > \lceil n \rceil$ is time-consuming. We can remove this drawback by selecting a subsequence from the sequence $a^2 - n$ such that the elements of this subsequence are multiplied together to yield a perfect square. Such a subsequence can be guaranteed by the notion of smooth numbers. The following lemma gives the guarantee of finding a subsequence with the above condition.

Lemma

If $m_1, m_2, \ldots, m_k$ are positive B-smooth integers, and if $k > \pi(B)$, then some non-empty sub-sequence $\left(m_{i_j}\right)$ has a product which is a perfect square. Here, $\pi(B)$ denotes the number of primes in the interval $[1, B]$ (Pomerance, 2008; Seibert, 2011).

Note that a factor base $\mathcal{B}$ is a non-empty set of positive prime integers and an integer x is called smooth over a factor base $\mathcal{B}$, if all primes factors of x are numbers of $\mathcal{B}$. Consider a factor base $\mathcal{B} = \left\{p_1, p_2, \ldots, p_{\pi(B)}\right\}$ and a set $M = \left\{m_1, m_2, \ldots, m_k\right\}$ of integers, which is smooth over the factor base $\mathcal{B}$, i.e., $m_i = p_1^{v_{i1}} p_2^{v_{i2}} \cdots p_{\pi(B)}^{v_{i\pi(B)}}$, for $i = 1, 2, \ldots, k$. Then, a subsequence $M^* = \left\{m_{i_1}, m_{i_2}, \ldots, m_{i_t}\right\}$ has the product a perfect square iff $\sum_{l=1}^{t} v_{i_l j} \equiv 0 \pmod{2}$, $\forall j, 1 \le j \le \pi(B)$, i.e., the problem of finding a subsequence M^* is now simplified to finding a set of exponent vectors such that their sum is the zero vector over modulo 2. Here $\left(v_{i_1}, v_{i_2}, \ldots, v_{i\pi(B)}\right)$ is the exponent vector of $m_{i_l}, \forall l$. Note that we can easily find a subsequence with the above condition by using the linear algebra techniques such as by Gaussian elimination method.

Thus, we describe the factoring algorithm as follows (Bimpikis & Jaiswal, 2005):

1. Choose a composite number n.
2. Choose a bound B.
3. Consider a factor base $\mathcal{B} = \left\{p_1, p_2, \ldots, p_{\pi(B)}\right\}$, i.e., find all the primes in $[1, B]$.
4. Consider a set $U = \{r_1, r_2, \ldots, r_k\}$ of integers such that
   i. $f(r_i) = r_i^2 \pmod{n}, \forall i$ is smooth over the factor base $\mathcal{B}$,
      i.e., $f(r_i) = r_i^2 \equiv p_1^{e_{i1}} p_2^{e_{i2}} \cdots p_{\pi(B)}^{e_{i\pi(B)}} \pmod{n}, \forall i$
   ii. $\sum_{i=1}^{k} e_{ij} = 2k_i, \forall\, i, 1 \le j \le \pi(B)$
      where e_{ij} is the exponent of p_j occurring in the factorization of $f(r_i)$ and $k_i \in \mathbb{Z}$
5. Compute the integers a and b, where $a = \prod_{i=1}^{n} r_i$ and $b = \prod_{i=1}^{\pi(B)} p_i^{e_i}$

Then, we have:

$$a^2 = \prod_{i=1}^{k} r_i^2 \equiv \left(p_1^{e_1} p_2^{e_2} \cdots p_{\pi(B)}^{e_{\pi(B)}}\right)^2 \equiv b^2 \pmod{n}$$

Therefore, we find two integers a and b such that $a^2 \equiv b^2 \pmod{n}$ with $a \not\equiv \pm b \pmod{n}$. Hence, the nontrivial factors of n can be determined.

For example, consider $n = 301$ and $B = 13$. Here, $\mathcal{B} = \{2, 3, 5, 7, 11, 13\}$. Now we perform the following computation:

$$19^2 \equiv 60 \pmod{301} = 2^2 \times 3 \times 5$$

$$20^2 \equiv 99 \pmod{301} = 3^2 \times 11$$

$$21^2 \equiv 140 \pmod{301} = 2^2 \times 5 \times 7$$

$$22^2 \equiv 183 \ (\mathrm{mod}\, 301) = 3 \times 61$$

$$23^2 \equiv 228 \ (\mathrm{mod}\, 301) = 2^2 \times 3 \times 19$$

$$24^2 \equiv 275 \ (\mathrm{mod}\, 301) = 5^2 \times 11$$

$$26^2 \equiv 74 \ (\mathrm{mod}\, 301) = 2 \times 37$$

$$25^2 \equiv 23 \ (\mathrm{mod}\, 301) = 1 \times 23$$

$$28^2 \equiv 182 \ (\mathrm{mod}\, 301) = 2 \times 7 \times 13$$

Here, for the set $M = \{19, 20, 21, 24, 28\}$, $f(r) = r^2 \ (\mathrm{mod}\, 301)$ is smooth over $\mathcal{B}$. Also, we obtain:

$20^2 \times 24^2 \equiv 3^2 \times 5^2 \times 11^2 (\mathrm{mod}\, 301)$ and hence $a = 480$ and $b = 165$. Therefore, the factors of 301 are $\gcd(480 - 165, 301) = 7$ and $\gcd(480 + 165, 301) = 43$.

Note that a and b can be found by adding the exponent vectors, i.e., solving the system of congruence:

$$3x_1 + 2x_3 + x_5 \equiv 0 \ (\mathrm{mod}\, 2)$$

$$x_1 + 2x_2 \equiv 0 \ (\mathrm{mod}\, 2)$$

$$x_3 + 2x_4 \equiv 0 \ (\mathrm{mod}\, 2)$$

$$x_3 \equiv 0 \ (\mathrm{mod}\, 2)$$

$$x_2 + x_4 \equiv 0 \ (\mathrm{mod}\, 2)$$

$$x_5 \equiv 0 \ (\mathrm{mod}\, 2)$$

One nontrivial solution of this system of congruence is $(0, 1, 0, 1, 0)$. Thus, we get the congruence $20^2 \times 24^2 \equiv 3^2 \times 5^2 \times 11^2 \ (\mathrm{mod}\, 301)$. Here, $U = \{20, 24\}$ is a subsequence of M.

5.3.2 Quadratic Sieve Method

Pomerance (1982) developed a new factoring algorithm called QS method. This method is similar to the random square method, but in this method, a sieving procedure is used to find the values $r^2 \bmod n$. The algorithm is defined as follows:

1. Choose a composite number n.
2. Choose a bound B.
3. Consider a factor base $\mathcal{B} = \{p_1, p_2, \ldots, p_{\pi(B)}\}$.
4. Choose an integers r such that $f(r) = r^2 - n$ is smooth over $\mathcal{B}$.
5. Find a subset $U = \{r_1, r_2, \ldots, r_m\}$ such that:

    i. $f(r_i) = r_i^2 - n \equiv p_1^{e_{i1}} p_2^{e_{i2}} \cdots p_{\pi(B)}^{e_{i\pi(B)}} \ (\mathrm{mod}\, n), \forall i$

    ii. $\displaystyle\sum_{i=1}^{k} e_{ij} = 2k_i, \forall i, 1 \le j \le \pi(B)$, for some $k_i \in \mathbb{Z}$

    where e_{ij} is the exponent of p_j occurring in the factorization of $f(r_i)$.

6. Compute the integers a and b such that $a = \prod_{i=1}^{\pi(B)} r_i$ and $b = \prod_{i=1}^{\pi(B)} p_i^{e_i}$

7. Then, we have:

$$a^2 = \prod_{i=1}^{k} r_i^2 = \prod_{i=1}^{k} r_i^2 - n \equiv \left(p_1^{e_1} \ p_2^{e_2} \cdots p_{\pi(B)}^{e_{\pi(B)}} \right)^2 \equiv b^2 \ (\text{mod } n)$$

which gives the nontrivial factors of n.

Note that this algorithm is similar to the previous algorithm; only the function $f(r_i) = r_i^2 \bmod n$ is replaced by $f(r_i) = r_i^2 - n$. This improves the running time of the algorithm, as for the function $f(r_i) = r_i^2 - n$, it is easy to check whether $f(r_i)$ for some r_i is smooth over a factor base on not. The process of finding all r_i for which $f(r_i)$ are smooth over a factor base is called Sieving process. Now, let $f(r_i) \equiv 0 (\bmod p_i)$ for some integer r_i, where $p_i \in B$. This implies that $r_i^2 - n \equiv 0 \bmod p_i$. So

$$\left(r_i + t p_i \right)^2 - n \equiv 0 (\bmod p_i), \text{ for } t \in \mathbb{Z}$$

i.e., $f(r_i + t p_i) \equiv 0 \ (\bmod p_i)$

Therefore, in the QS method, one can consider all primes p_i in B and for all r_i for which p_i divides $f(r_i)$, which is equivalent to finding the square roots of n modulo p_i, $\forall p_i \in B$. Because of the above relation (i.e., if $p_i \mid f(r_i)$ then $p_i \mid f(r_i + t p_i)$ for $t \in \mathbb{Z}$), the sieving process is simplified. This is more efficient than checking of the smoothness of $f(r_i)$ for a randomly selected r_i, as in that case, one has to verify that for $f(r_i)$ is divisible by all primes for p_i in the factor base B using the trial division method (which is the case in the random square method).

The expected running time of this method is asymptotically bounded by $O\left(e^{(1+\varepsilon)\sqrt{\log n \log \log n}} \right)$, for any $\varepsilon > 0$ (Bimpikis & Jaiswal, 2005). The details of this method can be found in Back and Shallit (1997) and Pomerance (1985). This method is considered as the fastest factorization method before the NFS method was proposed. For the numbers up to 110-digit long, the QS method is faster than the NFS method.

5.3.3 Number Field Sieve Method

The NFS method is considered as the fastest method for factoring a large number. Like the other factoring method, this method has also used the congruence $x^2 \equiv y^2 \bmod n$ to find the factors of n, but this time computation is done in a ring of algebraic integers. The NFS method was first designed for some integer of a special form, and this variant is called special number field sieve (SNF) method (Lenstra et al., 1993). Later in 1992, this algorithm is modified for arbitrary integer and named as general number field sieve (GNFS) or NFS method.

The GNFS method includes two parameters: the first parameter is a polynomial function $f : \mathbb{R} \to \mathbb{R}$ with integer coefficient and the second parameter is a number $m \in \mathbb{N}$ such that $f(m) \equiv 0 \bmod n$, where n is a composite number whose factor have to determine. Such f and m can be found easily if m is chosen first. If we choose m first, then we can express n as $n = a_k m^k + a_{k-1} m^{k-1} + \cdots + a_0$. Therefore, the function f can be defined as $f(x) = a_k x^k + a_{k-1} x^{k-1} + \cdots + a_0$ and which implies $f(m) \equiv 0 \bmod n$.

If $\theta \in \mathbb{C}$ is a root of the polynomial function f and k is the degree of f, then the space $\mathbb{Z}[\theta] = \{x : x = a_{k-1}\theta^{k-1} + a_{k-2}\theta^{k-2} + \cdots + a_0 \text{ for } \{a_i\} \subset \mathbb{Z}\}$ form a ring with respect to the polynomial multiplication. The basic idea of the NFS method is based on the following theorem of $\mathbb{Z}[\theta]$ (Case, 2003).

Theorem (Case, 2003)

Let $f(x)$ be a polynomial function with integer coefficients. If $\theta \in \mathbb{C}$ is a root of $f(x)$, and $m \in \mathbb{Z}_n$ is an integer such that $f(m) \equiv 0 \,(\mathrm{mod}\, n)$, then there exists a unique mapping $\phi : \mathbb{Z}[\theta] \rightarrow \mathbb{Z}_n$ satisfying:

   i. $\phi(ab) = \phi(a)\phi(b) \,\forall a,b \in \mathbb{Z}[\theta]$
   ii. $\phi(a+b) = \phi(a) + \phi(b) \,\forall a,b \in \mathbb{Z}[\theta]$
   iii. $\phi(1) \equiv 1 \,\mathrm{mod}\, n$
   iv. $\phi(\theta) \equiv m \,\mathrm{mod}\, n$

Note that the above conditions also imply that $\phi(za) = z\phi(a) \,\forall a \in \mathbb{Z}[\theta], z \in \mathbb{Z}$. Using the above theorem, we can construct a difference of square as follows.

Consider a finite set U of pairs of integer (a,b) such that

$$\prod\nolimits_{(a,b)\in U} (a+b\theta) = \beta^2 \text{ and } \prod\nolimits_{(a,b)\in U} (a+bm) = z^2$$

for $\beta \in \mathbb{Z}[\theta]$ and $z \in \mathbb{Z}$. Then using the homomorphism as defined in the above theorem, we can define $x = \phi(\beta)$ and $z = y$ for any integer x,y. The pair (x,y) constructs a difference of squares

$$x^2 = \phi\left(\Pi_{(a,b)\in U}(a+b\theta)\right)$$

$$= \Pi_{(a,b)\in U}\left(\phi(a+b\theta)\right)$$

$$= \Pi_{(a,b)\in U}(a+bm)$$

$$\equiv y^2 \,\mathrm{mod}\, n$$

The algorithm (Bimpikis & Jaiswal, 2005) is defined as follows:

1. Choose an integer $m \in \mathbb{Z}$ and find a polynomial function f such that $f(m) \equiv 0 \,\mathrm{mod}\, n$.
2. Consider a factor base $\mathcal{A} \in \mathbb{Z}[\theta]$ and $\mathcal{B} \in \mathbb{Z}$.
3. Choose two integers a,b such that $(a+b\theta)$ and $\phi(a+b\theta) = a + bm$ are smooth over $\mathcal{A}$ and $\mathcal{B}$, respectively.
4. Consider a subset U of pairs of such (a,b) found in Step (3) such that for $\beta \in \mathbb{Z}[\theta]$ and $z \in \mathbb{Z}$

$$\prod\nolimits_{(a,b)\in U} (a+b\theta) = \beta^2 \text{ and } \prod\nolimits_{(a,b)\in U} (a+bm) = z^2$$

Then, we have $x = \phi(\beta)$ and $z = y$ such that $x^2 \equiv y^2 \,\mathrm{mod}\, n$, which gives the nontrivial factors of n.

Note that a finite set $\mathcal{A} \subset \mathbb{Z}[\theta]$ is an algebraic factor base where each $a + b\theta$ satisfies the condition that $\nexists c, d \in \mathbb{Z}[\theta]$ such that $c \cdot d = a + b\theta, \forall (a, b)$ and an element $l \in \mathbb{Z}[\theta]$ is called smooth over $\mathcal{A}$ if $\exists W \subset \mathcal{A}$ such that $\prod_{(c,\, d) \in W} (c + d\theta) = l$. Using the following theorem (Case, 2003), an algebraic factor base $\mathcal{A}$ can be represented as a set of pairs of integers $\{(r, p)\}$.

Theorem

Let $f(x)$ be a polynomial function with integer coefficients and let $\theta \in \mathbb{C}$ be a root of $f(x)$. Then, the set of pairs $\{(r, p)\}$ where p is a prime integer and an integer $r \in \mathbb{Z}_n$ such that $f(r) \equiv 0 \bmod p$ is in bijective correspondence with the set of $a + b\theta \in \mathbb{Z}[\theta]$ (Case, 2003).

To find the square in $\mathbb{Z}[\theta]$ and in $\mathbb{Z}$, one has to find a pair of numbers (a, b) such that $a + b\theta$ and $a + bm$ are smooth in $\mathcal{A}$ and $\mathcal{B}$, respectively. This can be found in the following ways (Case, 2003):

i. Choose an integer $b \in \mathbb{Z}$ and an arbitrary positive integer N. Let a runs from $-N$ to N.

ii. Find two arrays in the following ways:

$$\begin{vmatrix} -N + b\theta \\ \vdots \\ N + b\theta \end{vmatrix} \begin{vmatrix} -N + bm \\ \vdots \\ N + bm \end{vmatrix}$$

iii. Note that q_i will divide $a + bm$ if $a \equiv -bm \bmod q_i$, for each $q_i \in \mathcal{B}$. Find all the values of a such that $a \equiv -bm \bmod q_i$, and for each value of a, list this factor of $a + bm$ in the sieve array. This process will be repeated for every $q_i \in \mathcal{B}$. List all the $a + bm$ in the sieve array that is completely factored by this method. These $a + bm$ are smooth in $\mathcal{B}$.

iv. A pair $(r_i, p_i) \in \mathcal{A}$ divides $a + b\theta$ iff $a \equiv -br_i \bmod p_i$. Find all the value of a such that $a \equiv -br_i \bmod p_i$. For each such a, list all of this (r_i, p_i) factor of $a + b\theta$ in the sieve array. At last, there will be a list of (r_i, p_i) factors for all $a + b\theta$ in the sieve array. Now, this list of factors is a complete factorization if $\Pi p_i = (-b)^d f(-a/b)$, and in that case, $a + b\theta$ is smooth over $\mathcal{A}$.

v. Compare the two arrays one can find (a, b) such that both $a + b\theta$ and $a + bm$ are smooth.

Again, the following two theorems give the guarantee of finding an element in $\mathbb{Z}[\theta]$ which is a perfect square:

Theorem (Case, 2003)

Let $l \in \mathbb{Z}[\theta]$ have the factorization $l = (a_1 + b_1\theta)^{e_1} (a_2 + b_2\theta)^{e_2} \cdots$ where for every i, $a_i + b_i\theta$ satisfies the criteria to be in an algebraic factor base. If l is a perfect square in $\mathbb{Z}[\theta]$, then $e_i \equiv 0 \bmod 2, \forall i$ (Case, 2003).

Theorem (Case, 2003)

Let U be a set of (a,b) pairs such that $\prod_{(a,b)\in U}(a+b\theta)$ is a perfect square in $\mathbb{Z}[\theta]$. Then for any (s,q) with $(s,q) \nmid a+b\theta$ for any $(a,b)\in U$,

$$\prod_{(a,b)\in U}\left(\frac{a+bs}{q}\right)=1$$

where q is prime and $s\in\mathbb{Z}_n$ with $f(s)\equiv 0\bmod q$ (Case, 2003).

For a prime q, the set Q of pairs of numbers (s,q) with $f(s)\equiv 0\bmod q$ is called quadratic character base and the element (s,q) is called quadratic character (Case, 2003). Note that each $(a,b)\in U$ can be represented as a row vector of order $1+k+l+u$ with the help of algebraic factor base A of k elements, factor base B of l elements and quadratic factor base Q of u elements. For details, one can see Case (2003). Also, it is proved that $\prod_{(a,b)\in U}(a_i+b_i\theta)$ and $\prod_{(a,b)\in U}(a_i+b_i\theta)$ both are perfect square in $\mathbb{Z}[\theta]$ and $\mathbb{Z}$, respectively, if and only if the sum of the vector representation of each $(a_i,b_i)\in U$ is equal to the zero vector modulo 2. Therefore, it is sufficient to check whether the sum of the vector representation of each $(a_i,b_i)\in U$ is congruence to $0\bmod 2$ or not. Once we get $\prod_{(a,b)\in U}(a_i+b_im)$ and $\prod_{(a,b)\in U}(a_i+b_i\theta)$ as a perfect square in $\mathbb{Z}$ and $\mathbb{Z}[\theta]$, respectively, then we can factor n with the help of ring homomorphism ϕ.

The expected running time of this algorithm is $O\!\left(e^{c+o(1)(\log n)^{1/3}(\log\log n)^{2/3}}\right)$, where c is a constant with value ≈ 1.92. The details of this method can be found in Bimpikis & Jaiswal, 2005, Lenstra (2000), etc.

5.4 Factorization Method Based on Square Root Approximation

Zalaket and Hajj-Boutros (2011) proposed a new integer factorization method based on the square root approximation. It consists of finding a reducible primorial $R_p=(p_i\times p_{i+1}\times\cdots\times p_j)$, where the $\gcd(R_p,n)$ gives the first two factors of n in a reduced time. This process is repeated for the obtained two factors until achieving all the prime factors of n.

The algorithm is defined as follows: let $n=AB$ be the composite numbers whose factors have to be determined. Here A and B are prime factors of n, and it is assumed that any prime factor B can be expressed as $B=A\dfrac{p^2}{q^2}+\dfrac{m}{q^2}$, where p and q are prime or relatively prime, m is an integer, and $\dfrac{m}{q^2}$ is small compared to $A\dfrac{p^2}{q^2}$. Then, $n=\dfrac{A^2p^2}{q^2}\left(1+\dfrac{m}{Ap^2}\right)$, which gives $\sqrt{n} = \dfrac{Ap}{q}\left(1 + \dfrac{m}{Ap^2}\right)^{\frac{1}{2}}$. As $\dfrac{m}{Ap^2} \ll 1$ so $\left(1 + \dfrac{m}{Ap^2}\right)^{\frac{1}{2}} \approx 1 + \dfrac{m}{2Ap^2}$ and hence $\sqrt{n} = \dfrac{Ap}{q} + \dfrac{m}{2pq}$.

Since $\dfrac{m}{2pq} < 1$, so the following two approximations can be considered:

i. The integer part of $\sqrt{n}$ is $\left[\sqrt{n}\right] \approx \dfrac{Ap}{q} \pm \varepsilon'$.

ii. The fractional part of $\sqrt{n}$ is $\left(\sqrt{n} - \left[\sqrt{n}\right]\right) \times pq \approx k \pm \varepsilon'$, where k is a positive integer or half-integer and $0 \leq \varepsilon < 1$ and $0 \leq \varepsilon' \ll 1$.

Therefore, one can find k by multiplying the fractional part $\sqrt{n} - \left[\sqrt{n}\right]$ by $r = pq$. As r is a small integer, so if k is found, then one can easily find p and q by using any of the factorization methods. Then, multiplying the integer part $\left[\sqrt{n}\right]$ by $\dfrac{p}{q}$ gives a number that is approximately equal to A. The algorithm is presented below. We consider the factorization algorithm from Zalaket and Hajj-Boutros (2011).

Algorithm

Input: n: an integer number to be factored
           P_r: a list of a subset of prime numbers
           Upr: a threshold indicating the highest $r = pq$ allowed
           Highstep: a threshold indicating the steps allowed to find a solution
Output: A and B, two integer factors of n.
begin
   Initialization;

   $\varepsilon := 0.02, r := 2, steps := 0;$

   $fx := \sqrt{n} - \left[\sqrt{n}\right];$

   Iterate until finding factors or until reaching the threshold specified for *steps*;
   repeat
       while $(r < upr$ AND $fx \times r \neq \left[fx \times r\right] \pm \varepsilon)$ **do**
           find a possible multiplier r of fx which generates its nearest integer;
           $steps := steps + 1$
           factor r into p and q;

           $A := \left[\sqrt{n}\right] \times \dfrac{p}{q};$

           if A is not a divisor of n **then**
               find in P_r the nearest value P_{ri} to A;

               $A := \gcd\left(P_{ri-2} \times P_{ri-1} \times P_{ri} \times P_{ri+1} \times P_{ri+2}, n\right);$

       until A is an integer divisor of n OR $steps > Highstep;$

        if *steps > Highstep* **then**

$$B := \frac{n}{A};$$

      else

            print ("*n* is prime or no solution is found for the chosen thresholds");

end

5.5 Prime Factorization Algorithm Without Using Any Approximation

Balasubramaniam, Muthukumar, and Othman (2012) pointed out some drawbacks of the above method and proposed a new factorization method without using any approximation. The algorithm is defined as follows:

1. Choose a composite number n.
2. If $n - 3 \equiv 0 \bmod 4$, then
    i. find l, m such that $4lm + 3l + m = N$ where $N = (n-3)/4$.
    ii. find $p - 1 = 4l$ and $q - 3 = 4m$.
3. If $n - 1 \equiv 0 \bmod 4$, then
    i. find l, m such that $4lm + l + m = N$ where $N = (n-1)/4$.
    ii. find $p - 1 = 4l$ and $q - 1 = 4m$.
4. If $n - 9 \equiv 0 \bmod 4$, then
    i. find l, m such that $4lm + l + m = N$ where $N = (n-9)/4$.
    ii. find $p - 3 = 4l$ and $q - 3 = 4m$.

For example, consider $n = 63{,}787$. Then $n - 3 \equiv 0 \bmod 4$, i.e., Step (2) is satisfied, where $N = 15{,}946$. We get $l = 70$ and $m = 56$ such that $4lm + 3l + m = N$. Hence, $p = 281$ and $q = 227$ are the prime factors of n.

5.6 Factoring Records

Although there are various integer factorization algorithms that have been proposed, so far there is no efficient integer factorization algorithm published. In recent years, the NFS method is considered as the fastest method for integer factorization, but it takes a long time to compute. As integer factorization is the core of many public key cryptosystems, so it is important to know the size of numbers which can be factored in current technology. Morrison and Brillhart (1975) categorized integer factorization record in two categories. In the first category, they consider the general composite number such as the RSA modulus (Lenstra, 2017), and in the second category, they consider the composite numbers with

TABLE 5.2

List of Some Factored RSA Moduli

Number	Digit	Year	Methods
RSA-100	100	1991	MPQS[a]
RSA-110	110	1992	NFS
RSA-120	120	1993	QS
RSA-129	129	1994	MPQS
RSA-130	130	1996	NFS
RSA-140	140	1999	NFS
RSA-155	155	1999	NFS
RSA-160	160	2003	NFS
RSA-576	174	2003	NFS
RSA-150	150	2004	NFS
RSA-200	200	2005	NFS
RSA-640	193	2005	NFS
RSA-768	**232**	**2009**	**NFS**
RSA-170	170	2009	NFS
RSA-180	180	2010	NFS
RSA-190	190	2010	NFS
RSA-704	212	2012	NFS
RSA-210	210	2013	NFS
RSA-220	220	2016	NFS

[a] Multiple Polynomial Quadratic Sieve (MPQS) is a variant of QS. Several polynomials of the form $ax^2 + 2bx + c$ are used, where a is square, $0 \leq b < a$ such that $b^2 \equiv n \bmod a$ and c is such that $b^2 - 4ac = n$. Here, n is a large composite number.

special form such as the Fermat number (i.e., $F_n = 2^{2^k} + 1, k \geq 0$) and Cunningham number (i.e., $C^{\pm}(b,k) = b^k \pm 1$, where b and k are integer, and b is not a perfect number) (Brillhart et al., 2002; Cunningham & Woodall, 1925; Lenstra, 2017).

In the first category, RSA-768 is the largest number, which is factored on December 2009 (Kleinjung et al., 2010). Table 5.2 lists some already factored RSA modulus along with the method.

In the second category, the number of the form $2^m - 1$, where $m \in$ $\{1007, 1009, 1081, 1111, 1129, 1151, 1153, 1159, 1177, 1193, 1199\}$ are factored in the year 2014 (Coppersmith, 1993; Kleinjung, Bos & Lenstra, 2014; Lenstra, 2017). Some of the numbers that are factored between 1990 and 2017 are given in Table 5.3. Note that this is not the complete list. For further details, refer to Brent (1996), Brent and Pollard (1981), Brent, Montgomery and te Riele (2000), and Brent (2000). The table also lists some special composite numbers along with their factoring method.

TABLE 5.3

List of Some Factored Composite Numbers of Special Types

Number	Digits	Year	Method
$2^{2^9}+1$	155	1990	NFS
$12^{151}-1$	163	1993	SNFS
$2^{2^{10}}+1$	309	1995	ECM
$10^{211}-1$	211	1999	SNFS
$2^{773}+1$	233	2000	SNFS
$2^{953}+1$	287	2002	NFS
$2^{809}-1$	244	2003	SNFS
$6^{353}+1$	275	2006	SNFS
$2^{1,039}-1$	313	2007	SNFS
$2^{1,061}-1$	320	2012	SNFS

5.7 Conclusion

The integer factorization problem is a core of many public key cryptosystems, and it is a challenging problem to find the factors of a large composite number. There are various methods that have been proposed for factoring a composite integer. In this chapter, we briefly discussed the various factorization methods such as Fermat's factorization method, Pollard's $p-1$ method, Pollard's ρ method, and QS and NFS methods. Recently, two new factorization methods have been proposed in the years 2011 and 2012: the first method is based on square root approximation and the other is developed to remove the drawback of the first one without any approximation.

References

Alkin, A. O. L. & Morain, F. (1993). Finding suitable curves for the elliptic curves method of factorization. *Mathematics of Computation, 60(201)*, 399–405. doi:10.1090/S0025-5718-1993-1140645-1.

Back, E. & Shallit, J. (1997). *Algorithmic Number Theory Volume I: Efficient Algorithm* (2nd Edition). London: MIT Press. ISBN-13:978–0262024051.

Balasubramaniam, P., Muthukumar, P. & Othman, W. A. B. M. (2012). Prime factorization without using any approximations. In *Proceeding P. Balasubramaniam and R. Uthayakumar (eds.) Communication in computer and information science-ICMMSC 2012*, CCIS 283, 537–541. doi:10.1007/978-3-642-28926-2_61.

Bimpikis, K. & Jaiswal, R. (2005) *Modern Factoring Algorithm*. San Diego: University of California. www.cs.columbia.edu.

Brent, R. P. (1980). An improved Monte Carlo Factorization Algorithm. *BIT, 20*, 176–184.

Brent, R. P. (1986). Some integer factorization algorithm using elliptic curves. *Australian Computer Science Communications, 8*, 149–163.

Brent, R. P. (1996). Factorization of the tenth and eleventh fermat numbers. The Australian National University. Technical report TS-CS-96-02, February 1996.

Brent, R. P. (2000). Recent progress and prospects of integer factorization algorithms. In *D.-Z. Du et al. (eds.) COCOON 2000* (LNCS 1858, pp. 3–22). Springer-Verlag.

Brent, R. P., Montgomery, P. L. & te Riele, H. J. (2000). *Factorization of Cunningham numbers with bases 13 to 99: Millennium edition.* PRG TR-14-00, 31 December 2000.

Brent, R. P. & Pollard, J. M. (1981). Factorization of the eighth Fermat numbers. *Mathematics of Computation, 36*, 627–630. doi: 10.2307/2007666.

Brillhart, J., Lehmer, D. H., Selfridge, J. L., Tuckerman, B. & Wagstaff Jr., S. S. (2002), Factorization of $b^n \pm 1$, b = 2, 3, 5, 6, 7, 10, 11, 12 up to high powers. *Contemporary Mathematics, 22*, 3rd Edition.

Brillhart, J. & Selfridge, J. (1967). Some factorizations of $2^n + 1$ and related results. *Mathematics of Computation 21(97)*, 87–96. doi: 10.2307/2003473.

Case, M. (2003). *A beginner's guide to the general number field sieve.* Oregon State University, ECE575 Data Security and Cryptography Project, 2003.

Coppersmith, D. (1993). Modifications to the number field sieve. *Journal of Cryptology, 6(3)*, 169–180. doi:10.1007/BF00198464.

Cunningham, A. J. C. & Woodall, H. J. (1925). Factorization of $y^n \mp 1$, y = 2, 3, 5, 6, 7, 10, 11, 12 up to high powers n. London: Frances Hodgson.

Kleinjung, T., Aoki, T. K., Franke, J., Lenstra, A. K., Thomé, E., Bos, J. W., Gaudry, P., Kruppa, A., Montgomery, P. L., Osvik, D. A., te Riele, H., Timofeev, A. & Zimmermann, P. (2010). Factorization of a 768-bit RSA modulus. In *Proceedings T. Rabin (ed.) Advances in Cryptology – CRYPTO 2010*, (LNSC 6223, pp. 333–350). Heidelberg: Springer.

Kleinjung, T., Bos, J. W. & Lenstra, A. K. (2014). Mersenne factorization factory. In *Proceedings P. Sarkar and T. Iwata (eds.), Advances in Cryptology – ASIACRYPT 2014, Part I* (LNCS 8873, pp. 358–377). Heidelberg: Springer.

Lehman, R. S. (1974). Factoring large integer. *Mathematics of Computation, 28*, 637–646. doi:10.2307/2005940.

Lenstra, A. K. (2000). Integer factorization. *Designs, Codes and Cryptography, 19*, 101–128. doi:10.1023/A:1008397921377.

Lenstra, A. K. (2017). General Purpose Integer Factoring. In J. Bos & A. Lenstra (Eds.), *Topics in Computational Number Theory Inspired by Peter L. Montgomery* (pp. 116–160). Cambridge: Cambridge University Press. doi: 10.1017/9781316271575.006.

Lenstra, A. K. & Lenstra Jr., H. W. (1993). The development of the number field sieve. *Lecture Notes in Mathematics, Vol. 1554.* Springer-Verlag. ISBN: 978-3-540-47892-8.

Lenstra, A. K., Lenstra Jr., H. W., Manasse, M. S. & Pollard, J. M. (1993). The number field sieve. In *Proceedings A. K. Lenstra and H. W. Lenstra (eds.), The Development of the Number Field Sieve* (Lecture Notes in Mathematics, 1554, pp. 11–42). Berlin, Heidelberg: Springer.

Lenstra Jr., H. W. (1987). Factoring integers with elliptic curves. *Annals of Mathematics, 126*, 649–673. doi: 10.2307/1971363.

Montgomery, P. L. (1987). Speeding the Pollard and elliptic curve methods for factorization. *Mathematics of Computation, 48*, 243–264. doi: 10.1090/S0025-5718-1987-0866116-7.

Montgomery, P. L. (1992). An FFT extension of the elliptic curve method of factorization. Ph.D. dissertation, Mathematics, University of California at Los Angeles.

Montgomery, P. L. (1994). A survey of integer factorization algorithms. *CWI Quarterly, 7*, 337–366.

Morrison, M. A. & Brillhart, J. (1975). A method of factoring and the factorization of F_7. *Mathematics of Computation, AMS, 29(129)*, 183–205. doi:10.1090/S0025-5718-1975-0371800-5.

Pollard, J. M. (1974). Theorems on factorization and primality testing. *Proceedings of the Cambridge Philosophical Society, 76*, 521–528.

Pollard, J. M. (1975). A Monte Carlo method for factorization. *BIT, 15*, 331–334.

Pomerance, C. (1982). Analysis and comparison of some integer factorization algorithms. In *Proceedings H. W. Lenstra Jr. and R. Tijdeman (eds.), Computational Methods in Number Theory, Part I*, Mathematical Centre Tracts, Amsterdam: Mathematisch Centrum, (154, pp. 89–139).

Pomerance, C. (1985). The quadratic sieve factoring algorithm. In *Proceedings T. Beth, N. Cot and I. Ingemarsson (eds.), Advances in Cryptology – EUROCRYPT 1984* (LNCS 209, pp. 169–182). Berlin: Springer.

Pomerance, C. (2008). Smooth number and quadratic sieve. *Algebraic Number Theory, 44,* 69–81.

Rivest, R. L., Shamir, A. & Adleman, L. (1978). A method for obtaining digital signature and public-key cryptography. *Communications of the Association for Computing Machinery, 21(2),* 120–126. doi:10.1145/359340.359342.

Seibert, C. (2011). Integer factorization using quadratic sieve. In *Proceedings of the 44th Annual Midwest Instruction and Computing Symposium 2011,* 8–9 April 2011. www.micsyposium.org.

Silverman, R. D. & Wagstaff, Jr., S. S. (1993). A practical analysis of the elliptic curve factoring algorithm. *Mathematics of Computation, 61,* 445–462. doi:10.2307/2152967.

Stinson, D. R. (2013). Cryptography Theory and Practice (3rd Edition). Chapman & Hall/CRC. ISBN-13:978-1-58488-508-5.

Zalaket, J. & Hajj-Boutros, J. (2011). Prime factorization using square root approximation, *Computer and Mathematics with Applications, 61,* 2463–2467. doi:10.1016/j.camwa.2011.02.027.

6

Symmetric Algorithms I

Faheem Syeed Masoodi
University of Kashmir

Mohammad Ubaidullah Bokhari
Aligarh Muslim University

CONTENTS

6.1 Introduction 80
    6.1.1 Background of Symmetric Key Cryptography 80
    6.1.2 Classification of Symmetric Key Cryptography 81
    6.1.3 Confusion and Diffusion 81
    6.1.4 Substitution Box (S-Box) 81
6.2 Block Ciphers 82
    6.2.1 Modes of Operation 82
        6.2.1.1 Electronic Code Book 82
        6.2.1.2 Cipher Block Chaining Mode 82
        6.2.1.3 Cipher Feedback Mode 83
        6.2.1.4 Output Feedback Mode 83
        6.2.1.5 The Counter Mode 83
    6.2.2 Pseudorandom Number Generation 83
6.3 Stream Ciphers 84
    6.3.1 Synchronous and Self-Synchronous Stream Ciphers 84
    6.3.2 Linear Feedback Shift Register 85
    6.3.3 Nonlinear Filter 86
    6.3.4 Clock Controlled Generators 87
6.4 Data Encryption Standard 87
    6.4.1 How DES Works? 88
    6.4.2 What Happens in Each Round? 88
    6.4.3 The Feistel Function (f) 88
6.5 Triple DES (3DES) 89
    6.5.1 How It Works? 89
6.6 Advanced Encryption Standard 90
    6.6.1 How It Works? 90
    6.6.2 What Happens in Each Round? 91
6.7 International Data Encryption Algorithm 91
    6.7.1 How It Works? 91
    6.7.2 What Happens in Each Round? 92
6.8 Word Auto Key Encryption 93
    6.8.1 How It Works? 93

6.9   Conclusion 94
References..94

6.1 Introduction

6.1.1 Background of Symmetric Key Cryptography

The primary objective of securing the data transmission is that it should be processed in such a manner that it gets difficult for an unintended receiver to be able to discern its true or intended meaning. This is what broadly cryptology does. The very term *cryptology* has a Greek etymology, stemming from the words *kryptos*, meaning "hidden", and *logos*, meaning "Study" (Katz, Menezes, Van Oorschot, & Vanstone, 1996).

Traditionally, cryptology would allude to the study of the standards and methods by which certain information could be hidden in ciphers and subsequently uncovered by authorized individuals using the private key. The practice has now evolved to encompass the entire matrix of key-driven conversion of data into forms that are either perplexing or computationally infeasible to copy or undo for unauthorized users. Cryptology is broadly classified into cryptography and cryptanalysis. The former deals with the design of cryptographic primitives, and the latter investigates breaking of cryptographic algorithms.

The science of encrypting a message in an incomprehensible form with the use of some dynamic key text and subsequent decryption of this incomprehensible text back into the actual message, with the objective of securing information from potential attackers, is an extended applied description of cryptology (Katz & Lindell, 2014). Put simply, the transformation of actual information (indented message) to ciphertext—which is coded in an incomprehensible form so as to keep the content hidden from unintended users—is termed as *encryption*, and the process of transforming the ciphertext into actual information is called *decryption*. The need for encryption and decryption comes into play because of the potentially insecure channel over which the sender transmits the message to the intended recipient. The objective of cryptography is not to hide the existence of the message, but rather to conceal its intended meaning (Mao, 2003).

Encryption is generally subdivided into two types. The first is a symmetric key or private key—in which identical key is employed for both the operations of encrypting and decrypting information; the second is an asymmetric key or public key introduced in the year 1976. This employs the practice of using unidentical keys for *encryption* and subsequent decryption. The knowledge of encryption key is implicitly or explicitly equivalent to knowing the decryption key. These algorithms are typically fast and much more efficient than asymmetric algorithms in both hardware and software (Paar & Pelzl, 2009).

Symmetric encryption gives a level of authentication since the information is encoded with one symmetric key; it cannot be decoded with some other key. In this manner, until the primitive is kept a secret between the two imparting parties who are involved in communication, both the sender and receiver can be sure that it is communicating with the intended partner as long as the decoded text is meaningful. The trade-off though with the symmetric key cryptography is the *key security*, given that an identical key is employed for both the encryption and decryption operations. It is imperative that the key is conveyed in a safe way prior to correspondence and known at both sender and receiver ends (Stinson, 2005).

6.1.2 Classification of Symmetric Key Cryptography

Secure file transfer protocols make use of symmetric key primitives to encrypt data. These primitives are categorized into two types: *block ciphers* and *stream ciphers*.

Block cipher, one of the popular methods of concealing information, works on the principle of dividing the input stream of plaintext into fixed-size strings called blocks, typically 64, 128, or 256 bits long. Stream ciphers usually act on individual elements of the message. A private key and a randomized seed value called initialization vector (IV) are passed as input to the stream cipher, and a sequence of bits known as keystream is generated as output. This keystream is then XORed with the plaintext to compute the ciphertext (Schneier, 2007).

6.1.3 Confusion and Diffusion

Shannon (1949) introduced the concept of two primitive operations that can lead to a specific end goal of accomplishing strong encryption. The two operations termed as *Confusion* and *Diffusion* were introduced in his work "Communication Theory of Secrecy Systems".

  i. **Confusion**: Shannon defined confusion as an encryption task in which the correlation within key and ciphertext is as intricate as attainable. The idea is to entangle key discovery even when a large amount of ciphertext is generated from plaintext using the same key. Typically change in a single bit of key should alter the entire stream of ciphertext.
  ii. **Diffusion**: Encryption activity in which the impact of input bits is dissipated over numerous output bits with the objective that any change in plaintext should alter entire ciphertext in a pseudorandom manner. The idea is to dissipate the statistical properties of plaintext into ciphertext.

In the design of symmetric encryption systems, Substitution and Permutation network is a typical component to accomplish both confusion and diffusion. The combination of these two basic functions in multiple rounds is employed to provide for a maximum level of confusion and diffusion and thereby achieve a reasonably fair level of secrecy.

6.1.4 Substitution Box (S-Box)

The S-boxes play a significant role in symmetric key cryptography and their primary goal is to render the necessary confusion. They work on the principle of mapping an input of size m bits into an output of size n bits, where n and m need not be equivalent. The idea is to complicate the connection between input and output bits. Generally, for an $m \times n$ S-Box construction, lookup table with a total of $2m$ strings of n bits each is employed (William, 2006). An important attribute associated with S-Box is its size and there is a trade-off in terms of size between implementation complexity and resistance to attacks. More often than not, larger S-boxes are cryptanalytically secure but challenging to design. The table may either be static in which case some pre-computed values are used or they may be dynamic in which case the values change during execution (Bokhari & Masoodi, 2012).

6.2 Block Ciphers

A block cipher works on the principle of dividing the input stream of plaintext into fixed-size strings called blocks. The conventional size of a block cipher block is 64, 128, or 256 bits. Block cipher remains one of the popular methods of concealing information (Lai & Massey, 1990). If the plaintext length is shorter than the size of the block, then the padding technique is used. This is to ensure that the data are appended to the actual plaintext, such that the resulting plaintext is partitioned into pieces that are multiple of block size. The resultant blocks are then encrypted one at a time, using a cryptographic key to produce the ciphertext and there is no correlation between the encrypting of one message block and another. Finally, the plaintext text blocks of specific length are converted into their corresponding ciphertext blocks of equal length.

6.2.1 Modes of Operation

A cryptographic mode refers to the way in which plaintext can be encrypted into ciphertext by combining the basic cipher, some sort of feedback, and some elementary operations (Dworkin, 2001). Multiple methods of encrypting a large set of plaintext with a block cipher have been proposed. In this chapter, we discuss the following five modes of operation.

6.2.1.1 Electronic Code Book

As the name suggests, this mode can be perceived as a large codebook that replaces each input by a distinct output. ECB is the most classical way of encryption (decryption) in which the message blocks are encrypted independently of one another with a key by the source. These message blocks are typically of size b bits and in case of message exceeding b bits, the entire message is divided into blocks and the last block is padded in case it is not multiple of b. Each such block is encrypted and transmitted individually to the decrypting device, which uses the same key to decrypt. The advantage with this mode is that synchronization is not mandatory, i.e., in case some blocks are lost during transmission, the received blocks can still be decrypted. Also if some errors occur during transmission, they will only affect the corresponding block and no other blocks. The drawback with this mode is that the identical message block will be encrypted to identical ciphertext as long as key remains the same, but with different keys, the output will be different.

6.2.1.2 Cipher Block Chaining Mode

The objective here is to overcome the limitations faced in ECB mode by chaining together the encryption of all blocks in a way that each cipher block is reliant on the respective plaintext block as well as all preceding plaintext blocks. The underlying idea is to feedback the previously generated ciphertext as input to the algorithm and XOR it with next message block, which is then encrypted to generate next ciphertext block and it continues till all the message blocks are encrypted. Pertinently no ciphertext for feedback is available here for the first message block. CBC mode takes care of this by adding an IV to the first message block. The initialization vector is a randomized seed value and plays a vital role in making CBC encryption random.

6.2.1.3 Cipher Feedback Mode

The CFB mode of operation utilizes the block cipher to implement stream cipher for the generation of a sequence of pseudorandom bits, and the ciphertext is produced by doing an XOR operation between output bits and plaintext. The first keystream block is generated by initially filling the input shift register with a seed value known as an initialization vector (IV) and for subsequent keystream blocks, the previously generated ciphertext blocks are supplied back as an input to the block cipher. The mode of operation is similar to OFB except here ciphertext is used as feedback instead of block cipher output. The primary advantage of this mode is that in case of short plaintext blocks, a variant of this mode can be used in which the high-order s bits of block cipher output are XORED with the plaintext to generate ciphertext. This ciphertext is then fed back to block cipher and current s MSB bits are shifted out and the process continues. On the flipside, any bit error that occurs during the course of communication is propagated.

6.2.1.4 Output Feedback Mode

The idea behind OFB mode of operation is to feed the output of block cipher back to it and unlike CFB mode, the keystream generated does not depend on plaintext or ciphertext. The input text is then mixed with the output keystream using an XOR operation to compute the corresponding ciphertext. Since no output keystream feedback is available for the first set of keystream bits, an initialization vector (IV) is required as the initial random n-bit input block. Initially, an IV is encrypted with a block cipher to generate the first set of keystream bits, and the next block of keystream is computed by feeding previously generated keystream back to the block cipher. OFB mode operates in a similar fashion to that of a stream cipher, and both encryption and decryption operations are the same as in the case of a stream cipher. Also like CBC mode, the use of IV makes OFB encryption nondeterministic. Since no chaining dependencies are there in OFB mode, the decryption of subsequent blocks is not affected even if bit errors occur during transmission.

6.2.1.5 The Counter Mode

The counter mode of operation also uses block cipher as a means of achieving a stream cipher function. Similar to OFB and CFB, keystream is generated in terms of blocks, but unlike both of these modes, no feedback is required in counter mode. In this mode, the underlying algorithm is fed with a counter value which counts from a starting value and takes a different value every time a new keystream block is generated. This keystream is then used to generate ciphertext by doing an XOR operation between input text and keystream. The counter, in this case, may be any function with a large period, but the most popular is one that simply increments by one. CTR mode does not suffer the limitations like error propagation and chaining dependencies which were present in the earlier discussed modes (Lipmaa, Rogaway, & Wagner, 2000). Another advantage of CTR mode is that it can significantly increase the speed by running two or more block ciphers in parallel.

6.2.2 Pseudorandom Number Generation

Random number generator is considered as one of the primary building blocks for cryptographic system essentially for the logic that they are quite unpredictable for potential attackers. Uniformly distributed random numbers are pivotal to cryptographic applications, and even though the values generated by random processes are very unpredictable, they lack uniform distribution (Random.org, 2018). Pseudorandom number generators (PRNGs) generate sequences commonly known as pseudorandom numbers by starting with a short random seed value and then recursively expand it into a large set of random-looking sequences. It is important to mention here that PRNGS are completely deterministic in nature, implying that if the generator is put back into the same state, the output sequence will be a replica of the earlier generated sequence (Van Tilborg, 2005). One of the primary requirements of these pseudorandom numbers is that they carry good statistical properties and a reasonable number of mathematical tests verify their statistical behavior. Owing to their good statistical properties, PRNGs are widely used in stream ciphers (Goldreich, 1998). A good PRNG must carry statistical properties that resemble the properties of a true random number generator, and the size of the seed value must be long enough to resist exhaustive key search attack.

6.3 Stream Ciphers

Stream ciphers usually act on individual elements of the message. A private key and a randomized seed value called initialization vector (IV) are passed as input to the stream cipher and a sequence of bits known as keystream is generated as output. This keystream is then XORed with the plaintext to compute the ciphertext. The aim is to construct a cryptographically secure primitive that is optimized for speed, easy to execute, no short cycles, and deduction of internal state from the output should be impractical. Stream ciphers on the basis of an internal state are commonly subdivided into synchronous and self-synchronous stream ciphers (Bokhari & Masoodi, 2010).

6.3.1 Synchronous and Self-Synchronous Stream Ciphers

Given how the keystream is generated, stream ciphers are generally categorized into two types. If the generated keystream is not dependent on plaintext and ciphertext, then the cryptosystem is termed as synchronous and if the keystream generation is dependent on plaintext and several bits of previously generated ciphertext then the cryptosystem is referred to as self-synchronized stream ciphers. Majority of the functional stream ciphers are synchronous in nature while as cipher feedback mode (CFB) is an example of self-synchronous. Figure 6.4 depicts the encryption and decryption of synchronous stream cipher, and if the dotted line is present, then it represents self-synchronous stream cipher.

In order to allow accurate decryption in synchronous stream ciphers, both parties need to be synchronized in terms of key and state and lost synchronization may be regained by using techniques like sufficient redundancy in plaintext, re-initialization, or using different stamps at fixed intervals within ciphertext, while as in the case of self-synchronous stream ciphers, lost synchronization is regained automatically and only a limited number of plaintext symbols are lost.

Also with synchronous stream ciphers, there is no error propagation since the modified bit has no effect on other ciphertext bits. In the case of self-synchronous stream ciphers, there is limited error propagation as the modified bit affects a certain number of other ciphertext bits.

6.3.2 Linear Feedback Shift Register

In stream cipher design, LFSRs have received a wide acknowledgment primarily due to their inherited characteristics including extensive period, desired statistical properties, and low implementation costs (Masoodi, Alam, & Bokhari, 2012). An LFSR is a shift register that consists of multiple cascaded storage units (*flip-flops*) and using feedback, cycles the bits on each clock tick. The aim is to iteratively generate a sequence of pseudorandom numbers by performing exclusive-OR (or exclusive-NOR) operation on chosen bits (taps) and blend them with input. An LFSR is said to have generated a maximal sequence (*m*-sequence), if it can generate every possible unique value (precluding an all 0's state in exclusive-OR and an all 1's state in exclusive-NOR) before arriving at its starting state.

LFSRs corresponding to primitive polynomials are capable of generating maximum length sequences, while as LFFSRs corresponding to irreducible polynomials do not generate *m*-sequence but the length of the generated bit stream is independent of the starting value LFSR and the third kind of LFSRs corresponding to reducible polynomials doesn't generate an *m*-sequence and also the length of the generated sequence is dependent on the starting values of the LFSR. Pertinently irreducible polynomials refer to the type of polynomials that cannot be factored and all primitive polynomials in addition to satisfying some mathematical conditions are also irreducible polynomials of degree n, where n represents the length of the register. An important property associated with Linear feedback shift registers is linear complexity and can be determined by the Berlekamp–Massey algorithm (Upadhyay, Sharma, & Sampalli, 2016).

Example 1

Let's consider two XOR-based LFSRs of degree 4 with different tap selection. The four flip-flops are driven by a common clock *CLK* and are denoted as q_0, q_1, q_2, and q_3. At every clock pulse, the value of the LFSR is shifted to right by one and the value at q_3 is the taken as the output bit. As shown in Tables 6.1 and 6.2, both the LFSRs are initialized with the same value of {1, 1, 1, 1} but after the third clock pulse, the sequences deviate.

TABLE 6.1

Sequence of States for LFSR 1

Clk	q_0	q_1	q_2	q_3	Val
0	1	1	1	1	F
1	0	1	1	1	E
2	1	0	1	1	D
3	0	1	0	1	A
4	1	0	1	0	S
5	1	1	0	1	B
6	0	1	1	0	6
7	0	0	1	1	C
8	1	0	0	1	9
9	0	1	0	0	2
10	0	0	1	0	4
11	0	0	0	1	8
12	1	0	0	0	1
13	1	1	0	0	3
14	1	1	1	0	7

TABLE 6.2

Sequence of States for LFSR 2

Clk	q_0	q_1	q_2	q_3	Val
0	1	1	1	1	F
1	0	1	1	1	E
2	0	0	1	1	C
3	0	0	0	1	8
4	1	0	0	0	1
5	0	1	0	0	2
6	0	0	1	0	4
7	1	0	0	1	9
8	1	1	0	0	3
9	0	1	1	0	6
10	1	0	1	1	D
11	0	1	0	1	A
12	1	0	1	0	5
13	1	1	0	1	B
14	1	1	1	0	7

(a)

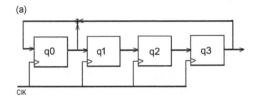

(b)

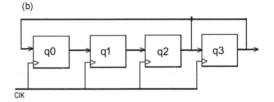

FIGURE 6.1

(a) LFSR with degree 4 and tap at q_0. (b) LFSR with degree 4 and tap at q_2.

This deviation happens primarily given the manner in which different taps have been selected to compute new LSB values (see Figure 6.1a,b).

An LFSR of degree 4 can generate a maximum of ($2^4 - 1 = 15$) different sequences before it starts repeating itself. In general, for an n-bit LFSR, the maximum number of sequences that can be generated is $2^n - 1$ and is referred to as the period of that LFSR. The decision of keeping a feedback enabled or disabled is defined by the feedback coefficients. If the coefficient value is "1", the feedback is active, and a value of "0" indicates feedback is disabled.

The corresponding sequence of states of both these LFSRs is given in Tables 6.1 and 6.2.

Cryptographically Linear feedback shift registers are insecure on the grounds that the design of an LFSR of degree n can effectively be found by examining 2^n successive bits of its output with the help of the Berlekamp–Massey calculation (Upadhyay, Sharma, & Sampalli, 2016). Despite the fact that the satisfying statistical properties are important, they are not adequate to declare that a stream cipher is not vulnerable to potential attacks.

6.3.3 Nonlinear Filter

In an attempt to defeat the issue of linearity in LFSRs, a number of alternatives have been proposed. *Nonlinear filter* is one such variation and is based on the principle of generating

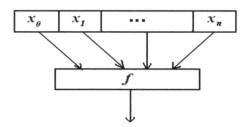

FIGURE 6.2
Nonlinear filter generator.

keystream as a nonlinear function f of the state of the LFSR. In contrast to combination generators, wherein multiple LFSRs are tied together in parallel fashion, the Nonlinear filter works on the principle of computing output by a nonlinear function of some taps of a single LFSR (see Figure 6.2) (Braeken & Lano, 2005, August). The nonlinear function f is known as the filter function.

The quandary with sequences generated by nonlinear feedback shift registers is that there may be bias (more 1's than 0's), smaller period than expected and the different period for variant initial values.

6.3.4 Clock Controlled Generators

The clock-controlled generators are an alternate way of generating keystream with good statistical properties. These kinds of generators are built using two registers, and the movement of data of one register is controlled by the output of another LFSR. The register responsible for controlling the clock is called as a control register, and denoted as CR and the register that generates keystream according to the output sequence of CR is called as generator register, and denoted as GR (Figure 6.3).

6.4 Data Encryption Standard

DES was introduced by IBM in 1970 and was based on a cryptographic cipher "Lucifer" developed by Horst Feistel. DES is a block cipher that takes a fixed-size string of plaintext as input and performs some operations on the input to generate the ciphertext of the same length (Thakur & Kumar, 2011). DES block size is 64 bits which undergoes 16-round structure to generate the ciphertext. DES operates at the key of size 56 bits out of which 8 bits are taken as parity bits leaving the actual key length as 48 bits only.

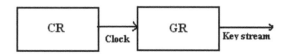

FIGURE 6.3
Clock-controlled generator.

6.4.1 How DES Works?

The input plain text undergoes an initial permutation in which the 64-bit plain text is interchanged (bit position interchange). The permuted text is then passed to a series of 16 rounds with different keys. The different keys generated for these 16 rounds are generated from the same 48 bits of the key, which is then permuted in a unique way to generate the 16 keys for 16 rounds. The output of the 16th round is then permuted again, i.e., a reverse of initial permutation is done to get the final result (ciphertext) (see Figure 6.4a).

6.4.2 What Happens in Each Round?

The input 64-bit plain text is divided into two blocks of 32 bits: right and left halves (see Figure 6.4b). The fundamental operations on the information are enveloped into what is alluded to as the cipher function and are marked as "*f*". This function takes two distinct inputs of size 32 bits (right half) and 48 bits (key) and yields a solitary 32-bit number. Different operations are performed on data and key but in a parallel fashion.

The 32-bit value generated by function *f* is then XORed with 32 bits (Left half) and is given as input (right half) to the next round. The left half for next round is simply the right half (unchanged 32-bit sequences) from the previous round.

6.4.3 The Feistel Function (*f*)

The Feistel function comprises of four stages and typically operates on 32 bits at one point of time as shown in Figure 6.4c.

1. **Expansion**: In this stage, the 32-bit block is widened to 48 bits with the help of an expansion permutation. The generated output comprises of 8 chunks with each chunk accommodating 6 bits. Out of these six bits, four bits (central bits) come

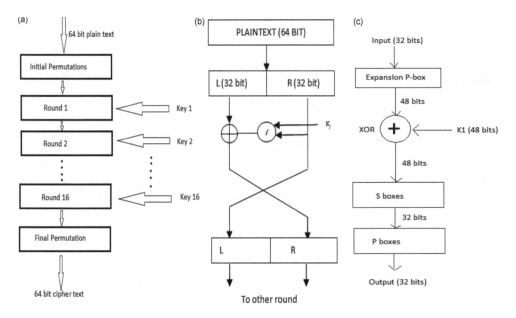

FIGURE 6.4
(a) DES structure. (b) A single round. (c) The Feistel (*F*) function.

from corresponding input bits and the other two bits (first and last) are taken from the last bit of the previous chunk and the first bit of the next chunk, respectively.

2. **Key mixing**: An XOR operation is performed on the output from the expansion stage and the sub-key that is derived from the main key. There are a total of 16 sub-keys of length 48-bit each, and every round uses one such sub-key.

3. **Substitution**: The output from key mixing stage is passed on to a substitution box and a total of 6×4 substitution boxes are used. Each S-Box is given a six-bit input, and a corresponding four-bit output is computed with the help of a lookup table. The S-Box is very critical to the security of DES, as it plays a very vital role in the nonlinearity of this primitive.

4. **Permutation**: Permutation forms the last stage, and the idea here is to dissipate the output of each S-Box across four different S-Boxes in the next round. A permutation box (P-Box) is used to rearrange the output generated by Substitution phase.

6.5 Triple DES (3DES)

One of the major weaknesses found in Data Encryption Standard (DES) is in its key size. With a key size of 56 bits, an adversary can analyze all possible combinations within a very short period of time. Multiple solutions have been proposed to overcome this problem of DES, and one such popular solution is triple DES (3-DES) (William, 2006).

6.5.1 How It Works?

3-DES increases the key size by repeating the DES algorithm thrice with three different keys of DES keys, which leads to an aggregate key size of 168 bits. The triple data encryption algorithm employs three different 56-bit DEA keys (k_1, k_2, and k_3) in an EDE mode. The basic idea here is to use key k_1 to encrypt (E) the message followed by a decryption (D) using k_2 and finally one more encryption (E) using key k_3 (Smid, 2000). The three keys used may either be completely independent of each other or else the first and third keys may be the same and the second key is different. Based on the configuration used, we may either have a key space of 168 or 112 bits, respectively.

Let's assume (k_1, k_2, and k_3) represent the three keys, also known as a key bundle. If $E_k(x)$ and $D_k(x)$ denote the encryption and decryption operations, respectively, x is the input message and C represents the corresponding ciphertext. Then, the encryption and decryption operations are defined as follows:

$$\text{Encryption:} \quad C = E_{k_3}\left(D_{k_2}\left(E_{k_1}(x)\right)\right)$$

$$\text{Decryption:} \quad x = D_{k_3}\left(E_{k_2}\left(D_{k_1}(C)\right)\right)$$

The input message x is first encrypted with a function $E_{k_1}(x)$ using key k_1, and the output is decrypted using key k_2 and then again encrypted using key k_3 and the decryption operation is exactly the reverse of encryption operation. 3DES with mutually independent three keys are considered to be much secure in comparison to DES and provides a security of 168-bits against brute force attack.

6.6 Advanced Encryption Standard

AES is a symmetric block cipher that won an open submission call issued by NIST in January 1997 (Smid, 2000). A total of 15 candidate ciphers were accepted in the first round and five finalists were chosen. A subset of Rijndael (Daemon & Rijmen, 2002) primitive submitted by Daemen & Rijmen was found to be the most fitting candidate and therefore selected to be the Advanced Encryption Standard. The block length and the key length of Rijndael are at least 128 bits but in compliance to the adopted FIPS standard, NIST set the length of block to 128 bits and three variant keys of length 128 bits or 192 bits or 256 bits—thus making up three variants: AES(128), AES(192), and AES(256).

6.6.1 How It Works?

The algorithm consists of 10 rounds (12 for AES(192) and 14 in the case of AES(256)). Every round having its own key obtained from the original one by a key generator. The plaintext block of 128 bits forms the input for the first round leaving and output of 128 bits which forms the input for the next round, and so on. Each of the rounds consists of four basic steps as follows:

1. The ByteSub
2. The ShiftRow
3. The MixColumn
4. Add RoundKey

We have chosen to show the structure for AES-128. The structure for the other two versions is the same, but with a more number of rounds. To start with, both the plaintext block as well as the key is partitioned into smaller 16-bit blocks.

The steps can be enumerated as follows:

- Expand 16-byte key into key blocks: Input of the algorithm is the key and the plain text block. Here the size of the key is 16 bytes. These 16 bytes are expanded into 11 two-dimensional arrays consisting of four rows and columns. The first array is used in the initialization process, and other 10 arrays are used in each round.
- Convert 16-bytes plain text block into a state: Here, a 16-byte input message block is transcribed into a state. A state is a two-dimensional 4×4 array. The input text block is copied into the state in the column order. That means first 4 bytes of input message get copied into the first column of state and second 4 bytes of input in the second column and so on.
- XORing state with a key block: In this step, first 16 bytes of the extended key and 16-byte state array are XORed. Thus, every byte in the state array is replaced by the XOR of itself and the corresponding byte in the expanded key.

This follows 10 rounds with identical steps, which at the end result in a 128-bit ciphertext block.

6.6.2 What Happens in Each Round?

The following four steps are carried out in each round:

i. Byte-Substitution: Here a state array is substituted byte-by-byte by looking up into the S-box. Only one S-box is used irrespective of multiple S-boxes used in DES.

ii. Shift Rows: Each row in state array is shifted to the left. The first row is shifted by 0 bytes, the second row 1 byte and so on till the fourth row.

Original array				Modified array			
1	5	9	13	1	5	9	13
2	6	10	14	6	10	14	2
4	7	11	15	11	15	3	7
4	8	12	16	16	4	8	12

iii. Mix Columns: In this step, matrix multiplication is done. This is done to create new state array for the next round. Every column of state array is multiplied with a 4×4 matrix to compute a new state array. Remember this step is not performed toward the final step as we do not need new state array.

iv. Add Round Key: Here, the 16-byte state matrix is XORed with 16 bytes of the sub-key. If this is the last round, then these 16 bytes are considered as ciphertext. Otherwise, these 16 bytes are used in another round.

6.7 International Data Encryption Algorithm

IDEA (Lai & Massey, 1990) was introduced by Lai and Massey in 1991. The primitive is actually a modification of Proposed Encryption Standard that was proposed in 1990 to counter the limitations of DES.

6.7.1 How It Works?

IDEA operates on blocks of size 64 bit, and each such block is partitioned into four 16 bit sub-blocks. The encryption operation comprises a total of nine rounds on plaintext blocks, and during round one, all four of these sub-blocks are operated upon by some arithmetic and logical operations. The output of each round is fed as input to the next round, and the same sequence of operations is performed up to round eight. Round nine, also called as the OUTPUT TRANSFORMATION phase, is the final round, and only arithmetic operations are performed in this round. The output of round nine is the resultant ciphertext of the corresponding plaintext.

The 128-bit key of IDEA is divided into eight 16-bit sub-keys and six of these sub-keys are used in round one and the remaining two are used in round two. For subsequent rounds, a left cyclic shift of 25 bits on the 128-bit key results in another 128-bit block, which is then partitioned to form the keys. This process is repeated for all the eight plus one rounds. In the entire encryption process, a total of 52 keys are used, six sub-keys per round for the first eight rounds, and the final four sub-keys for the output transformation phase. The decryption process remains the same as the encryption process except that the sub-keys are used in an inverted manner at each round to get back the original plain text.

6.7.2 What Happens in Each Round?

The entire encryption algorithm can be broken up into 14 steps. For the eight complete rounds, the 64-bit plaintext is divided into four 16-bit sub-blocks M_1, M_2, M_3, and M_4. Each complete round requires six sub-keys. The 128-bit key is split into eight 16-bit blocks, which become eight sub-keys. The first six are used in round one and the remaining two in round two. Each round uses each of the three algebraic operations: bitwise XOR, addition modulo 2^{16}, and multiplication modulo $2^{16} + 1$. Table 6.3 illustrates the 14 steps of a complete round (^ indicates bitwise XOR, * indicates multiplication modulo $2^{16} + 1$, + indicates addition modulo 2^{16} and R_i stores the results of various steps).

For every round except the final transformation, a swap occurs and the input to the next round is: result of step 11 || result of step 13 || result of step 12 || result of step 14, which then becomes $M_1 \| M_2 \| M_3 \| M_4$, the input for the next round.

After round 8, as shown in Table 6.4, a ninth "half-round" final/OUTPUT transformation occurs.

The concatenation of the blocks is the output.

Decryption: The computational process used for decryption of the ciphertext is essentially the same as that used for encryption of the plaintext. The only difference compared with encryption is that during decryption, different 16-bit key sub-blocks are generated. More precisely, each of the 52 16-bit key sub-blocks used for decryption is the inverse of the key sub-block used during encryption in respect of the applied algebraic group operation. Additionally, the key sub-blocks must be used in the reverse order during decryption in order to reverse the encryption process.

TABLE 6.3

Fourteen Steps of One Complete Round

Step No.	Corresponding Operation	Step No.	Corresponding Operation
1	$R_1 = M_1 * K_1$	8	$R_8 = R_6 + R_7$
2	$R_2 = M_2 + K_2$	9	$R_9 = R_8 * K_6$
3	$R_3 = M_3 + K_3$	10	$R_{10} = R_7 + R_9$
4	$R_4 = M_4 * K_4$	11	$R_{11} = R_1 \wedge R_9$
5	$R_5 = R_1 \wedge R_3$	12	$R_{12} = R_3 \wedge R_9$
6	$R_6 = R_2 \wedge R_4$	13	$R_{13} = R_2 \wedge R_{10}$
7	$R_7 = R_5 * K_5$	14	$R_{14} = R_4 \wedge R_{10}$

TABLE 6.4

Four Steps of One Complete Round

Step No.	Corresponding Operation	Step No.	Corresponding Operation
1	$R_1 = M_1 * K_1$	3	$R_3 = M_3 + K_3$
2	$R_2 = M_2 + K_2$	4	$R_4 = M_4 * K_4$

The fundamental criteria for the development of IDEA were military strength for all security requirements and easy hardware and software implementation. The algorithm is used widely and is used in various banking and industry applications. Due to its strength against cryptanalytic attacks and due to its inclusion in several popular cryptographic packages, IDEA is widely used (Meier, 1993, May).

6.8 Word Auto Key Encryption

David Wheeler in 1993 introduced a new cryptographic primitive named *Word Auto Key Encryption* (WAKE) (Pudovkina, 2001). It is a self-synchronous stream cipher that operates in CFB and essentially relies on three primitive operations: addition, shifting, and XOR. The keystream is generated in terms of blocks of size 256 and the preceding ciphertext block is used to generate the new block of keystream.

6.8.1 How It Works?

An S-Box "T" with 2^n entries is central to the design of WAKE stream cipher. Each entry in this table is a $4n$-bit word with a special attribute that the most significant n of $4n$ is a permutation of all the available n-bit words and rest of the $3n$ words are generated randomly. The initial values of the $4n$-bits $\{a_0, b_0, c_0, d_0\}$ are set up with the key. At any point of time t, the new values of $\{a, b, c, d\}$ are computed using an invertible transformation function S as illustrated here:

$$a_{\text{new}} = S\left(a_{\text{prev}}, y_{\text{prev}}, T\right)$$

$$b_{\text{new}} = S\left(b_{\text{prev}}, a_{\text{new}}, T\right)$$

$$c_{\text{new}} = S\left(c_{\text{prev}}, b_{\text{new}}, T\right)$$

$$d_{\text{new}} = S\left(d_{\text{prev}}, c_{\text{new}}, T\right)$$

where Y and T represent the ciphertext and S-BOX, respectively, and S is the invertible transformation function defined as

$$S(x, y, T) = (x + y) \gg n \oplus T\left[(x + y) \bmod 2^n\right];$$

where $\gg$ and $\oplus$ are a shift and XOR operations, respectively.

The d_{new} represents the final output, and ciphertext is produced by doing an XOR between plaintext and d_{new}. Similarly, at the decryption end, plaintext can be produced by doing an XOR between ciphertext and d_{new}.

The cipher security relies on a repeated table using a large state space. The table and the initial constant are generated from the secret key. Since the block size is too large, it is not suitable for applying in real-time communications.

6.9 Conclusion

This chapter introduced the basic concepts of symmetric key cryptography. At first, the classification of 11symmetric encryption is given—block and stream cipher, followed by an explanation of various modes of operation for block ciphers. A detailed description of various stream cipher building blocks is presented and the focus of the discussion has been the different properties associated with random number generation. Linear feedback shift registers role as m-sequence generator in stream cipher design and various alternatives to LFSR are presented. Finally, a detailed description of design and functioning of various prominent cryptographic primitives including DES, 3DES, Advanced Encryption System (AES), International Data Encryption algorithm (IDEA) and WAKE is given at the end of the chapter.

References

Bokhari, M. U., & Masoodi, F. (2010, February). Comparative analysis of structures and attacks on various stream ciphers. In *Proceedings of the 4th National Conference*, India.

Bokhari, M. U., & Masoodi, F. (2012, October). BOKHARI: A new software oriented stream cipher: A proposal. In *Information and Communication Technologies (WICT), 2012 World Congresson* (pp. 128–131). IEEE, India.

Braeken, A., & Lano, J. (2005, August). On the (im) possibility of practical and secure nonlinear filters and combiners. In *International Workshop on Selected Areas in Cryptography* (pp. 159–174). Springer, Berlin, Heidelberg.

Bucerzan, D., Crăciun, M., Chiș, V., & Rațiu, C. (2010). Stream ciphers analysis methods. *International Journal of Computers Communications & Control*, 5(4), 483–489.

Daemon, J., & Rijmen, V. (2002). The advanced encryption standard: Rijndael, *The Design of Rijndael: AES - The Advanced Encryption Standard*, 1st Edition. Springer, Berlin, Heidelberg, pp. 151–161.

Diffie, W., & Hellman, M. (1976). New directions in cryptography. *IEEE transactions on Information Theory*, 22(6), 644–654.

Dworkin, M. (2001). Recommendation for block cipher modes of operation, methods and techniques (No. NIST-SP-800–38A). National Inst of Standards and Technology Gaithersburg Md Computer Security Div.

Goldreich, O. (1998). *Modern Cryptography, Probabilistic Proofs and Pseudorandomness* (Vol. 17). Springer, Berlin, Heidelberg.

Katz, J., & Lindell, Y. (2014). *Introduction to Modern Cryptography*. CRC press, Boca Raton, FL.

Katz, J., Menezes, A. J., Van Oorschot, P. C., & Vanstone, S. A. (1996). *Handbook of Applied Cryptography*. CRC press, Boca Raton, FL.

Lai, X., & Massey, J. L. (1990, May). A proposal for a new block encryption standard. In *Workshop on the Theory and Application of Cryptographic Techniques* (pp. 389–404). Springer, Berlin, Heidelberg.

Lipmaa, H., Rogaway, P., & Wagner, D. (2000, October). CTR-mode encryption. In *First NIST Workshop on Modes of Operation*. Vol 39, Citeseer. MD.

Mao, W. (2003). Modern Cryptography: Theory and Practice. India: Pearson Education.

Masoodi, F., Alam, S., & Bokhari, M. U. (2012). An analysis of linear feedback shift registers in stream ciphers. *International Journal of Computer Applications*, 46(17), 46–49.

Meier, W. (1993, May). On the security of the IDEA block cipher. In *Workshop on the Theory and Application of Cryptographic Techniques* (pp. 371–385). Springer, Berlin, Heidelberg.

Paar, C., & Pelzl, J. (2009). *Understanding Cryptography: A Textbook for Students and Practitioners*. Springer, Berlin, Heidelberg.

Pudovkina, M. (2001). Analysis of chosen plaintext attacks on the WAKE Stream Cipher. *IACR Cryptology ePrint Archive, 2001*, 65.

Random.org. (2018) Available at www.random.org/randomness/. Accessed on 26 Feb, 2018.

Schneier, B. (2007). *Applied Cryptography: Protocols, Algorithms, and Source Code in C*. John Wiley & Sons, New Delhi.

Shannon, C. E. (1949). Communication theory of secrecy systems. *Bell Labs Technical Journal, 28*(4), 656–715.

Smid, M. E. (2000). From DES to AES. www.nist gov/aes.

Stinson, D. R. (2005). *Cryptography: Theory and Practice*. CRC press, Boca Raton, FL.

Thakur, J., & Kumar, N. (2011). DES, AES and Blowfish: Symmetric key cryptography algorithms simulation based performance analysis. *International journal of emerging technology and advanced engineering, 1*(2), 6–12.

Upadhyay, D., Sharma, P., & Sampalli, S. (2016). Enhancement of GSM stream Cipher security using variable taps mechanism and nonlinear combination functions on Linear feedback shift registers. In *Proceedings of First International Conference on Information and Communication Technology for Intelligent Systems:* Volume 2 (pp. 175–183). Springer, Cham.

Van Tilborg, H. C. A. (2005). *Encyclopedia of Cryptography and Security*. Springer, New York.

William, S. (2006). *Cryptography and Network Security: Principles and Practices*. Pearson Education India, New Delhi.

7
Symmetric Algorithms II

Vivek Kapoor
Devi Ahilya University

Shubhamoy Dey
Indian Institute of Management Indore

CONTENTS

7.1 Introduction 98
    7.1.1 Terminology and Background 99
    7.1.2 Key Range and Key Size 100
    7.1.3 Possible Types of Attacks...101
7.2 Blowfish Algorithm 101
    7.2.1 Background and History 101
    7.2.2 How Blowfish Algorithm Works?...102
        7.2.2.1 Operation...102
    7.2.3 Strength and Weakness of Algorithm 103
7.3 CAST-128 (CAST5) Algorithm..104
    7.3.1 Background and History 104
        7.3.1.1 CAST-128 Purpose...104
    7.3.2 How CAST-128 (CAST5) Algorithm Works? 105
        7.3.2.1 Mathematics Used..106
    7.3.3 Advantages and Disadvantages 106
7.4 RC Algorithms..106
    7.4.1 Background and History 106
    7.4.2 How RC5 Algorithm Works?..106
        7.4.2.1 Sub-Key Generation Process 107
    7.4.3 Variants of RC2, i.e., RC4, RC5, and RC6 Algorithms 108
7.5 Serpent Algorithm 108
    7.5.1 General Description..108
        7.5.1.1 Serpent Structure...108
7.6 Twofish Algorithm 109
    7.6.1 General Description..109
        7.6.1.1 Twofish Design Goals 110
    7.6.2 Security Attacks on Twofish Algorithm 112
    7.6.3 Performance of Twofish Algorithm...112
7.7 Algebraically Homomorphic Scheme (AHS) 112
    7.7.1 Classical Homomorphic Encryption Systems 113
    7.7.2 Applications and Properties of Homomorphic Encryption Schemes............113
        7.7.2.1 Bitcoin Split-Key Vanity Mining 113
    7.7.3 Some Properties of Homomorphic Encryption Schemes 113

7.8   Lightweight Cryptography 113
       7.8.1   Performance Metrics 114
       7.8.2   Lightweight Cryptographic Primitives 115
               7.8.2.1   Lightweight Block Ciphers 115
               7.8.2.2   Lightweight Hash Functions.. 115
       7.8.3   Design Considerations 116
7.9   Symmetric Searchable Encryption (SSE)... 116
       7.9.1   How Do Searchable Encryption Work?..................................... 116
7.10   Conclusion 117
References... 117

7.1 Introduction

Privacy is also an imperative aspect of computing security. Computers do not formulate privacy issues. But high-speed processing and data storage capability have affected privacy in the cyber world. Since privacy is a part of confidentiality, it is an aspect of computer security. As confidentiality, integrity and availability can conflict, so too can privacy and other aspects of security. Privacy is a broad topic affected by computing but just not a security topic. In this chapter, we look at various symmetric key security algorithms (Kahate, 2013).

Cryptography—secret writing—is the strongest tool for scheming alongside many kinds of security coercion. Well-disguised data cannot be read, tailored, or fabricated with no trouble. Cryptography is entrenched in superior mathematics: group and field theory, computational convolution, and even real psychoanalysis, not to mention probability and statistics "Stallings (2017)". Fortunately, it is not essential to appreciate the underlying mathematics to be clever to use cryptography.

We start this chapter by exploratory what symmetric key encryption does and how it works. We introduce the basic values of encryption with two simple encryption methods: substitution and transposition. Next, we discover how they can be expanded and enhanced to create stronger, more urbane defense. Because weak or flawed encryption makes available only illusion of defense, we also look at how encryption can be unsuccessful. We examine techniques used to break through the defensive scheme and disclose the original text. Few very popular algorithms are in use today: Blow Fish, CAST, RC5, Serpent, Two Fish, and Lightweight Cryptography. We look at them in some point to see how these algorithms can be used in building blocks by means of protocols and structures to carry out other computing tasks, such as signing credentials and swap over sensitive data (Figure 7.1).

Let us for a short time appraise symmetric key cryptography. It is also recognized as private key cryptography. Here only one key is used for mutually encryption and decryption of plain text. Both parties must agree on the key before the start of exchange of messages and key should be kept secret. On sender's side, we convert plain text into ciphertext by applying the key. At receiver's side, the same key is applied to convert or decrypt the ciphertext, thus extracting the original message from it. In practice, there are few problems with a symmetric key. We will revise them quickly.

The main problem is of key exchange. The second problem is more serious than this. Since the same key is used, one key is used between a sender and a receiver. Hence, as the number of pairs of sender and receiver increases, the number of keys also increases between them as to provide confidentiality between each pair of sender and receiver.

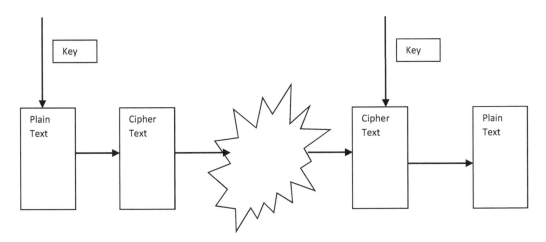

FIGURE 7.1
Symmetric key cryptography.

Since the Internet has millions of merchants selling their product to billions of customers, using this scheme is not feasible because every pair of communicating parties will need a separate key (Menezes, 2012).

Regardless of all the drawbacks, symmetric key has several advantages also, and it is widely used in practice. These drawbacks can be overcome using intelligent solutions.

These chapters present a deeper study of symmetric key encryption techniques and algorithms, counting their mathematical bases, the apparatus that can make them work, and their confines. Most client of cryptography will never discover their own algorithms, just as most users of electricity do not construct their own power generators. Still, a deeper acquaintance of how cryptography works can help you use it efficiently, just as a deeper acquaintance of energy issues helps you appreciate the environmental and cost trade-offs among different energy sources. This chapter offers you a rudimentary perceptive of what the symmetric key cryptography and their algorithms are.

7.1.1 Terminology and Background

Cryptography has an elongated and striking history. The most absolute non-technical explanation of the area under discussion is Khan's *The Codebreakers*. This book traces cryptography from its near-the-beginning and imperfect use by the Egyptians several 4,000 years ago, to the 20th century where it played a very important role in the result of both world wars. Khan's book covers those features of history which were most important to the expansion of the subject. The prime practitioners were those who were connected with the military, diplomatic service, and government in general. Cryptography was used as a tool to defend national secrets and strategies.

The propagation of computers and communication systems in the 1960s brought with it a command from the private sector for means to defend information in digital form and to offer security services. Commencement with a work of Feistel Information Processing Standard for encrypting uncategorized in sequence, DES (Data Encryption Standard), is the most well cryptographic means in the account. It vestiges the average means for protected electronic commerce for a group of financial establishment around the globe.

Cryptography (surreptitious writing) is the strongest instrument for scheming flanking many kinds of security bullying. Well-disguised data cannot understand writing, customized or made-up, with no trouble. Cryptography is entrenched in advanced mathematics. Providentially, it is not obligatory to appreciate the original arithmetic to be capable to use cryptography. We commence this chapter by giving the impression of being at what encryption does and its modus operandi. We discover how it can be prolonged and enhanced to create a stronger, more complicated defense, because weak and defective encryption provides only a false impression of defense. DES, AES, Blowfish, RC5, CAST Algorithm, and Twofish algorithm are some of the well-known algorithms that are in use today. We glance at them in a few features to see how they work and can be used as building blocks to hold out computing responsibilities such as swap over sensitive data, signing, etc. Most client of cryptography will, by no resources, find out their individual algorithms (Pfleeger, Pfleeger, & Margulies, 2016).

Think about the steps involved in sending messages from sender S to a recipient R. If S entrusts the message to T, who transports it to R, then T becomes communication intermediate. If an outsider O desires to right to use the message, then we call O as interceptor or trespasser.

Encryption is the practice of brainwashing a communication so that its connotation cannot be forecast. Decryption is the undo process, i.e., alter an encrypted communication back into its usual form. Substitute terms used are encode or decode or encipher and decipher as an alternative of encrypt or decrypt. A scheme of encryption and decryption is called the cryptosystem.

The original outline of a message is identified as plain text, and encrypted form is acknowledged as ciphertext. The crypto arrangement involves a set of systems for how to encrypt a plain text and how to decrypt the ciphertext. The encryption and decryption systems are called algorithms. These algorithms frequently use a device called a key, denoted by K such the ensuing ciphertext depends on plain text message, algorithm, and the key value. We write this as $C = E(K, P)$. E is a set of encryption algorithms, and the key K selects one specific algorithm from the set. In the case of symmetric key encryption and decryption, keys are alike, so $P = De(K, Ee(K, P))$. Here, De and Ee are the mirror image course (Figure 7.2).

A key gives suppleness in using an encryption system. We can create dissimilar encryption of one plain text message by varying the key. Key also offers supplementary security, i.e., if the encryption algorithm falls into the interceptor's hands, potential communication can still be reserved undisclosed because interceptor does not be acquainted with the key significance.

7.1.2 Key Range and Key Size

If encrypted messages are assault, then encryption/decryption algorithms are identified to everyone. One can access encrypted message by a variety of other ways. Thus, only the

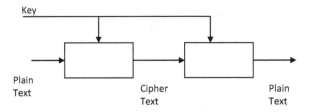

FIGURE 7.2
Symmetric cryptosystem.

value of key remains a face up to an attacker. We can think brute force attack, which works on the belief of trying every possible key in the range, in anticipation of you get the right key. It frequently takes a small amount of time to check a key. In the most excellent case, attacker found the key in the first effort, and in the most horrible case, it found a key in the last effort. Mathematics tells us that key can be brought into being about after half possible values in the key range are tartan. Thus, the solution to this difficulty lies in expanding or increasing the key array to a dimension which necessitates the attacker to work for more than the time you necessitate keeping message a secret.

In computational terms, the idea of a key range leads to key size. We gauge key size in bits and represent it in binary (twofold) number scheme. In order to thwart brute force attack, key size should be such that the attacker cannot crack it within a specified period of time given the computational possessions it has got.

At the simplest stage, the key size can be just 1 bit. Possible values can be 0 or 1. If key size is 2, then likely values are 00, 01, 10, and 11. This works on the values of binary (twofold) figures, where additional bit two times the number of promising states. Thus, with supplementary bit, attacker has to carry out two times the number of maneuver as weigh against to previous key dimension. We can assume that a large key size is considered to be safe. As computing power augment, these numbers may change. A key size of 128 bits that may be considered safe at one point of time may not be measured safe after a few years.

7.1.3 Possible Types of Attacks

A cryptanalyst's objective is to break an encryption. Thus, the cryptanalyst makes efforts to infer the original meaning of a ciphertext message. It attempts the following things:

- Smash a single message.
- Be familiar with patterns in encrypted messages.
- Infer the key, to break succeeding messages easily.
- Find fault in the completion of encryption.
- Find general fault in an encryption algorithm, without interrupting any messages.

7.2 Blowfish Algorithm

7.2.1 Background and History

Blowfish algorithm is a symmetric key block code, developed in 1993. It is used in a huge way in encryption and decryption services. Blowfish is a symmetric key algorithm that can be used as an exchange for DES or International Data Encryption Algorithm. Blowfish was premeditated by Bruce Schneier as a speedy, open alternate to handy symmetric key encryption. It uses an uneven span key, ranging from 32 to 448 bits which is ideal for light and heavy encryption depending upon its use. As then it has been look at by a long way, and it is slowly but surely gaining improvement as a strong encryption symmetric key algorithm. Blowfish is not patented and licensed without charge, and is at no cost for all consumers and all types of exercise. While no effectual cryptanalysis of Blowfish has been set up so far, supplementary interest is now approved to block ciphers with a bigger block size, such as Advance Encryption Standard or Twofish Algorithm. Schneier affirmed that "Blowfish is not patented, and will be available free of charge to the users. This symmetric key algorithm is positioned in the community realm, and can be used by anyone free of cost."

7.2.2 How Blowfish Algorithm Works?

According to the inventor, Blowfish was intended with the following objectives in mind.

- Quick: Blowfish encryption rate on 32-bit microprocessors is 26 clock cycles per byte.
- Dense: Blowfish uses only prehistoric operations, such as addition, Exclusive-OR, and table lookup, making its design and completion easy.
- Safe: Blowfish has a variable key length up to a maximum of 448 bits long, building it equally supple and protected.
- Blowfish is appropriate for appliance where key remnants unvarying for a long time, but not where key changes regularly.

7.2.2.1 Operation

Blowfish is a 64-bit block cipher with an erratic key length. It consists of two components:

1. Key expansion: This practice converts the key up to 448 bits long to sub-keys totaling 4,168 bits.

      The encryption of a 64-bit block input plain text PT is as follows (Figure 7.3):

    1. Divide PT into two blocks: Right Plain Text (RPT) and Left Pain Text (LPT), of equal sizes.

    2. For i = 1 to 16

        LPT = LPT Exclusive-OR $P[i]$

        RPT = F(LPT) Exclusive-OR RPT

        Swap LPT, RPT   Next i

    3. Swap LPT, RPT (i.e., undo the last swap)

    4. RPT = RPT Exclusive-OR P17

    5. LPT = LPT Exclusive-OR P18

    6. Mingle LPT and RPT into PT

2. Data Encryption: This method involves the iteration of an effortless function 16 times. Every round controls a key needy permutation and data needy changeover.

Key generation in Blowfish algorithm is initiated by using hexadecimal digits of P-array and Sand boxes copied from $P[i]$, which holds no clear blueprint. The undisclosed key is then Exclusive ORed with the P-entries in tidy (rotating the key if necessary). The algorithm then encrypts a 64 bits plain text block. Then, ciphertext block $P1$ and $P2$ are swapped with each other. Encryption of resultant block is done consecutively for a succeeding time with the new-fangled sub-keys, and $P3$ and $P4$ are swap by the fresh ciphertext. It carries on swapping the whole P-array and the entire the Sand-box admission.

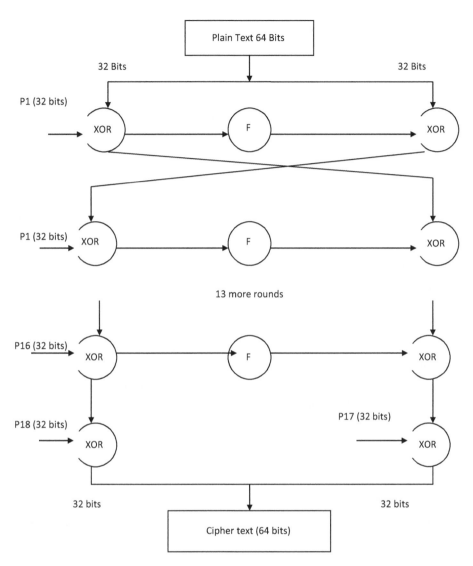

FIGURE 7.3
Blowfish encryption scheme.

Thus, Blowfish encryption algorithm sprints 521 times to create the entire sub-keys—a proposed 4 kB of data is managed.

7.2.3 Strength and Weakness of Algorithm

The following are the potencies of this algorithm:

- Blowfish's pace makes it an excellent alternative for relevance that encrypts an intermediate sum of data, such as archetypal of network communications (e-mail, file transfers).
- No assault on Blowfish has branded that effort on the full 16-round official version (definite attack recuperates some information from an account with up to 14 rounds).

- Blowfish is a symmetric key algorithm being used heavily; it is finest in its class by using the facility of altering keys.
- The algorithm is somewhat slow as compared to other algorithms in its class. This is due to the fact that the new key generation requires pre-processing of 4 kbps of data every time.
- Due to the sluggish feature of the algorithm, its use is confined to some purpose only. This sluggish feature is not bane in other purpose, but a lead for example password hashing process damaged in open BSD exploit procedure from Blowfish that employ its lethargic key process. This is useful in resistance subsequently to a dictionary attack.
- Blowfish algorithm uses a very less memory of 4 kbps. Hence, it can be effectively used in smaller less capacity computers such as palm tops. But it cannot be used as an encryption algorithm in smart cards.
- Blowfish can be used freely by anyone as it is unpatented. This feature adds to its recognition.
- Put into practice in SSL and other security suites (Sahu & Ansari, 2017).

The following are the flaw of this algorithm:

- Squat key-agility and/or high-memory involvedness make Blowfish not viable in aberrant background.
- Minute (64-bit) block size formulates it unsure of yourself for submission that encrypts a bulky sum of data with the matching key (such as data archival and file scheme encryption).

7.3 CAST-128 (CAST5) Algorithm

CAST algorithm is a symmetric key algorithm widely in use. It is a block code with uniform alteration. It is used broadly in numerous security protocols and is widely used to encrypt a large amount of data.

7.3.1 Background and History

CAST-128 is a block cipher algorithm used in a group of products like GPG and Pretty Good Privacy. CAST-128 was agreed for use by the Communication Security Organization of Canada. It was created in 1996 by Carlisle Adams and Stafford Tavares using the CAST design. The CAST name is based on the premature of its inventors.

7.3.1.1 CAST-128 Purpose

CAST-128 is not designed for a precise purpose. It can be used extensively for small, large, and any type of data. It is used for general-purpose communication wherever a free, physically powerful cryptography algorithm is desirable. Though CAST-128 is available freely for business and nonbusiness uses, Entrust Incorporation had totally designed a CAST algorithm, and hence, it holds the propriety rights of it (Adams, 1997).

7.3.2 How CAST-128 (CAST5) Algorithm Works?

In CAST-128, there are either 12 or 16 rounds. It is a block code in which input is 64-bit plain text block. It has a varying key range from 40 to 128 bits with only 8-bit addition. If the key size is more than 80 bits, then the use of 16 rounds is advisable by the makers of this algorithm. The algorithm uses 8- by 32-bit Sand boxes based on bent functions, subject to key alternation, modular arithmetic, and Exclusive-OR function. Figure 7.4 shows three

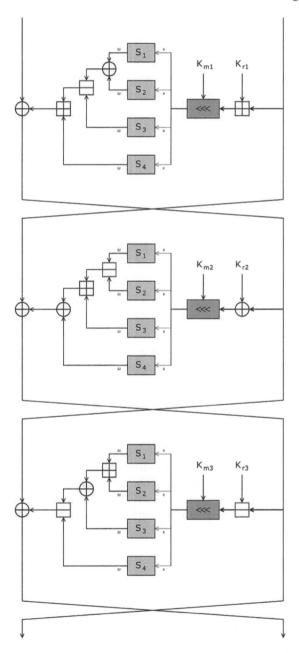

FIGURE 7.4
CAST-128 algorithm.

irregular classes of round occupation, which are similar in pattern and are diverse merely in the selection of the accurate action (addition, subtraction or Exclusive-OR) at a range of points. Also, it's based on the Feistel Cipher arrangement.

7.3.2.1 Mathematics Used

1. (key schedule) Compute 16 pairs of sub-keys $\{K_{m_i}, K_{n_i}\}$ from K
2. $(L_0, R_0) \leftarrow (m_1 \ldots m_{64})$. (Split the plaintext into left and right 32-bit halves $L_0 = m_1 \ldots m_{32}$ and $R_0 = m_{33} \ldots m_{64}$.)
3. (16 rounds) for i from 1 to 16, compute L_i and R_i as follows: $L_i = R_i - 1$ $R_i = L_i - 1$ $\wedge f(R_i - 1, K_{m_i}, K_{n_i})$ (f is of Type 1, Type 2, or Type 3, depending on i.)
4. $c_1 \ldots c_{64} \leftarrow (R_{16}, L_{16})$. (Exchange final blocks L_{16}, R_{16}, and concatenate to form the ciphertext.)

7.3.3 Advantages and Disadvantages

Since it is based on the Feistel Cipher structure, it has a benefit that encryption and decryption procedures are very comparable. Even though it is impossible to differentiate in a number of cases, it necessitates merely a rotate of the key schedule. Consequently, the size of the code necessary to put into practice such a cipher is almost divided in two.

7.4 RC Algorithms

The series of RC algorithms was developed by R. Rivest in addition to MD5 algorithm series. It is widely used in a lot of networking acquiescence because of their optimistic speed and variable key-length capability. There are quite a lot of discrepancies of RC algorithms together with: RC2, RC4, RC5, and RC6.

7.4.1 Background and History

It is a proprietary cipher owned by RSADSI. It is designed by Ronald Rivest (of RSA fame) used in various RSADSI products. We can differ in the key size/data size/number of rounds. It has a very spotless and straightforward design. It is trouble-free to implement on various CPUs and yet still look upon as secure. Its nominal version is RC5-32/12/16, i.e., 32-bit words so encrypts 64-bit data blocks using 12 rounds with 16-byte (128-bit) secret key.

7.4.2 How RC5 Algorithm Works?

In RC5, word size (i.e., plain text size), number of rounds, and number of 8-bit bytes of the key can be of variable length. Word size could be 16, 32, 64 bits, etc. The quantity of rounds could be in between 0 and 255, and the number of 8-bit bytes of the key could be in between 0 and 255. Once determined, these values remain identical for a particular cryptographic procedure.

Initially, RC5 is quite simple to comprehend. There is one original operation consisting of two steps and then number of rounds. The number of rounds can be in between 0 and 255. If we start with a plain text block of 64 bits, then in first two steps plain text block is alienated into two 32-bit blocks A and B. Then, the first two sub-keys are added to A and B. The consequence C and D marks the end of one-time operation. After this round begins, each round consists of bitwise Exclusive-OR, left circular swing, and a calculation with the next sub-key. Here the output of one block is fed back as the contribution of another block, making hard to decipher the ciphertext. Mathematical illustration of RC5 encryption is as follows (Kahate, 2013):

$A = A + S[0]$

$B = B + S[1]$

For $I = 1$ to r

$\quad A = ((A \text{ XOR } B) <<< B) + S[2i]$

$\quad B = ((B \text{ XOR } A) <<< A) + S[2i + 1]$

Next i

Mathematical illustration of RC5 decryption is as follows:

For $I = 1$ to r to 1 step-1 (i.e., decrement I each time by 1)

$\quad A = ((B - S[2i + 1]) >>> A) \text{ XOR } B$

$\quad B = ((A - S[2i + 1]) >>> B) \text{ XOR } B$

Next i

$B = B - S[1]$

$A = A - S[0]$

7.4.2.1 Sub-Key Generation Process

Mathematical representation of RC5 sub-key generation is as follows:

$S[0] = P, P = $ B7E15163 in hexadecimal, $Q = $ 9E3779B9 in hexadecimal

For $I = 1$ to $2(r + 1) - 1$

$\quad S[i] = (S[I - 1] + Q) \bmod 2^{32}$

Mathematical representation of sub-key mixing is as follows:

$i = j = 0$

$A = B = 0$

Do $3n$ times (where n is the maximum of $2(r + 1)$ and c)

$\quad A = S[i] = A + B) <<< 3$

$\quad B = L[i] = (L[i] + A + b) <<< (A + B)$

$\quad i = (i + 1) \bmod 2(r + 1)$

$\quad j = (j + 1) \bmod c$

End do

7.4.3 Variants of RC2, i.e., RC4, RC5, and RC6 Algorithms

Description	RC2	RC4	RC5	RC6
Timeline	1987	1987	1994	1998
Type of Algorithm	Block Cipher	Stream Cipher	Block Cipher	Stream Cipher
Key Size	40–64	1–256	0–2,040	128, 192, 256
Use	Variable key-size block cipher that was intended as a "drop-in" replacement for DES	It is a changing key, stream cipher. It is used in Secure Socket layer-protected communication. It is fast and completely secure	It is a variable key block code. It has both changing block and key span	Finalist in AES Competition. It has an input of 128–256 bits plain text block. It covers all features of AES

7.5 Serpent Algorithm

Serpent is a 128-bit block cipher planned by Ross Anderson, Eli Biham, and Lars Knudsen as a claimant for the Advanced Encryption Standard. It was a finalist in the AES struggle. The winner, Rijndael, received 86 votes at the preceding AES symposium, while Serpent got 59 votes, Twofish 31 votes, RC6 23 votes, and MARS 13 votes. So, NIST's alternative of Rijndael as the AES was not astonishing, and they had to be satisfied with silvery in the 'encryption olympics'. Serpent and Rijndael are fairly comparable; the main dissimilarity is that Rijndael is quicker (having a smaller number of rounds), but Serpent is additional protected (Nazlee, Hussin, & Ali, 2009).

7.5.1 General Description

Serpent is designed to offer client with the maximum realistic point of declaration that no shortcut assault will originate. It also used two times as many rounds as are adequate to stop all presently known shortcut attack. We thought this to be a cautious practice for a cipher that capacity has an overhaul existence of a century or more. In spite of this demanding blueprint limitation, Serpent is much quicker than DES. Its design ropes a very proficient bit-slice execution, and the best ever version at the time of the contest ran at over 45 Mbit/s on a 200 MHz Pentium (compared with about 15 Mbit/s for DES).

7.5.1.1 Serpent Structure

It Pad shorter keys to 256 bits. Use to initialize $w - 8 \ldots w - 1$. Use a linear recurrence relation to derive 132 more words $w_0 \ldots w_{131}$. Put four words at a time through the S-boxes, just like the cipher itself. There are eight Sand boxes. Each round uses 32 copies of matching Sand box. Round i uses Sand box S_i mod 8. Sand boxes are resultant from DES S-boxes, so no one thinks that they're cooked. Each S-box is 4-bits to 4-bits, transformation converse Sand boxes used for decryption. 32-bit words $X_{0\ldots3}$ are mixed:

- Rotate X_0 and X_2 (different amounts)
- Mix X_0 and X_2 into X_1 and X_3 (different ways)
- Rotate X_1 and X_3 (different amounts)

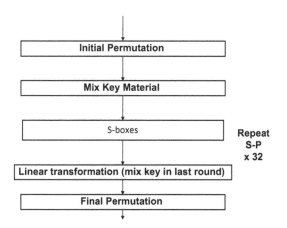

FIGURE 7.5
Serpent structure.

- Mix X_1 and X_3 into X_0 and X_2 (different ways)
- Rotate X_0 and X_2 (different amounts).

Very simple to parallelize, proficient on superscalar machines like Pentium Pro (Figure 7.5).

7.6 Twofish Algorithm

It is a symmetric key block cipher code. It was invented by Counterpane Labs. Twofish algorithm is absolutely free for use for all types of purpose as it is not patented and its source code is not copyrighted. It was one of the five AES finalists.

7.6.1 General Description

Here the input is 128-bit plain text block. The algorithm allows the use of variable keys up to 256 bits span. It has 16 rounds Feistel network with a bijective F function completed by four key-dependent 8- by 8-bit S and boxes, a fixed 4-by-4 utmost space divisible matrix over GF(2^8), a not actual Hadamard alter, bitwise alternation, and a carefully intended key calendar.

Pentium pro encrypts with 17.8 clock cycles/byte, and 8-bit smart card encrypts at the rate of 1,660 clock cycles/byte. Twofish can be used using 14,000 hardware gates. The use of round function and key timetable helps us to analyze a wide assessment of trade-offs considering thrust, code size, key set-up duration, hardware gate calculation, and memory. An extensive study of Twofish algorithm shows that the finest hit breaks 5 rounds with $2^{22.5}$ chosen plaintexts and 2^{51} pains (Aparna, Solomon, Harini, & Indhumathi, 2016)

General parameters of twofish algorithms are as follows:

- Size of Input Block: 128-bit block
- Key size: Variable from 128- to 256-bit key
- Total number of rounds: 16
- Works in all possible cases

- Data rate encryption: 18 clocks/byte on a Pentium or 16.1 clocks/byte on a Pentium Pro
- At period crypt analyzed
- Not patented
- Not copyrighted
- Absolutely free of charge.

7.6.1.1 Twofish Design Goals

Twofish was intended to meet NIST's blueprint criterion for AES. Purposely, they are as follows:

- A 128-bit plain text block.
- Changing keys from 128 to 256 bits.
- Key strength.
- Compatible on various Intel processors and other wide diverse hardware platforms.
- Flexibility in design, e.g., key range or length can be increased, compatible in a diverse variety of environment, can be implemented as a stream cipher, appropriate for the hash function, and Message Authentic Code.
- Less complex design in order to have less effort in its analysis and encryption/ decryption services.

Furthermore, they are obligatory to the following presentation criteria on their propose:

- Any key up to 256 bits length can be used.
- Time to execute the algorithm can be reduced by using 128-bit key in order to encrypt 32 blocks on various processors.
- Data can be encrypted with a reduction of 5,000 clock cycles per block on various processors.
- It can be executed effectively on 8-, 16-, 32-, 64-bit processors.
- It does not consist of any overheads that create problems in hardware.
- A combination of various efficiencies is seen with various key sizes.
- Algorithm in its optimized form can be executed with only 64 bytes of RAM.
- It encrypts data in a slighter quantity than 10 ms on a product 8-bit microprocessor.
- Algorithm can be executed in hardware by means of fewer than 20,000 gates.

Figure 7.6 demonstrates a general idea of the Twofish block cipher. Twofish employs a 16-round Feistel-like structure with additional whitening of the putin and putout. The solitary non-Feistel elements are the 1-bit rotates. The regular change can be moved into the *F* function to generate a pure Feistel arrangement, but this demands an extra turning round of the words just previous to the putin whitening pace. Input plain text block is cut into four 32-bit words. In the input whitening pace, these are XORed with four keywords. It is then put into a loop of 16 rounds.

In every encompassing, the two words on the left are used as input to the *g* functions. (One of them is swapped by 8 bits first.) The *g* function consists of 32-bit general key-dependent Sand boxes and linear MDS matrix. The consequences of the two *g* functions

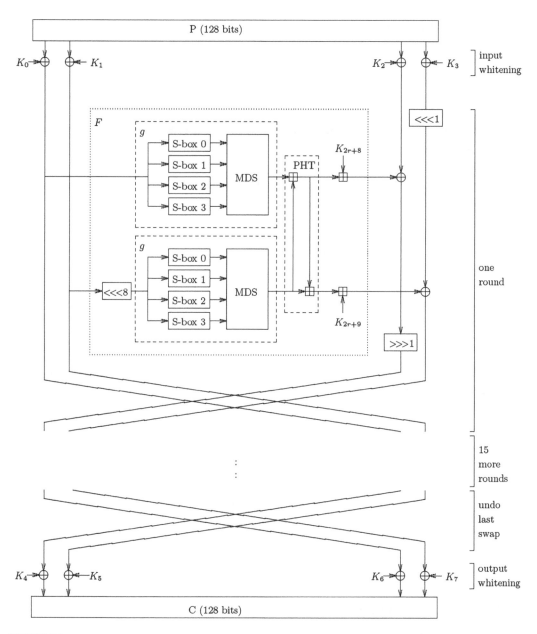

FIGURE 7.6
Twofish algorithm.

are common using a Pseudo Hadamard Transform (PHT), and two keywords are extra. These two consequences are then XORed into the words on the right (one of which is left shift by 1 bit first, the further is rotated right afterward). The left and right are divided into two parts and swapped for further rounds. At the end of all 16 rounds, four words are Exclusive ORed with four keywords to generate the final ciphertext. Thus, 16 bytes of plaintext $p_0 \ldots p_{15}$ are the first tear into four words $P_0 \ldots P_3$ of 32 bits apiece by means of the little-endian gathering.

7.6.2 Security Attacks on Twofish Algorithm

The developers of Twofish have worn out over 1000 man-hours cryptanalyzing Twofish. These are the most significant findings of their cryptanalysis of Twofish:

- A victorious chosen-key attack neighboring to Twofish requires to prefer 160 bits of a pair of keys and necessitates 234 work, 232 chosen-plaintext inquiry, and 212 adaptive chosen-plaintext inquiry so that 10 rounds Twofish can be out of order.
- The meet-in-the-middle attack on usual Twofish requires 4 rounds, 256 known plaintexts, 2,225 memory, and 2,232 work.
- The doing well disparity attack on typical Twofish can break 5 rounds with 2,232 work and 241 chosen-plaintext stipulation.
- There is also an unbeaten meet-in-the-middle attack on 11 rounds Twofish with permanent S-boxes, no 1-bit rotations and no whitening which entail 2,225 memory, 256 known plaintexts, and 2,232 work. The differential attack on this nine rounds Twofish requirements 241 memory, 241 chosen plaintexts, and 2,254 work.
- The related-key hit alongside 10-round Twofish with no whitening needs 2,155 related-key queries, 2,187 work, and for every one of the 2,155 keys it needs 212 adaptive chosen plaintexts and chosen 232 plaintexts.
- The mostly capable attack adjacent to Twofish is the brute force attack, as for 128-bit key, it needs 2,128 involvedness, for the 192-bit key, it requires 2,192 difficulties, and for the 256-bit key, the complication is 2,256.

7.6.3 Performance of Twofish Algorithm

Maybe one of the algorithm's most interesting features is that it allows dissimilar implementations to get better the qualified performance of the algorithm. It is seen that Algorithm has performed better on the following parameters: coding pace, key bargain, hardware gate tally, and memory allocation. The algorithm performs better in all these parameters.

7.7 Algebraically Homomorphic Scheme (AHS)

In this type of encryption process, we can make a calculation on the cipher block, which confronts the process as if it has occurred on the input Plaintext block. The code of homomorphic encryption approves the computation on the ciphertext.

High computing resources are needed for a homomorphic encryption system. For example, cloud computing can stomach the complexity of this encryption system such as corporation calculates stock exchange rate, regression values, and exchange rate of various currencies without helpful to the unencrypted in order to cater those services. It is also used in secure systems such as Black Box Voting, online gambling, clash gratis hash functions, and secure multi-party computation.

Homomorphic encryption systems provide feeble security. It is defenseless against various types of security attacks.

7.7.1 Classical Homomorphic Encryption Systems

Some of the classical homomorphic encryption systems are Goldwasser–Micali scheme, Benaloh's scheme, Naccache–Stern scheme, Okamoto–Uchiyama scheme, Paillier scheme, Damgard–Jurik scheme, Galbraith scheme, and Castagnos scheme.

7.7.2 Applications and Properties of Homomorphic Encryption Schemes

An intrinsic negative aspect of homomorphic cryptosystems is that attacks on these systems might perhaps make use of their additional structural information. For instance, using plain RSA for signing, the multiplication of two signatures gives up a valid signature of the product of the two equivalent messages. Even though there are many ways to keep away from such attacks, for example, by application of hash functions, the use of idleness, or probabilistic schemes, this possible weakness shows us the way to the question why homomorphic schemes should be used as an alternative of conventional cryptosystems under certain circumstances. The main reason for the notice in homomorphic cryptosystems is its wide application span. There are hypothetical as well as realistic applications in diverse areas of cryptography. Some of the objectives of Homomorphic Encryption Schemes are protection of mobile agents, multiparty computation, secret sharing scheme, threshold schemes, zero-knowledge proofs, watermarking and fingerprinting schemes, commitment schemes, and data aggregation in wireless sensor networks (Potey, Dhote, & Sharma, 2016).

7.7.2.1 Bitcoin Split-Key Vanity Mining

In the case of bitcoin currency, its addresses are hashes or message digests derived from the public keys of ECDSA pairs which have homomorphic properties for various computations.

In the case of communication between Harry and Eve, Harry possesses private-public key pair (h, H). It declares its public key as H. Eve possesses private-public key pair (e, E). It declares its public key as E. Now Hash of (H + E) results in a favorable vanity address. He sells e and E to Harry; H, E, and e are visibly known to all and sundry; and hence, one can confirm the address and Hash (H + E) is preferred.

Harry computes mutual private key (h + e) and is used as private key for the corresponding public key (H + E). Thus, in the place of AND operator, OR operator can be used.

7.7.3 Some Properties of Homomorphic Encryption Schemes

Homomorphic encryption systems have some smart mathematical properties. In the following, we talk about some of these characteristics: re-randomizable encryption/re-encryption, random self-reducibility, demonstrable encryptions/fair encryptions, etc.

7.8 Lightweight Cryptography

There are quite a few up-and-coming areas in which highly controlled devices are interconnected, working in recital to achieve some task. Examples of these areas include automotive systems, sensor networks, healthcare, distributed control systems, the Internet of Things (IoT), cyber-physical systems, and the smart grid. Security and privacy can be very significant in all of these areas. Because the mainstreams of modern cryptographic algorithms were intended for desktop/server environments, many of these algorithms cannot be put into practice in the controlled devices used by these applications. When existing standard algorithms can be engineered to fit into the partial resources of controlled environments, their performance may not be satisfactory. For these reasons, a lightweight cryptography project was started to examine the issues and then build up a strategy for the consistency of lightweight cryptographic algorithms (McKay, Bassham, Turan, & Mouha, 2017).

Lightweight cryptography is a subfield of cryptography that aspires to offer explanation tailored for resource-constrained devices. There has been a major amount of work done by the academic community linked to lightweight cryptography; this includes competent implementations of conventional cryptography standards, and the design and analysis of new-fangled lightweight algorithms and protocols.

Lightweight cryptography targets an ample diversity of devices that can be implemented on a wide spectrum of hardware and software. On the high end of the device, spectra are servers and desktop computers followed by tablets and smart phones. Conventional cryptographic algorithms usually execute well in these devices; therefore, these platforms do not necessitate lightweight algorithms. On the lower end of the spectrum are devices such as embedded systems, RFID devices, and sensor networks. Lightweight cryptography is mainly paying attention to the highly constrained devices that can be brought into being at this end of the spectrum.

7.8.1 Performance Metrics

In the cryptographic algorithm design, there is a trade-off between performance and assets required for a given security level. Performance can be articulated in terms such as power and energy consumption, latency, and throughput. In software, this is replicated in the register, RAM and ROM usage. Resource requirements are sometimes referred to as costs, as adding more gates or memory tends to augment the production cost of a device. Power and energy consumption are pertinent metrics due to the nature of many constrained devices. Power may be of particular significance in devices that produce power from their surroundings.

Energy consumption (i.e., power consumption over a certain time period) is particularly significant in battery-operated devices that have a preset amount of stored energy. The batteries in some devices may be not easy or impossible to recharge or restore once arrange. It should also be noted that power consumption depends on many features other than the algorithm used, such as the threshold voltage, the clock frequency, and the technology used for execution.

Latency is, in particular, relevant for certain real-time submission, for example, automotive applications where very fast answer times for components such as steering, airbags, or brakes are required. It can be defined as the measure of time between the early request of an operation and turn out the output. For example, the latency of an encryption operation is the time involving the initial request for the encryption of a plaintext and the reply that returns the equivalent ciphertext.

Throughput is the rate at which new outputs (e.g., authentication tags or ciphertext) are produced. Unlike conventional algorithms, high throughput may not be a design aim in lightweight designs. However, reasonable throughput is still compulsory in most applications.

7.8.2 Lightweight Cryptographic Primitives

Over the previous decade, a number of lightweight cryptographic primitives, including block ciphers, hash functions, message authentication codes, and stream ciphers, have been proposed, which offer recital advantages over conservative cryptographic standards. These primitives are different from conventional algorithms with the postulation that lightweight primitives are not intended for a wide range of applications, and may impose limits on the power of the attacker. For example, the amount of data available to the attacker under a single key may be incomplete. However, it should be distinguished that this does not mean that the lightweight algorithms are feeble—rather, the idea is to use progress that results in designs with a better equilibrium between security, performance, and resource requirements for specific resource-constrained environments.

7.8.2.1 Lightweight Block Ciphers

Lesser block sizes: In order to save memory in lightweight cryptography, lighter block sizes from 64 to 80 bits in place of 18 bits are used. This helps in reduction of limits in the number of plain text blocks to be encrypted and saves memory.

Lesser key spans: It uses lesser key spans such as less than 96 bits for better performance.

Simpler rounds: Procedures in lightweight cryptography are comparatively simpler than used in conventional cryptography techniques. Lightweight cryptography uses lighter Sand boxes of size 4 bits as compared to 8-bit Sand boxes used in conventional cryptographic procedures. This reduces the use of resources.

Simpler key agenda: Relatively simpler key generation module results in less usage of memory, power consumptions, latency, and other resources, etc. Thus, in lightweight cryptographic procedures, simple sub-keys are generated, but they are susceptible to various attacks.

Minimal implementations: There are several modes of process and protocols that require only the encryption function of a block cipher. Some claim may require a device to only support one of the encryption or decryption operations. Implementing only the essential functions of a cipher may necessitate fewer possessions than implementing the full cipher.

7.8.2.2 Lightweight Hash Functions

Conventional hash functions may not be appropriate for constrained environments, mostly due to their large internal state sizes and high power consumption supplies. This has led to the development of lightweight hash functions, such as PHOTON, Quark, SPONGENT, and Lesamnta-LW. The expected usage of conventional and lightweight hash functions is different in various aspects.

Smaller internal state and output sizes. Large output sizes are significant for applications that require collision resistance of hash functions. For applications that do not need collision resistance, smaller internal states and output sizes might be used. When a collision-resistant hash function is required, it may be suitable that this hash function has the same security against preimage, second-preimage, and collision attacks. This may decrease the size of the internal state.

Smaller message size. Conventional hash functions are probable to support inputs with very large sizes (around 264 bits). In most of the target protocols for lightweight hash functions, typical input sizes are much lesser (e.g., at most 256 bits). Hash functions that are optimized for short messages may consequently be more suitable for lightweight applications.

7.8.3 Design Considerations

While specific requirements diverge by application, there are more than a few generally desired properties that will be used to assess designs: security strength, flexibility, low overhead for multiple functions, ciphertext expansion, side channel and fault attacks, limits on the number of plaintext-ciphertext pairs, and related-key attacks.

It may not be likely to satisfy all properties, in scrupulous when this increases the resources further than what is available for a given application. Still, any algorithm chosen for the portfolio must offer adequate security. In particular, the security alongside key-recovery attacks should be at least 112 bits.

7.9 Symmetric Searchable Encryption (SSE)

Searchable encryption scheme is a cryptographic system that permits to investigate precise information in an encrypted at ease. This results that a party to bond out the storage space of his data to one more party in a classified mode, while upholding the talent to selectively hunt in excess of it.

Assume that Alice wants to forward the mail to Bob containing the term "office" as she is gone on a holiday. Now how would the service contributor know which mail to send to Bob if it is not able to interpret the data? Again it is a security apprehension to convey the data without end-to-end encryption. An explanation of this dilemma is to use searchable encryption. Here Alice chooses to encrypt the message and generates an assortment of tags or looks for a keyword that can exclusively recognize the message for a detailed search. Now their tags are add on to the encrypted message so that the service offer can perform the task of forwarding the mail without knowing the details of the mail (Curtmola, Garay, Kamara, & Ostrovsky, 2006).

7.9.1 How Do Searchable Encryption Work?

Searchable exception mechanism by revealing a segment of information of a context that can be used to recognize a context. For example, if there is a post about the electoral debate in BBC, transcript of the debate can be easily identified by "electoral debate" in "BBC" on these keywords. They, along with date, may be enough to find that precise manuscript among thousand others.

So searchable encryption facilitates recognition of an encrypted content based on some segments of information available about the content without revealing the content itself. In this example, those keywords would be able to discover the transcript of electoral debate but won't be enough to get the real content. So the content owner would encrypt the data with a private key and share the keyword encrypted with the public key of the search provider, or a shared secret key that is shared with the provider as well. Now once the search query comes, the provider would open the envelop of the keyword collection and matches the search, once the match is established, it would establish the association between the content owner and the requester. Then, the requester and content owner can confer the terms of accessing the private content without the attendance of the search provider.

Another example: Suppose Alice wants to forward all the mails marked "Urgent" to john while she is away on her vacation. Now Bob sends an encrypted e-mail to Alice encrypted with Alice's public Key. But the gateway provider has no way of knowing whether to route the mail to John if is it marked "Urgent".

Then, John extracts the keyword urgent, performs searchable encryption or public key encryption with a keyword search on the extracted keyword, and sends the data to the gateway. The gateway can now decrypt the keyword list and route the message to John.

7.10 Conclusion

Traditionally, the focal point of cryptology has been on the exercise of symmetric encryption to make available privacy. In this chapter, we contemplate on the utilization of symmetric encryption to present confidentiality. Computers carry out encryption fairly with no trouble and rapid. Symmetric key encryption is individual quite well-liked over a quite number of time. Currently, it is failing to asymmetric key cryptography. In this chapter, we had glanced in the lead diverse feature and algorithms of symmetric key cryptography. We had also argued a number of symmetric key algorithms in the facet. We went up to the previous bit of clarification, which could make the reader appreciate about the operational character tics of these algorithms. After analysis of this chapter, reader will have absolute thoughtful of small details of the algorithms. This matter remains significant in itself. Indulgent of the above matter argue in the above chapter helps in the expansion of public key encryption.

References

Adams, C. (1997). The CAST-128 Encryption Algorithm, 01–15. doi:10.17487/rfc2144.

Aparna, K., Solomon, J., Harini, M., Indhumathi, V. (2016). A study of Twofish algorithm. *International Journal of Engineering Development and Research, 04*(21), 148–150. Retrieved May 26, 2018, from www.ijedr.org/papers/IJEDR1602023.pdf.

Curtmola, R., Garay, J., Kamara, S., & Ostrovsky, R. (2011). Searchable symmetric encryption: Improved definitions and efficient constructions. *Journal of Computer Security, 19*(5), 895–934. doi:10.3233/jcs-2011-0426.

Kahate, A. (2013). *Cryptography and Network Security*. New Delhi: McGraw Hill Education.

Mckay, K.A., Bassham, L., Turan, M.S., & Mouha, N. (2017). Report on lightweight cryptography. *National Institute of Standards and Technology Internal Report 8114*, 01–21. doi:10.6028/nist.ir.8114.

Menezes, B. (2012). *Network Security and Cryptography*. Belmont, CA: Wadsworth Publishing Co.

Nazlee, A.M., Hussin, F.A., & Ali, N.B. (2009). Serpent encryption algorithm implementation on Compute Unified Device Architecture (CUDA). *2009 IEEE Student Conference on Research and Development (SCOReD), 72*, 74–81. doi:10.1109/scored.2009.5443190.

Pfleeger, C.P., Pfleeger, S.L., & Margulies, J. (2016). *Security in Computing*. Bejing: Publishing House of Electronics Industries.

Potey, M.M., Dhote, C.A., & Sharma, D.H. (2016). Homomorphic encryption for security of cloud data. *Procedia Computer Science, 79*, 175–181. doi:10.1016/j.procs.2016.03.023.

Sahu, R., & Ansari, M.S. (2017). Securing messages from brute force attack by combined approach of honey encryption and blowfish. *International Research Journal of Engineering and Technology, 04*(09), 1019–1023. Retrieved April 23, 2018, from www.irjet.net/archives/V4/i9/IRJET-V4I9179.pdf.

Stallings, W. (2017). *Cryptography and Network Security Principles and Practices*. Boston, MA: Pearson.

8

Asymmetric Cryptography

Rajiv Ranjan, Abir Mukherjee, and Pankaj Rai
Birsa Institute of Technology Sindri

Khaleel Ahmad
Maulana Azad National Urdu University

CONTENTS

8.1 Introduction 119
8.2 Preliminaries 121
    8.2.1 Divisibility and the Euclidean Algorithm 121
    8.2.2 Algebra 122
    8.2.3 Modular Arithmetic 123
    8.2.4 Basic Number Theory 123
8.3 The RSA Algorithm 124
8.4 Rabin Cryptosystem 126
8.5 ElGamal Public-Key Algorithm 128
8.6 Knapsack Algorithm 130
8.7 Chaos-Based Asymmetric Cryptography 132
8.8 Conclusion 136
References 136

8.1 Introduction

Asymmetric or public-key cryptography was first publicly introduced by W. Diffie, M. Hellman, and R. Merkle in 1976. In the iconic paper "New Directions in Cryptography," they introduced the basic structure of public-key cryptosystems and also demonstrated how key exchange can be performed securely using a public channel (Diffie & Hellman, 1976; Goldwasser, 1997).

Any general cryptographic system involves three parties: a sender S, a receiver R, and an eavesdropper O. Privacy is achieved in a cryptographic system when S and R can communicate without O being able to extract any usable information about the communication. In order to avoid the need of a secure communication channel, it is generally assumed that the eavesdropper is able to intercept the encrypted messages.

Asymmetric cryptosystems are characterized by their use of two different keys—the public key $\mathbf{K_{\{PU\}}}$ and the private key $\mathbf{K_{\{PR\}}}$. In order to communicate securely using public-key cryptography, both the sender S and the receiver R generate pairs of distinct keys $\{K_{\{PU\}}, K_{\{PR\}}\}$. The public keys are stored in some public files or are exchanged, while the private keys are kept as secret with its user. Now, in order for S to send a message $\mathbf{M}$ to

$\mathcal{R}$, it uses the public key $K_{\{PU,R\}}$ of receiver $\mathcal{R}$, to encrypt M using the encryption system **Enc**(M, $K_{\{PU,R\}}$), generating the cipher text **C**. This encryption system is designed to act like a trapdoor one-way function. In a trapdoor one-way function $F(x)$, it is easy to computer $y = F(x)$ but is computationally infeasible to obtain the inverse $x = F^{-1}(y)$, unless a trapdoor is used. The decryption system **Dec**(C, $K_{\{PR,R\}}$) on the receiving end is designed to be a trapdoor for the one-way encryption system, provided that the private key is available with the receiver. Hence, it acts as an inverse to the encryption system. Only the intended receiver $\mathcal{R}$ is able to easily decrypt the encrypted message C, and thus secure communication takes place. Since the public key can be used by anyone to send encrypted messages to a particular receiver, public-key cryptosystems are also called multiple-access ciphers. The general structure of public-key cryptosystems is depicted in Figure 8.1.

A public-key cryptosystem can be formally defined as a pair of two invertible transformations **Enc**(X, $K_{\{PU\}}$) and **Dec**(Y, $K_{\{PR\}}$), where X belongs to the message space $\mathcal{M}$, and $K_{\{PU\}}$ and $K_{\{PR\}}$ belong to the key space $\mathcal{K}$ containing all the keys of a given size **n**. Every asymmetric cryptosystem is required to satisfy a few properties in order to provide the required security. These properties are the following:

- Every pair of encryption and decryption systems, for a given pair of public and private keys, must be the inverse of each other.

- It must be computationally easy to compute **Enc**(X, $K_{\{PU\}}$) and **Dec**(Y, $K_{\{PR\}}$) when using the public and private keys, respectively.

- It must be computationally infeasible for an adversary to derive the private key $K_{\{PR\}}$ from the public key $K_{\{PU\}}$.

- It must be computationally easy to generate a pair of keys ($K_{\{PU\}}$ and $K_{\{PR\}}$) by a communicating party.

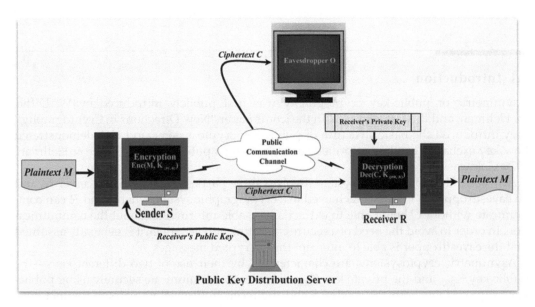

FIGURE 8.1
General structure of asymmetric (public - key) cryptosystems.

Asymmetric cryptosystems can be applied to various problems of security. The three broad categories of applications are confidentiality, key exchange, and digital signatures. Key exchange schemes are used to exchange symmetric keys between two users over an insecure channel. Digital signature schemes are complementary in terms of their usage of keys—the private key is used to encrypt a message and the public key for decryption. When a message is encrypted using a sender's private key, it generates a ciphertext that acts as a digital signature of the sender. Whenever someone successfully decrypts the message using the sender's public key, it is immediately established that the message was generated by the sender. Thus, possessing the ciphertext generated using the sender's private key solves the problem of non-repudiation. The user can no longer deny sending the message, since it is the only party possessing the secret key, which was used for encrypting the received message.

In the following sections, we first discuss some preliminaries that will be required to understand the asymmetric algorithms in a formal manner. Then, we proceed to discuss some of the most important public-key cryptographic systems, their properties, and applications.

8.2 Preliminaries

8.2.1 Divisibility and the Euclidean Algorithm

An integer a is said to divide an integer b, if $a = q \cdot b$ for some integer q, and is expressed as $a|b$. If p is an integer and q is a positive integer, then the **modulus** operation, denoted by p **mod** q, is defined to be the remainder obtained when p is divided by q. A prime number is a number that has no divisors other than 1 and itself; a number is called composite if it has at least one nontrivial divisor. The **greatest common divisor** of two integers a and b, denoted here by **gcd(a, b)**, is the largest integer d that divides both a and b.

The Euclidean algorithm is a fast algorithm which can be used to find the gcd(a, b) of two integers a and b. If two numbers a and b have gcd(a, b) = 1, then they are called relatively prime numbers. By using the Euclidean algorithm, it is possible to express the gcd(a, b) of two numbers a and b as a linear combination of a and b with integer coefficients. The process is called the extended Euclidean algorithm.

Algorithm 1: The Euclidean Algorithm

```
Algorithm: The Euclidean Algorithm

Input: Positive integers a, b;

procedure gcd(a , b):
{
 x := a;
 y := b;
 while(y != 0){
  r := x mod y;
  x := y;
  y := r;
 }
 return x;
}
```

8.2.2 Algebra

In this subsection, we describe the basic abstract algebraic structures such as groups, rings, and fields.

Definition 1: A binary operation ● on a set S describes a mapping between each ordered pairs (a, b) of the elements of set S to a unique element of the same set S.

Definition 2: A **group** $\mathcal{G}$, denoted by $(\mathcal{G}, ●)$, is a nonempty set of elements together with a binary operation ● that satisfies a set of axioms. The axioms that define a group are the following:

- Closure: If $p \in \mathcal{G}$ and $q \in \mathcal{G}$, then $(p ● q) \in \mathcal{G}$.
- Associativity: $(p ● q) ● r = p ● (q ● r)$, $\forall\ p, q, r \in \mathcal{G}$.
- Identity element: Every group must have a unique element $e \in \mathcal{G}$, such that $p ● e = e ● q = p$. This element e is called the identity element of the group.
- Inverse element: For every element $p \in \mathcal{G}$, there must exist a unique element q, such that $p ● q = q ● p = e$. This element is called the inverse element of a, and is denoted by p^{-1}.

For a group $\mathcal{G}$ to be termed as commutative, also called as an Abelian group, it must satisfy one additional axiom:

- Commutativity: $p ● q = q ● p$, $\forall\ p, q \in \mathcal{G}$.

An example of an Abelian group is the set of integers $\mathbb{Z}$ along with the binary operation of addition $(+)$.

A *semigroup* is a set $\mathcal{G}$ having a binary operation ● and satisfying only the group axioms of closure and associativity. A group having a finite number of elements is called a *finite group*; otherwise, it is called an *infinite group*. A group is said to have the order n, denoted $|\mathcal{G}| = n$, if the number of elements in the group is n. A nonempty subset of $\mathcal{G}$, satisfying all the group axioms under the same binary operation, is called a *subgroup* of $\mathcal{G}$.

Definition 3: A **ring** $\mathcal{R}$, denoted by $(\mathcal{R}, \oplus, \odot)$, is a set of at least two elements, having two binary operations $\oplus$ and $\odot$ called addition and multiplication, respectively, where $\mathcal{R}$ is an Abelian group with binary operation $\oplus$, a semigroup with the operation $\odot$ and the operation $\odot$ is distributive over the binary operation $\oplus$. If a ring satisfies the commutative property, then it is called a commutative ring.

Definition 4: An **integral domain** is a commutative ring that has an identity element and satisfies the following property:

- If $p, q \in \mathcal{R}$ and $p \odot q = 0$, then either $p = 0$ or $q = 0$.

Definition 5: A **field** $\mathcal{F}$, denoted by $(\mathcal{F}, \oplus, \odot)$, is an integral domain that has a multiplicative inverse for every element $p \in \mathcal{F}$, except the **0** elements (i.e., additive identity). A multiplicative inverse of an element $p \in \mathcal{F}$ is an element $p^{-1} \in \mathcal{F}$, such that $p \odot p^{-1} = p^{-1} \odot p = \mathbf{1}$. A field having a finite number of elements is called a finite field, or a Galois field. All finite fields must have an order that is a power of a prime number, i.e., the order must be p^n where

p is a prime and n is a positive integer. These finite fields are generally denoted by GF(p^n), for some integer n.

8.2.3 Modular Arithmetic

If two integers p and q have the same remainder or residue, when divided by n, then they are said to **congruent modulo n** and written as $p \equiv q$ (mod n). The set of integers, having the same remainder modulo n, forms the **residue class mod n**. The set of residue classes (mod n), denoted by $\mathbb{Z}_n$, is the set of integers from zero to $n - 1$.

$$\mathbb{Z}_n = \{0, 1, 2, 3, 4, \ldots, n-3, n-2, n-1\}$$

A multiplicative group of $\mathbb{Z}_n$ is the group $\mathbb{Z}_n^*$ such that if $a \in \mathbb{Z}_n^*$, then $a \in \mathbb{Z}_n$, and $\gcd(a, n) = 1$. A quadratic residue modulo n in a multiplicative group $\mathbb{Z}_n^*$ is the integer $a \in \mathbb{Z}_n^*$, such that $x^2 \equiv a$ (mod n) where $x \in \mathbb{Z}_n^*$, if it exists. If no such x exists, then a is called the quadratic non-residue modulo n. The set of all quadratic residues modulo n is denoted by Q_n. Conversely, given $a \in \mathbb{Z}_n^*$, if some $x \in \mathbb{Z}_n^*$ satisfies $x^2 \equiv a$ (mod n), then x is called the square root of a modulo n.

In order to find the square roots of a modulo p, where p is an odd prime, we first probabilistically compute $b \in \mathbb{Z}_p$, such that $b^2 - 4a$ is a quadratic non-residue modulo p. A polynomial $f(x) = x^2 - bx + a \in \mathbb{Z}_p[x]$ is considered. Then, the value of $r = x^{(p+1)/2}$ (mod f) is computed. The square roots of a modulo p are thus obtained as $(r, -r)$.

8.2.4 Basic Number Theory

Some basic results in number theory are presented here, without their corresponding proofs and are accepted to be true.

Definition 6: Euler's totient function, written as $\phi(x)$, is defined as the number of positive integers less than x, which are relatively prime to x.

If p is a prime number, then $\phi(p) = p - 1$.

Theorem 1

Let p be a prime and a be a positive integer, such that $\gcd(p, a) = 1$, then they satisfy the following congruence relation:

$$a^{p-1} \equiv 1 \pmod{p}$$

This theorem is called **Fermat's Little Theorem**.

Theorem 2

Let a and n be two integers such that $\gcd(a, n) = 1$, then **Euler's theorem** states that these two integers satisfy the following:

$$a^{\phi(n)} \equiv 1 \pmod{n}$$

Here, a is called the primitive root of n, having the order of $\phi(n)$. The different powers of the primitive root, up to its order $\phi(n)$, produce distinct integers modulo n that are relatively prime to n.

Analogous to the logarithm function, the **discrete logarithm** for a given integer m is defined, for a given prime p and its primitive root a, to be the exponent i such that $m \equiv a^i \pmod{p}$ is satisfied. It is denoted by $\mathrm{dlog}_{(a,p)}(m)$ and is known to be computationally as hard as the factoring problem. It has widespread application in asymmetric cryptography.

In the next section, we begin our discussion of public-key cryptography algorithms, starting with one of the most influential public-key algorithms—the RSA.

8.3 The RSA Algorithm

The RSA algorithm, named after its creators Ron Rivest, Adi Shamir, and Leonard Adleman and published in 1977, was one of the first major breakthroughs in asymmetric cryptography. It was developed with the intention to solve the problem of both privacy and digital signatures, and has been the most widely used public-key algorithm since. Although not proven mathematically, the recovery of the plaintext, by a passive adversary, from the ciphertext produced by the RSA cryptosystem, is considered to be computationally as hard as the factoring problem (Rivest, Shamir, & Adleman, 1978).

The RSA system has two different parts—key generation algorithm and encryption/decryption algorithm. In the key generation algorithm, two distinct keys, called the public key and the private key, are generated. The public key of the receiver is used to encrypt the message, while the private key is used by the receiver to decrypt the message.

The key generation algorithm is described in Algorithm 4, while the encryption and decryption algorithms are described in Algorithms 2 and 3, respectively. For the purposes of privacy, the public key of the receiver is used for encryption, whereas for digital signatures, the private key is used for encrypting the signature.

Algorithm 2: Encryption Algorithm

```
Algorithm: Encryption for RSA Cryptosystem

Input: Receiver's Public Key {N, e}, Input Message M;
Output: Encrypted Message;

procedure Encryption({N, e}, M):
{
   M' := Represent M as an integer in range [1,N-1];
   C := (M')^e mod N;
   return C;
}
```

Algorithm 3: Decryption Algorithm

```
Algorithm: Decryption Algorithm For RSA Cryptosystem

Input: PrivateKey {d, N}, Encrypted Message C;
Output: Decrypted Message

procedure Decryption({d, N}, C ):
{
  M' := C^d mod N;
  return M';
}
```

The primes p and q selected for key generation are generally of the very large order of at least 200 decimal digits, with a very small value of $gcd(p - 1, q - 1)$, and having large prime factors for $p - 1$ and $q - 1$.

Algorithm 4: Key Generation Algorithm

```
Algorithm: Key Generation For RSA Cryptosystem

Input: Null
Ouput: Public Key, Private Key

procedure KeyGeneration( ):
{
 p := Select a large random prime;
 q := Select another large random prime != p;
 N := p * q;
 phi := (p - 1)(q - 1);
 e := Select random integer such that 1<e<phi, and gcd
(phi,e)==1;
 d := Multiplicative inverse of e mod phi;
 return PublicKey{N, e}, PrivateKey{d, N};
}
```

The following example demonstrates how a simple message can be encrypted and decrypted, for secure communication, using the RSA cryptosystem.

Example 1

Let the two communicating parties be Bob (*B*) and Alice (*A*). In this scenario, *B* intends to send a secure message *M* = "HELLOWORLD" to *A*, using the RSA cryptosystem.

First, the message *M* is encoded with a proper coding. Here we code each letter using a two-digit integer, starting with $A = 01$ and incrementing the code letter by letter up to $Z = 26$. Each block consists of two letters or four integers. So, the message *M* is encoded and split into blocks as follows:

$$M = \quad 0805 \quad 1212 \quad 1523 \quad 1518 \quad 1204$$

Selecting two primes $p = 53$ and $q = 61$, we get $N = p \times q = 3233$. Calculate $\phi(N) = (p - 1)(q - 1) = 3120$. Next, select the value of e such that $gcd(e, \phi(N)) = 1$.

Let $e = 47$, then using the extended Euclidean algorithm, we compute d such that $ed \equiv 1 \pmod{\phi(N)}$ is satisfied. Thus, we get $d = 863$.

Thus, $K_{\{PR, A\}} = \{N, d\} = \{3233, 863\}$ and $K_{\{PU, A\}} = \{N, e\} = \{3233, 47\}$ become the private and public keys of the receiver Alice (A), respectively.

Encrypting the first block $M_1 = 0805$, as follows:

$$C_1 \equiv (M_1)^e \pmod{N} \equiv (0805)^{47} \pmod{3233} \equiv 2462$$

Similarly, encrypting the other blocks, we obtain the ciphertext as follows:

$$C = 2462 \quad 0150 \quad 1386 \quad 1027 \quad 2550$$

Now, A receives this encrypted message C from B, and then uses her private key $N = 3233$ and $d = 863$ to decrypt C, block-by-block.

Decrypting the first block $C_1 = 2462$, it is as follows:

$$M_1 \equiv (C_1)^d \pmod{N} \equiv (2462)^{863} \pmod{3233} \equiv 0805$$

Similarly, Alice is able to decrypt the entire ciphertext C and recover the original message M. This is how secure communication takes place using the RSA cryptosystem.

Attacks on the RSA cryptosystem can be categorized into two types—mathematical attacks and implementation-based attacks. Mathematical attacks are attacks that try to break down the RSA algorithm by finding loop-holes in its structure. The most elementary attack is trying to factorize N, as by knowing the factors of N, one can derive d by calculating the inverse of e modulo $\phi(N)$. Since factoring is exponential in time complexity, it computationally infeasible for large prime factors. Another type of mathematical attack is based on the misuse of the RSA algorithm. Implementation-based attacks work by exploiting implementation faults and shortcomings. The most famous implementation-based attack is the one devised by Kocher, called the timing attack. The timing attack works by analyzing the time differences in encrypting different messages, and generating the key bit-by-bit. Adding appropriate time delay, to mask the timing information, can be used to prevent timing attacks (Boneh, 1999; Stallings, 2017).

With increasing computational power and gradual improvements in factoring, the RSA has been successfully broken for lower sized keys. Yet, no general vulnerability has been found for the RSA cryptosystem. The RSA cryptosystem provides very high security for large key sizes, generally of the order of 1024 bits. This has led to widespread use of this asymmetric algorithm, and it might very well continue to enjoy this popularity even further into the future.

8.4 Rabin Cryptosystem

The Rabin cryptosystem was published by Michael O. Rabin in 1979, and was the first asymmetric algorithm that was provably secure. It was proved that the problem of recovery of plaintext from a ciphertext by a passive adversary (i.e., having only the ability to listen to the communication) was as hard as the factoring problem. One disadvantage of this cryptosystem is the fact that the decryption algorithm produces four possible outputs, and hence there is a need for a mechanism to identify the correct plaintext from the four plaintexts that are generated. It can be used as a general-purpose asymmetric algorithm (Rabin, 1979; Katz & Lindell, 2007).

Algorithm 5: Key Generation

```
Algorithm: Key Generation For Rabin Cryptosystem

Input: Null
Ouput: Public Key, Private Key

procedure KeyGeneration( ):
{
  p := Select a large random prime;
  q := Select another large random prime != p;
  N := p * q;
  return PublicKey{N}, PrivateKey{p, q, N};
}
```

Algorithm 6: Encryption Algorithm

```
Algorithm: Encryption for Rabin Cryptosystem

Input: Receiver's Public Key {N}, Input Message M;
Output: Encrypted Message;

procedure Encryption({N}, M):
{
  M' := Represent M as an integer in range [1,N-1];
  C := (M')² mod N;
  return C;
}
```

In order to demonstrate how the Rabin cryptosystem works for secure transmission of a message, an example is discussed below.

Example 2

Let's assume a sender S and a receiver R intend to communicate using the Rabin cryptosystem. Let the message be $M = 83$. Let, $p = 43$, and $q = 47$, and hence $N = 43 \times 47 = 2021$. Thus, R generates its public key $K_{\{PU, R\}} = \{N\} = \{2021\}$ and its private key $K_{\{PR, R\}} = \{p, q\} = \{43, 47\}$. S obtains R's public key, and encrypts the message, to obtain the ciphertext C, as follows:

$$C \equiv M^2 \pmod{N}$$

or

$$C \equiv 83^2 \pmod{2021} \equiv 826$$

S sends C as the ciphertext to R.

On receiving the encrypted message, R decrypts the message using the decryption algorithm. The square roots $\pm r$ and $\pm s$ are obtained which satisfy $r^2 \equiv C \pmod{p} \equiv 826$ (mod 43) and $s^2 \equiv 826 \pmod{47}$. On computing, we find $r = 3$ and $s = 11$. Using the extended Euclidean algorithm, two coefficients $c = -12$ and $d = 11$ are obtained such that $cp + dq = 1$. Next, two integers x and y are computed such that

$$X = (rdq + scp) \bmod N = (1551 - 5676) \bmod 2021 = 1938$$

$$Y = (rdq - scp)\bmod N = (1551 + 5676)\bmod 2021 = 1164$$

Finally, the square roots of C are obtained modulo N as M_1, M_2, M_3, and M_4 which form the four prospective decrypted messages.

$$M_1 = X \bmod N = 1938;\ M_2 = -X \bmod N = 83;\ M_3 = Y \bmod N = 1164;\ M_4 = -Y \bmod N = 857$$

From the above results, it can be clearly seen that $M_2 = 83$ ($=M$) is the correct decryption of the ciphertext C.

Algorithm 7: Decryption Algorithm

```
Algorithm: Decryption Algorithm For Rabin Cryptosystem

Input: PrivateKey {p, q, N}, Encrypted Message C;
Output: Probable Decrypted Messages

procedure Decryption({p, q, N}, C ):
{
   {a, -a} := Square roots of (C mod p);
   {b, -b} := Square roots of (C mod q);
   {c, d} := Integers such that cp + dq = 1,use Extended
Euclidean Algorithm;
   u := (a.d.q + b.c.p) mod N;   v := (a.d.q - b.c.p) mod N;
   M₁ := u mod N;   M₂ := -u mod N;
   M₃ := v mod N;   M₄ := -v mod N;
   return {M₁, M₂, M₃, M₄};
}
```

For automating the process of recognizing the correct message from the outputs of the decryption algorithm, proper redundancy should be appended to the plaintext M. If the decrypted message contains the redundancy, then it is accepted. If none of the four outputs contains the redundancy, then the ciphertext can be rejected. The type of attacks and their resolution are similar to those of RSA. Redundancy also helps prevent adaptive chosen-ciphertext attacks. The encryption algorithm of the Rabin cryptosystem is extremely fast, whereas the decryption is of the order of the RSA algorithm. However, this algorithm is susceptible to chosen plaintext attacks.

8.5 ElGamal Public-Key Algorithm

Another popularly used asymmetric algorithm is the ElGamal Public-Key algorithm. It was developed by Taher El Gamal in 1984, and has found widespread use in many hybrid algorithms, where it is used alongside some symmetric algorithm. Its security against a passive adversary is known to be equivalent to the Diffie–Hellman problem (El Gamal, 1985; Schneier, 1995).

The ElGamal cryptosystem has gained widespread usage in digital signature schemes, e-mail security, key exchange, etc. The encryption algorithm is probabilistic in nature, which leads to the generation of different ciphertexts for the same plaintext during different trials. The following example demonstrates how the ElGamal cryptosystem can be used for privacy.

Example 3

Suppose Bob and Alice want to communicate again, but this time using the ElGamal asymmetric cryptosystem. Bob (*B*) wants to send Alice (*A*) a message $M = 15$. To generate her public/private key pair, *A* chooses a finite field GF(*q*) by choosing a prime *q*.

Algorithm 8: Key Generation Algorithm

Algorithm: Key Generation For ElGamal Cryptosystem

Input: Null
Ouput: Public Key, Private Key

procedure KeyGeneration( ):
{
  q := Select a large random prime;
  p := Select primitive root of multiplicative group $\mathbf{Z}_q$;
  a := Select random integer m, such 1<a<(q - 1);
  Y := p^a;
  return PublicKey{q, p, Y}, PrivateKey{q, a};
}

Let $q = 23$ be the prime with $\alpha = 7$ be the chosen primitive root. Alice selects $X = 14 \in \mathbb{Z}_q$, computing $Y = 7^{14} \pmod q = 2$, and thus generating the public key {23, 7, 2}. The value $X = 14$ is kept as her private key. Bob obtains Alice's public key, and creates a one-time key K by choosing a random integer $m \in \mathbb{Z}_q$, say $m = 13$, and computes $K = (Y)^m \bmod q = (2)^{13} \bmod 23 = 4$. Now, *B* encrypts the message M by generating an integer pair {C_1, C_2} as follows:

$$C_1 = \alpha^m \bmod q = 7^{13} \bmod 23 = 20; C_2 = (K \times M) \bmod q = (4 \times 15) \bmod 23 = 14$$

This forms the ciphertext $C = \{C_1, C_2\} = \{20, 14\}$, and *B* sends C to *A*.

Algorithm 9: Encryption Algorithm

Algorithm: Encryption for ElGamal Cryptosystem

Input: Receiver's Public Key {q, p, Y}, Input Message M;
Output: Encrypted Message;

procedure Encryption({q, p, Y}, M):
{
  M' := Represent M as an integer in range [1,q-1];
  r := Random integer, such that 0<r<q;
  K := Y^r mod q;
  C_1 := p^r;
  C_2 := (K.M') mod q;
  return {C_1, C_2};
}

Algorithm 10: Decryption Algorithm

Algorithm: Decryption Algorithm For ElGamal Cryptosystem

Input: PrivateKey {q, a}, Encrypted Message {C1, C2} ;
Output: Decrypted Message

procedure Decryption({q, a},{C1, C2}):
{
  K' := $(C_1)^d$ mod q;
  M' := $(C_2.K^{-1})$ mod q;
  return M';
}

Alice on receiving the ciphertext C containing $\{C_1, C_2\}$ decrypts the encrypted message by first calculating $K = (C_1)^x$ mod q, and then recovering M by computing $M = (C_2 \times K^{-1})$ mod q. So, K is calculated to be $K = (20)^{14}$ mod $23 = 4$, and then M is obtained as $M = (14 \times 6)$ mod $23 = 15$. Thus, the message M gets securely shared from Bob to Alice using the ElGamal asymmetric cryptosystem.

Although this algorithm has widespread use, it is susceptible to chosen-ciphertext attacks, which can be prevented by using some padding scheme. Also using the one-time key K multiple times creates vulnerabilities. Another disadvantage of this cryptosystem is the process of message expansion, in which the encrypted message has twice the size of the plaintext message M.

8.6 Knapsack Algorithm

The knapsack problem, also known as the subset sum problem, is an NP-complete problem in which the objective is to fit a knapsack of a given capacity from a set of objects, each having different weights. Formally, in a set-theoretic description, the problem can be described as finding the linear combination of the given values in a set, such that the sum is equal to a required amount. The best-known algorithm for this general subset sum problem is exponential in time complexity.

Although this general knapsack problem is computationally infeasible for a large set of objects, there exist special cases of this problem which can be computed easily. The knapsack-based schemes work by creating a transformation between a special case of a subset sum problem and the general subset sum problem for a fixed knapsack size.

The first knapsack-based asymmetric algorithm was the Merkle–Hellman cryptosystem (Merkle & Hellman, 1978; Shamir, 1982). It used a super-increasing set as its special instance of the knapsack problem, since it is easily solvable. Although it was proven to be insecure, its importance lies in the fact that it was the very first asymmetric encryption scheme. Many knapsack-based encryption schemes were developed following Merkle–Hellman, but all of them were proven to be insecure with the exception of one—the Chor–Rivest knapsack cryptosystem (Chor & Rivest, 1985; Menezes, van Oorschot, & Vanstone 2001).

Algorithm 11: Encryption Process

```
Algorithm: Encryption for Chor-Rivest Cryptosystem

Input: Public Key {[l_0, l_1, ...,l_{p-1}], p, h}, Input Message M;
Output: Encrypted Message;

procedure Encryption({[l_0, l_1, ...,l_{p-1}], p, h}, M):
{
    k := h;
    C := 0;
    for( i=1 to i=p){
        if(M >= ^{(p - i)}C_k){
            b := 1;
            M := M - ^{(p - i)}C_k;
            k := k - 1;
        } else {
            b := 0;
        }
        C := (C + (b*l_i)) mod (p^h - 1);
    }
    return C;
}
```

Algorithm 12: Encryption Process

```
Algorithm: Decryption Algorithm For Chor-Rivest Cryptosystem

Input: PrivateKey {fx, gx, prm[...], k, p, h}, Encrypted Message C ;
Output: Decrypted Message

procedure Decryption({fx, gx, prm[...], k, p, h},C):
{
    b := (C - (h*k)) mod (p_h - 1);
    ux := (gx)^b mod fx;
    sx := ux + fx, monic polynomial over Z_p with degree = h;
    Factorize sx into linear factors, so that sx :=(x+d_0)(x+d_1)...(x+d_h);
    m[..] := (m_0, m_1, ..., m_{p-1}), a binary vector all initalized to 0;
    invprm[...] := Compute inverse permutation of prm[...];
    for( i = 1 to i = h){
        d := d_i;
        m[invprm[d]] := 1;
    }
    M' := 0;
    b := h;
    for ( i = 1 to i = p){
        if( m[i] == 1){
            M' := M' + ^{(p - i)}C_b;
            b := b - 1;
        }
    }
    return M';
}
```

Algorithm 13: Key Generation

```
Algorithm: Key Generation For Chor-Rivest Cryptosystem

Input: Null
Ouput: Public Key, Private Key

procedure KeyGeneration( ):
{
 p := Select a large random prime;
 h := Select integer h (<=p);
 Fq := Select finite field GF(q), where q = p^h ;
 fx := Select random irreducible monic polynomial of degree h over Zp;
 gx := Primitive element of Fq;
 for( i=0 to i=p){
   ai = loggx(x + i);
 }
 k := Select non-negative integer less than p^h - 1;
 prm[...] := Generate random permutation of {0, 1, ..., p - 1};
 for( i=0 to i=p-1){
   li = ( aprm[i] + k) mod (p^h - 1);
 }
 return PublicKey{[l0, l1, ..., lp-1], p, h}, PrivateKey{fx, gx, prm, k, p, h};
}
```

Before this scheme was developed, all knapsack-based schemes used modular multiplication as their disguising transformation. On the contrary, this algorithm makes use of discrete logarithms over $GF(p^h)$ for transforming its generalized knapsack problem.

Apart from specialized attacks, where extra information is available to the adversary, the Chor–Rivest cryptosystem is accepted to be secure. The parameters must be chosen carefully for this scheme to work efficiently. The encryption process is very fast, the decryption is of the order of the RSA algorithm, while the computationally expensive part is the key generation process.

A large amount of computation involved in this Chor–Rivest knapsack-based asymmetric cryptosystem has restricted its use mostly to experimental purposes. If discrete logarithm computation becomes feasible someday, then this cryptosystem will become extremely useful, although consequently the number-theoretic algorithms like ElGamal cryptosystem will be rendered useless.

8.7 Chaos-Based Asymmetric Cryptography

A **dynamical system** is a system whose state evolves over the course of time. A state is a set of variables that completely defines the system. The set of all possible states forms the state-space. Dynamical systems can be both linear and nonlinear in nature. Continuous linear dynamical systems are called flows, while the discrete linear dynamical systems are called maps. Poincaré maps are used as a mapping between N-dimensional flows and $(N-1)$-dimensional maps. Examples of dynamical systems include systems describing the motion of undamped pendulums, population models, etc.

One-dimensional discrete dynamical systems are the simplest kind of systems for studying chaotic behavior. One-dimensional maps are dependent only on the present state of the dynamical system. The evolution function is used to get the next state of the system. One-dimensional discrete dynamical systems are usually denoted by

$$X_{n+1} = f(X_n)$$

Given an initial state X_0 of the system, future states can be obtained by repeatedly applying the evolution function. Thus, a discrete dynamical system is said to be iterated in this manner. An orbit of a state X_i of a given map $X_{n+1} = f(X_n)$ is the union of the set of states that the system went through up to the given state X_i, and the set of states that are obtained by continuously applying the evolution function starting from the given state. The set containing the previous states of a given state is called the backward orbit, while the set containing the future states is called the forward orbit. Non-invertible maps do not have backward orbits.

A fixed point of a map is a state which has the same next state as the present state, such that $f(X_i) = X_i$. If a map has a state X_p, such that on applying the evolution function on that state for n iterations, the system reaches the same state X_p, then the point X_p is called a periodic point of period n. A fixed point can be attracting or repelling in nature depending on whether the nearby orbits converge to or diverge from the point. Attracting fixed points and periodic points are called attractors.

Stability of a fixed point or periodic point is determined by the stability coefficient. A small change or perturbation is applied to a fixed periodic point of a dynamical system, and the corresponding change is observed in the system. The stability coefficient is obtained by analyzing this change in the system. If the stability coefficient is not equal to one, then the fixed point or the periodic point is hyperbolic—stability coefficient having a value less than one implies that the point is stable and for coefficient with a value greater than one, it is unstable. Nothing can be said about stability, if the stability coefficient is equal to one.

Nonlinear dynamical systems have evolution functions that are nonlinearly dependent on the input variables. These nonlinear dynamical systems exhibit deterministic chaotic behavior if they are deterministic in nature, have irregular and aperiodic behavior, and are sensitive to initial conditions. This sensitivity to initial conditions and irregular aperiodic behavior makes the chaotic systems mimic random behavior although being deterministic in true nature. This makes it suitable for use in cryptographic systems (Kocarev & Lian, 2011).

Chaotic maps form the core of chaos-based cryptography. The properties of chaotic dynamical systems parallel many properties of cryptographic systems. The ergodicity and sensitivity properties of chaotic systems are similar to the confusion and diffusion properties of cryptographic systems. The deterministic yet irregular behavior works similar to the deterministic pseudo-randomness of cryptographic systems. Despite the similarities, chaotic systems have many differences from cryptographic systems, as such chaotic systems work on subsets of real numbers contrary to cryptographic systems that work on finite sets of integers. Thus, initially many unsuccessful attempts were made to create cryptographic systems using chaotic systems, before finally secure systems were created based on chaos. One such successful attempt in creating an asymmetric cryptographic system involved integer based implementation using Chebyshev polynomials.

A **Chebyshev polynomial** is a map of degree r defined on the set of real numbers or its subset. It can be recursively defined as follows:

$$\mathbf{T}_{r+1}(\mathbf{x}) = 2x\mathbf{T}_r(\mathbf{x}) - \mathbf{T}_{r-1}(\mathbf{x})$$

with $\mathbf{T}_0 = 1$ and $\mathbf{T}_1 = \mathbf{x}$. The above expression can be transformed into an expression containing matrix algebra, which leads to the easy calculation of the Chebyshev polynomials of degree r. The matrix equation relating T_r and T_{r+1} to T_0 and T_1 is as follows:

$$\begin{bmatrix} \mathbf{T}_r \\ \mathbf{T}_r + 1 \end{bmatrix} = \mathbf{A}^r \cdot \begin{bmatrix} T_0 \\ T_1 \end{bmatrix}$$

Where,

$$\mathbf{A} = \begin{bmatrix} 0 & 1 \\ -1 & 2x \end{bmatrix}.$$

One important property of Chebyshev polynomials is that they commute under composition, i.e., $\mathbf{T}_r(\mathbf{T}_s(x)) = \mathbf{T}_s(\mathbf{T}_r(x))$ is satisfied. The asymmetric scheme discussed here is a generalization of the RSA algorithm using a modified form of Chebyshev polynomials. It is a secure algorithm, in the same way the RSA algorithm is, based on the computational intractability of the integer factorization problem (Kocarev, 2001).

Chebyshev map is modified to be restricted to the set $\mathbb{Z}_N$, such that $\mathbf{T}_r:\{0, 1, ..., N - 1\} \rightarrow \{0, 1, ..., N - 1\}$. The map is defined using modular arithmetic as follows:

$$\mathbf{MT}_r(x) = \mathbf{T}_r(x)(\bmod N)$$

where $\mathbf{MT}_r$ is the modified Chebyshev map, and x and N are integers. The modified polynomials still commute under composition, which is essential for modifying the RSA algorithm to use Chebyshev polynomials.

As in the case of the unmodified RSA algorithm, this algorithm also consists of two parts—key generation and encryption/decryption.

Let the sender S and receiver $\mathcal{R}$ be two parties who wish to communicate securely using this public-key cryptosystem. The receiver $\mathcal{R}$ first generates a pair of distinct keys—public key $\mathbf{K}_{\{PU,R\}}$ and private key $\mathbf{K}_{\{PR,R\}}$, using the key generation algorithm. As in other asymmetric schemes, the public key is shared with the sender while the private key is kept secret. The sender uses the public key $\mathbf{K}_{\{PU,R\}}$ and the encryption algorithm to generate the ciphertext $\mathbf{C}$ of the given input message $\mathbf{M}$. The receiver $\mathcal{R}$ on receiving the encrypted message $\mathbf{C}$ uses the secret private key $\mathbf{K}_{\{PR,R\}}$ to decrypt the message. In this particular chaos-based scheme, the modular exponentiation, used for encryption and decryption, is replaced by the modified Chebyshev polynomial.

The following example demonstrates the use of modified Chebyshev polynomials in the RSA cryptosystem to encrypt a given message.

Example 4

Bob (B) wants to send Alice (A) a message $M = 67$ using chaos-based implementation of RSA cryptosystem. In order to generate her public and private keys, Alice selects two primes $p = 17$ and $q = 23$. Thus, $N = p \times q = 391$ is computed. Using p and q, the value of $\phi(N)$ is calculated as $\phi(N) = (p - 1)(q - 1) = 352$.

Let the encryption exponent be $e = 59$, such that $\gcd(\phi(N), e) = 1$. Using the extended Euclidean algorithm, Alice computes the decryption exponent $d = 179$, satisfying the congruence $ed \equiv 1 \pmod{N}$. Thus, Alice's public key is $K_{\{PU, A\}} = \{391, 59\}$, and her private key is $K_{\{PR, A\}} = \{391, 179\}$.

Now, B obtains A's public key and then computes the modified Chebyshev polynomial $MT_r(x)$ for $r = e = 59$ and $x = M = 67$. This generates the ciphertext $C = 31$, which B sends to A.

On receiving the encrypted message C, Alice decrypts the ciphertext by using her private key $K_{\{PR, A\}}$. To perform the decryption, A calculates $MT_r(x)$ for $r = d = 179$ and $x = C = 31$, and obtains the original message $M = 67$. Thus, Alice is able to securely receive and recover the message M sent by Bob, using chaos-based implementation of RSA cryptosystem (Kocarev, Makraduli, & Amato, 2005, Kocarev & Tasev, 2003).

Despite the recent advances in chaos-based cryptographic algorithms, they still haven't gained much popularity as mainstream cryptographic algorithms. This can be attributed to the fact that most chaos-based cryptographic algorithms work on real numbers leading to practical implementation difficulties.

Algorithm 14: Key Generation Algorithm

```
Algorithm: Key Generation For Chaos-Based RSA

Input: Null
Ouput: Public Key, Private Key

procedure KeyGeneration( ):
{
 p := Select a large random prime;
 q := Select another large random prime != p;
 N := p * q;
 phi := (p - 1)(q - 1);
 e := Select random integer such that 1<e<phi, and gcd
(phi,e)==1;
 d := Multiplicative inverse of e mod phi;
 return PublicKey{N, e}, PrivateKey{d, N};
}
```

Algorithm 15: Encryption Algorithm

```
Algorithm: Encryption for Chaos-Based RSA

Input: Receiver's Public Key {N, e}, Input Message M
Output: Encrypted Message

procedure Encryption({N, e}, M):
{
   Let MT_r(x) := T_r(x) mod N, the modified Chebyshev
Polynomial of degree r;
   M' := Represent M as an integer in range [1,N-1];
   C := MT_d(M');
   return C;
}
```

Algorithm 16: Decryption Algorithm

Algorithm: Decryption Algorithm For Chaos-Based RSA

Input: PrivateKey {d, N}, Encrypted Message C;
Output: Decrypted Message

procedure Decryption({d, N}, C):
{
   Let $MT_r(x) := T_r(x)$ mod N, the modified Chebyshev
Polynomial of degree r;
   $M' := MT_d(C)$ mod N;
   return M';
}

Also, cryptographic analysis methods for chaos-based systems have not been developed properly, due to which it is very difficult to assess chaos-based methods. There is continuous ongoing development in chaos-based methods, including hardware realization of the methods, development of newer algorithms, and creation of new methods of performance analysis. It is quite hopeful that in near future chaos-based methods might become mainstream asymmetric cryptosystems, and possibly gain widespread usage.

8.8 Conclusion

We now conclude our discussion on public-key algorithms. We have described and analyzed five important asymmetric cryptosystems, and how they fit into the general framework of public-key cryptography, as was proposed by Diffie and Hellman. As final remarks, it can be said that asymmetric cryptography revolutionized the modern digital world by providing a very secure way of communication, solving important problems like non-repudiation, authentication, and key distribution; and leading to creation of digital systems like e-currency, zero-knowledge proofs, secure e-commerce, etc., which were almost impossible to realize previously.

References

Boneh, D. (1999). Twenty years of attacks on the RSA cryptosystem. *Notices of the AMS, 46*, 203–213.

Chor, B., & Rivest, R. L. (1985). A knapsack type public key cryptosystem based on arithmetic in finite fields (preliminary draft). In *Workshop on the Theory and Application of Cryptographic Techniques* (Vol. 196, pp. 54–65). Springer. doi:10.1007/3-540-39568-7_6.

Diffie, W., & Hellman, M. (1976). New directions in cryptography. *IEEE Transactions on Information Theory, 22*(6), 644–654. doi:10.1109/tit.1976.1055638.

El Gamal, T. (1985). A public key cryptosystem and a signature scheme based on discrete logarithms. In *Workshop on the Theory and Application of Cryptographic Techniques* (Vol. 196, pp. 10–18). Springer. doi:10.1007/3-540-39568-7_2.

Goldwasser, S. (1997). New directions in cryptography: Twenty some years later (or cryptography and complexity theory: A match made in heaven). *Proceedings 38th Annual Symposium on Foundations of Computer Science* (pp. 314–324). doi:10.1109/sfcs.1997.646120.

Katz, J., & Lindell, Y. (2007). *Introduction to Modern Cryptography*. Chapman & Hall (CRC Press). ISBN 13: 978-1-58488-551-1.

Kocarev, L. (2001). Chaos-based cryptography: A brief overview. *IEEE Circuits and Systems Magazine*, 1(3), 6–21. doi:10.1109/7384.963463.

Kocarev, L., & Lian, S. (2011). Chaos-Based Cryptography (1st ed.). Springer-Verlag, Berlin Heidelberg. doi:10.1007/978-3-642-20542-2.

Kocarev, L., Makraduli, J., & Amato, P. (2005). Public-key encryption based on Chebyshev polynomials. *Circuits, Systems and Signal Processing*, 24(5), 497–517. doi:10.1007/s00034-005-2403-x.

Kocarev, L., & Tasev, Z. (2003). Public-key encryption based on Chebyshev maps. In *Proceedings of the 2003 International Symposium on Circuits and Systems* (pp. III-28–III-31). IEEE. doi:10.1109/ISCAS.2003.1204947.

Menezes, A., van Oorschot, P., Vanstone, S. (2001). *Handbook of Applied Cryptography* (5th ed.). CRC Press. ISBN: 0-8493-8523-7.

Merkle, R., & Hellman, M. (1978). Hiding information and signatures in trapdoor knapsacks. *IEEE Transactions on Information Theory*, 24(5), 525–530. doi:10.1109/tit.1978.1055927.

Rabin, M. O. (1979). *Digitalized Signatures and Public-Key Functions as Intractable as Factorization* (pp. 1–16, Rep. No. MIT/LCS/TR-212). Massachusetts Institute of Technology, Cambridge.

Rivest, R. L., Shamir, A., & Adleman, L. (1978). A Method for Obtaining Digital Signatures and Public-Key Cryptosystems. doi:10.21236/ada606588.

Schneier, B. (1995). *Applied Cryptography: Protocols, Algorithms, and Source Code in C* (2nd ed.). John Wiley & Sons. ISBN: 0-471-11709-9.

Shamir, A. (1982). A polynomial time algorithm for breaking the basic Merkle–Hellman cryptosystem. In *Proceedings of the 23rd Annual Symposium on Foundations of Computer Science* (pp. 145–152). IEEE Computer Society. doi:10.1109/SFCS.1982.55.

Stallings, W. (2017). *Cryptography and Network Security: Principles and Practice*. Pearson Education, Boston.

9

Post-Quantum Cryptography

Amandeep Singh Bhatia and Ajay Kumar

Thapar Institute of Engineering & Technology

CONTENTS

9.1   Introduction 139
9.2   Preliminaries 141
9.3   Code-Based Cryptography 142
    9.3.1   McEliece Cryptosystem 142
        9.3.1.1   Disadvantages 145
        9.3.1.2   Security 145
    9.3.2   Niederreiter Cryptosystem 145
9.4   Lattice-Based Cryptography 146
    9.4.1   NTRU Algorithm 146
    9.4.2   Goldreich–Goldwasser–Halevi (GGH) 148
9.5   Multivariate Cryptography 150
    9.5.1   OV Signature Scheme 151
    9.5.2   Rainbow Signature Scheme 151
    9.5.3   Quartz Signature Scheme (2, 129, 103, 3, 4) 152
9.6   Hash-Based Cryptography 153
    9.6.1   Lamport–Diffie One-Time Signature Scheme (LD-OTS) 153
    9.6.2   Winternitz One-Time Signature Scheme (W-OTS) 154
9.7   Conclusion 155
References 155

9.1 Introduction

Quantum computing is a winsome field that deals with theoretical computational systems (i.e., quantum computers) combining visionary ideas of computer science, physics, and mathematics. It concerns with the behavior and nature of energy at the quantum level to improve the efficiency of computation. Initially, the idea of quantum computing is proposed after performing a quantum mechanics simulation on a classical computer (Feynman, 1982). Up until then, quantum computing was thought to be only a theoretical possibility, but research over the last three decades has evolved such as to make quantum computing applications a realistic possibility.

Over the last three decades, public-key cryptosystems (Diffie–Hellman key exchange, the RSA cryptosystem, the digital signature algorithm [DSA], and elliptic curve cryptosystems

[ECC]) have become a crucial component of cyber security. In this regard, security depends on the difficulty of a definite number of theoretic problems (integer factorization or the discrete log problem). (Shor, 1994) designed an algorithm to calculate factors of a large number n with space complexity $O(\log n)$ and runs in $O\big((\log n)^2 * \log \log n\big)$ on a quantum computer, and then perform $O(\log n)$ post-processing time on a classical computer, which could be applied for cracking the RSA algorithm at Bell Laboratories (US). (Grover, 1996) designed a searching algorithm with great quadratic gain for finding an element in an unstructured set of size n in $\sqrt{n}$ operations approximately. Furthermore, (Kwiat et al., 2000) implemented Grover's algorithm at Los Alamos National Laboratory using conventional optical interferometers. Quantum parallelism is one of the main features of quantum computation which allows quantum algorithms to gain speedup as compared to classical algorithms.

Through the impetus provided by Shor's algorithm (Shor, 1994), quantum computational complexity is an exhilarating area that transcends the boundaries of quantum physics and theoretical computer science. This algorithm is well known in the field of cryptography given its potential application in cracking various cryptosystems, such as the RSA algorithm and elliptic curve cryptography (Chen et al., 2016). These all public-key cryptosystems can be attacked in polynomial time. Although Grover's algorithm can also be steered to some attacks, it can be evaded by modifying the parameters. Table 9.1 represents the present status of several cryptosystems (Bernstein, 2009).

Post-quantum cryptography offers secure alternatives. The goal of post-quantum cryptography is to develop cryptographic systems that are secure against both quantum and classical computers, and compatible with existing communication protocols and networks. Apart from RSA, DSA, and ECDSA, there are other important classes of cryptographic systems which include *Code-based, Lattice-based, Hash-based and Multivariate quadratic equations*. In fact, nobody has been able to apply Shor's algorithm to these classes of cryptographic systems.

There are still many challenges regarding the successful execution of post-quantum cryptographic algorithms; for example, there is a need to enhance the usability and effectiveness of post-quantum cryptosystems, and time is needed to make sureness in post-quantum cryptosystems. By 2020, if the quantum computers come into action, these challenges need to be addressed because we are not yet organized to shift to an era of post-quantum cryptography. The following are the details of several cryptographic systems that have altogether withstand every attack.

TABLE 9.1

Impact of Quantum Computing on Cryptographic Algorithms

Cryptosystem	Broken by Quantum Algorithms?
Diffie–Hellman key exchange (Diffie and Hellman, 1976)	Broken
RSA public-key encryption (Rivest et al., 1978)	Broken
Algebraically homomorphic (Rivest et al., 1978)	Broken
Elliptic curve cryptography (Koblitz, 1987)	Broken
Buchmann–Williams key exchange (Buchmann and Williams, 1988)	Broken
McEliece public-key encryption (McEliece, 1978)	Not broken yet
NTRU public-key encryption (Hoffstein et al., 1998)	Not broken yet
Lattice-based public-key encryption (Cai and Cusick, 1998)	Not broken yet

9.2 Preliminaries

In this section, some preliminaries and basic notations are given, which will be used throughout the chapter.

- *Linear code*: Linear code C of length n and dimension k over a field F is a k-dimensional subspace of the vector space F_q^n with q elements, a set of n-dimensional vectors can be referred to as a $[n, k]$ code and elements of bits such that $F = GF(2) = \{0, 1\}$. If the minimum Hamming distance of the code is d, then the code is called a $[n, k, d]$ code.

- *Hamming distance*: A Hamming distance $d_H(x, y)$ is the number of positions in which two code words (x, y) differ (Löndahl, 2015). Let C be a $[n, k]$ linear code over F_q^n and $x = (x_1, x_2, \ldots, x_n)$, $y = (y_1, y_2, \ldots, y_n)$ are two code words.

$$d_H(x, y) = |i : x_i \neq y_i, 1 \leq i \leq n| \tag{9.1}$$

- *Hamming weight*: A Hamming weight $wt_H(x)$ is defined as the number of nonzero positions in the code word x (Löndahl, 2015). Let C be a $[n, k]$ linear code over F_q^n and $x = (x_1, x_2, \ldots, x_n)$ is a code word, such that

$$wt_H(x) = |i : x_i \neq 0, 1 \leq i \leq n| \tag{9.2}$$

- *Generator matrix*: A generator matrix for C is a $k \times n$ matrix G having the vectors of $V = (v_1, v_2, \ldots, v_k)$ as rows, which forms a basis of C such that

$$C = \left\{ mG : m \in F_q^n \right\}, G = \begin{bmatrix} v_1 \\ v_2 \\ \ldots \\ v_k \end{bmatrix} \tag{9.3}$$

The matrix G generates the code as a linear map: for each message $m \in F_q^n$, we obtain the corresponding code word mG (Löndahl, 2015).

- *Parity matrix*: A $(n-k) \times n$ generator matrix H is called a parity-check matrix for code word C (Löndahl, 2015), which is described by

$$C = \left\{ m \in F_q^n : mH^T = 0 \right\} \tag{9.4}$$

- *Lattice (L)*: It is defined as a set of all integer combinations of linearly independent vectors $(a_1, a_2, \ldots, a_n)$ of length n in R^n, which are called basis in the lattice.

$$L(a_1, a_2, \ldots, a_n) = \sum_{i=1}^{n} x_i a_i \mid x_i \in Z, \text{ for } 1 \leq i \leq n \tag{9.5}$$

It can be represented in a matrix form such that $A = (a_1, a_2, \ldots, a_n) \in R^{n \times n}$, where columns act as a basis vector (Peikert, 2016).

- *Polynomial ring*: Consider a commutative ring R, then the polynomial ring $R[x] = \left\{ a_n x^n + a_{n-1} x^{n-1} + \cdots + a_1 x + a_0 \right\}$, where x is the set of polynomials with $a_0, a_1, \ldots, a_n$ coefficients and $n \in Z$ is called the ring of polynomials over R (Buchmann, 2013).

9.3 Code-Based Cryptography

Code-based cryptography usually refers to cryptographic algorithms using an error-correcting code C (McEliece, 1978). Furthermore, it adds an intended error to a word and calculates a syndrome relative to its parity-check matrix of code. The process of code-based cryptography is a commutation between efficiency and security. There are several codes for which efficient decoders are known.

9.3.1 McEliece Cryptosystem

McEliece cryptosystem is the most successful cryptosystem, based on extended Golay code [24, 12, 8]. Originally, Golay codes were invented in the early 1950s and have experienced incredible responses in the last few years. McEliece (1978) proposed an asymmetric encryption cryptosystem based on Goppa codes, which remains unbroken, even after 15 years of adaptation of its proposal security parameters (Sendrier, 1998). McEliece cryptosystem is based on linear error-correcting code for creating public and private keys. The secret key can be drawn from the various alternate codes. Several versions of McEliece cryptosystem were proposed using various secret codes such as Reed–Solomon codes, concatenated codes, and Goppa codes. The original McEliece cryptosystem algorithm is given below. McEliece public-key cryptosystem is illustrated in Figure 9.1 (see also Table 9.2).

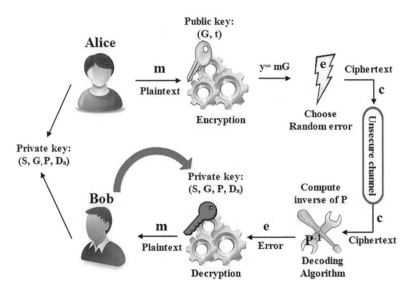

FIGURE 9.1
McEliece public-key cryptosystem.

TABLE 9.2

McEliece Public-Key Cryptosystem

Key Creation

- Select a random $[n, k, 2t + 1]$ linear code C and decoding algorithm (D_a) suitable for rectifying up to t errors efficiently
- Generate a $(k \times n)$ generator matrix (G) for C
- Select a random binary non-singular matrix (S) $(k \times k)$
- Select a random permutation matrix (P) $(n \times n)$
- Compute the $(k \times n)$ matrix $G' = SGP$
- The private key: (S, G, P, D_a) and public key: (G', t)

Encryption

- Choose a n-length vector $e \in \{0, 1\}^n$ randomly that contains exactly t ones
- Encrypt the message $m \in \{0, 1\}^k$ of length k and calculate the ciphertext (c) as

$$c = mG' + e$$

Decryption

- Compute the inverse of the permutation matrix P^{-1}
- Calculate $cP^{-1} = mSG + eP^{-1}$, and then employ decoding algorithm (D_a) to decrypt $c_1 = cP^{-1}$ to m'
- Compute the message $m = m'S^{-1}$

For instance, consider a (7, 4) Hamming code (n, k) and can generate a linear code (C) of length $n = 7$ and dimension $k = 4$. The generator matrix G for C can be computed in standard form $G = [I_k \mid Z]$, where I_k is the $(k \times k)$ identity matrix and Z is $(k \times r)$ such that r denotes redundant bits calculated as $r = n - k$:

$$G = \begin{bmatrix} 1 & 0 & 0 & 0 & 0 & 1 & 1 \\ 0 & 1 & 0 & 0 & 1 & 1 & 0 \\ 0 & 0 & 1 & 0 & 1 & 0 & 1 \\ 0 & 0 & 0 & 1 & 1 & 1 & 1 \end{bmatrix} \tag{9.6}$$

where generator matrix G is 4×7. Now, we have selected a random matrix S (non-singular) 4×4 as

$$S = \begin{bmatrix} 0 & 1 & 1 & 1 \\ 1 & 0 & 1 & 0 \\ 1 & 1 & 1 & 1 \\ 1 & 1 & 0 & 0 \end{bmatrix} \tag{9.7}$$

In the next step, we have chosen a random 7×7 permutation matrix as follows:

$$P = \begin{bmatrix} 0 & 1 & 0 & 0 & 0 & 0 & 0 \\ 0 & 0 & 1 & 0 & 0 & 0 & 0 \\ 1 & 0 & 0 & 0 & 0 & 0 & 0 \\ 0 & 0 & 0 & 0 & 0 & 1 & 0 \\ 0 & 0 & 0 & 0 & 1 & 0 & 0 \\ 0 & 0 & 0 & 1 & 0 & 0 & 0 \\ 0 & 0 & 0 & 0 & 0 & 0 & 1 \end{bmatrix} \tag{9.8}$$

Furthermore, we have computed the public generator matrix G' along with $t = 2$ for encryption such that

$$G' = SGP = \begin{bmatrix} 1 & 0 & 1 & 0 & 1 & 1 & 0 \\ 1 & 1 & 0 & 1 & 1 & 0 & 0 \\ 1 & 1 & 1 & 1 & 1 & 1 & 1 \\ 0 & 1 & 1 & 0 & 1 & 0 & 1 \end{bmatrix} \tag{9.9}$$

Thus, the private key is formed as (S, G, P, D) and public key (G', t). Suppose we wish to send a message $m = \begin{matrix} 1 & 1 & 0 & 0 \end{matrix}$ to another party. The matrix is computed as follows:

$$mG' = 0 \quad 1 \quad 1 \quad 1 \quad 0 \quad 1 \quad 0$$

In the next step, a random error vector $(e) = 0 \quad 1 \quad 1 \quad 0 \quad 0 \quad 0 \quad 0$ of length n contains exactly $t = 2$ ones is chosen and the ciphertext $c = mG' + e$ is formed as follows:

$$c = 0 \quad 0 \quad 0 \quad 1 \quad 0 \quad 1 \quad 0$$

On decryption side, the other party tries to retrieve the original message m from the ciphertext c. First, we have computed the inverse of the permutation matrix and applied the decoding algorithm to retrieve m such that

$$cP^{-1} = mG'P^{-1} + eP^{-1}$$

$$c_1 = mSG + e_1$$

Here, $c_1 = cP^{-1} = 0 \quad 0 \quad 0 \quad 1 \quad 0 \quad 1 \quad 0$ and found that error is in the first and second bits after calculating mSG such that

$$r = mSG = 1 \quad 1 \quad 0 \quad 1 \quad 0 \quad 1 \quad 0$$

By using the Gauss–Jordan elimination for $\left[G^T \mid r^T \right]$, we have computed that $mS = 1 \quad 1 \quad 0 \quad 1$. In the end, by multiplying with the inverse of S matrix, the original message (m) is retrieved as follows:

$$m = 1 \quad 1 \quad 0 \quad 1 \cdot S^{-1} = 1 \quad 1 \quad 0 \quad 1 \cdot \begin{bmatrix} 1 & 0 & 1 & 0 \\ 1 & 0 & 1 & 1 \\ 1 & 1 & 1 & 0 \\ 1 & 1 & 0 & 1 \end{bmatrix}$$

$$= 1 \quad 1 \quad 0 \quad 0$$

9.3.1.1 Disadvantages

- Large key size: Public key is sizeable. It consists of 2^{19} bits which can affect its implementation.
- Low transmission rate.
- It is asymmetric in nature, i.e., the encryption does not get commute with decryption.
- Increase in bandwidth size due to lengthy encrypted message than the plaintext, which makes it vulnerable to errors.

9.3.1.2 Security

The McEliece public-key cryptosystem is considered to be quite secure. Its security depends upon the thorough search on the key and decoding algorithm. The following are the two types of attacks on McEliece cryptosystem.

9.3.1.2.1 Decoding Attacks

This is due to decoding of the ciphertext. If the attack becomes successful, then it is easy to recover the plaintext. It has been investigated that decoding of ciphertext needs 2^{64} operations for a code of 1,024 length, 524 size, and 50 error-correcting ability (Loidreau, 2000). Thus, the overall cost of decoding attack is based on the three parameters for code C: its size, length of code, and error-correcting ability. Consequently, the original McEliece cryptosystem will be insecure as the computing power of system increases.

9.3.1.2.2 Structural Attacks

It is an effort to reconstruct the actual form of the code from the generated public key. In case of successful structural attack, the private key is recovered and the whole cryptosystem is broken. There is a need to scrutinize 2^{466} codes for the original McEliece cryptosystem of length $n = 1,024$, $k = 524$, and $t = 50$. (Loidreau and Sendrier, 2001) attacked the structure of McEliece encryption and disclosed the part of "weak" structure generator matrix G. Till now, the structural attacks have been attempted on its different variants.

9.3.2 Niederreiter Cryptosystem

The concept of knapsack-type cryptosystem is proposed by (Niederreiter, 1986) based on Reed–Solomon codes. Its security is equivalent to McEliece cryptosystem. It is different from McEliece cryptosystem in description of codes, i.e., Niederreiter uses parity-check matrix, whereas others use generator matrix. Later, it has been proved that Niederreiter cryptosystem is insecure using Reed–Solomon codes as well as Goppa codes (Sidelnikov et al., 1992). Many researchers proposed various cryptosystems as a modification to McEliece cryptosystem to shrink the size of its public key. However, they are proved to be insecure and ineffective as compared to original McEliece cryptosystem (Table 9.3).

It is noted that plaintext m is illustrated as an error of C instead of original word as in the case of McEliece cryptosystem. It is helpful to design digital signature scheme. Moreover, its decryption part can be executed more efficiently than original McEliece cryptosystem. Thus, McEliece and Niederreiter cryptosystems are dual to each other using linear codes, i.e., parity-check matrix can be computed from the generator matrix and vice versa. As a result, if an attacker can decipher the code in McEliece, then he can easily decipher the code in Niederreiter cryptosystem and conversely.

TABLE 9.3

Niederreiter Cryptosystem

Key Generation

- Select a $[n, k, 2t + 1]$ linear code C at random and efficient decoding algorithm (D) capable of correcting up to t errors
- Generate a parity-check matrix (H) $(n - k) \times n$ for C
- Select a $(n - k) \times (n - k)$ binary non-singular matrix (S) randomly
- Select a permutation matrix (P) $(n \times n)$ at random
- Compute the $(n - k) \times n$ matrix $H' = SHP$
- Private key: (S, H, P, D) and public key: (H', t)

Encryption

- Encrypt the message $m \in \{0, 1\}^k$ of length k and weight t, and compute the ciphertext (c) as

$$c = mH'^T$$

Decryption

- Compute the inverse of non-singular matrix S^{-1}
- Calculate $S^{-1}c^T = HPm^T$.
- Find a vector (v) such that $Hv^T = HPm^T$ using linear algebra
- As mP^T has weight t, then, in order to find an error mP^T, apply the efficient decoding algorithm (D) on v and retrieve the message m

---

9.4 Lattice-Based Cryptography

The construction of lattice-based cryptosystems holds strong security proofs on the basis of worst-case resistance of lattice problems which offers efficient execution and simplicity. Moreover, such cryptosystems are reliable and secure against the attacks of quantum computers. Basically, a lattice representing arbitrary basis is given as an input and expected the output to be the shortest nonzero vector. The concept of lattice-based cryptosystem for the shortest vector problems is proposed by (Lenstra and Lenstra, 1993). It runs in polynomial time. It is the most extensively studied algorithm for lattice problems, but later on, its various extensions have been introduced. We described the main lattice-based cryptosystems that have been introduced so far. We start with the NTRU cryptosystem, which is the most well-known, practically implemented, lattice-based encryption scheme till now.

9.4.1 NTRU Algorithm

NTRUEncrypt is a public-key cryptosystem developed by three mathematicians: Hoffstein, Pipher, and Silverman in 1998. It is ring-based cryptosystem. The lattices used by NTRU algorithm are closed under linear transformation and are q-ary lattices, i.e., they contain sublattice also. Since the proposal of NTRU algorithm, many researchers tried to improve its security, efficiency, and speedup process. Moreover, several attacks have been introduced, which are focused on finding the secret key instead of recovering a message. NTRU is short for Nth-degree truncated polynomial ring R, and it can also signify Number Theory Research Unit (Liu, 2015). Since the introduction of original NTRU algorithm, its several representations have been proposed on the basis of suitable selection of parameters. The working of NTRU cryptosystem is as follows (Table 9.4).

TABLE 9.4

NTRU Cryptosystem

Public Parameters

- Operations are formed on objects in a polynomial ring (R) that is truncated having coefficients to a certain degree such that $R = Z[X]/(X^N - 1)$. Select two moduli p and q relatively prime in R. The moduli p is smaller than q, and q is smaller than N such that $\gcd(p, q) = 1$. Note that the symbol * denotes the multiplication in R
- Additional parameter d_f is a set of polynomials in R having exactly $d - 1$'s and $(d + 1)$ 1's; d_g and d_r are sets of polynomials in R having exactly similar $d - 1$'s and d 1's and d_m is a set of polynomials in R_p such that coefficients are in between $-1/2(p + 1)$ and $1/2(p + 1)$

Key Creation

- Choose a random $f \in d_f$ and invertible in $R = Z[X]/(X^N - 1) \bmod p$ and q. In case, if it does not satisfy, then new random f is chosen
- Compute the inverse of $f \bmod p$ and $f \bmod q$ such that $f_q^{-1} \cdot f \equiv 1 \bmod q$ and $f_p^{-1} * f \equiv 1 \bmod p$
- Hence, the pair $\left(f, f_q^{-1}\right)$ is a private key
- For the public key h, compute the polynomial $h = f_q^{-1} * g \pmod q$

Encryption

- Choose a random polynomial $r \in d_r$
- Select a message $m \in d_m$
- Compute the ciphertext (c): $c = pr * h + m \pmod q$

Decryption

- Compute a polynomial a using private key polynomial such that $a = f * c \pmod q$
- Compute $b = a \pmod q$ and lessen each of the coefficients of $b \bmod p$
- Compute $z = f_p^{-1} * b \pmod p$ by using a private key polynomial f_p^{-1} to retrieve the original message m

It must be noted that if the chosen public parameters (n, p, q, d) of NTRU satisfy that $q > (6d + 1)p$, it assures that decryption will never be unsuccessful (Malekian and Zakerolhosseini, 2010). The advantages of NTRU cryptosystem are as follows: it offers efficient implementation of encryption and decryption in software and hardware (Hoffstein et al., 1998; Bu and Zhou, 2009). The keys are generated much faster. It takes less memory to implement and can be effective for small memory devices such as smart cards (integrated circuit) and mobile phones.

To demonstrate the NTRU cryptosystem, we considered the following parameters: $N = 7$, $p = 3$, $q = 41$.

We have selected random polynomial $f \in d_f$ such that it is invertible in p and q. Then, randomly selected polynomial $g \in d_g$.

$$f(x) = -x^6 - x^5 + x^4 + x^2 + 1, \quad g(x) = x^5 - x^4 + x^3 - x^2$$

In the next step, we have computed the inverse of $f(x)$ to modulo q, modulo p and named it as f_q^{-1} and f_p^{-1}.

$$f_q^{-1}(x) = 38x^6 + 16x^5 + 38x^4 + 35x^3 + 22x^2 + 22x + 35,$$

$$f_p^{-1}(x) = 2x^6 + 2x^4 + x^3 + 2x^2 + 2x + 1$$

Here, the pair $\left(f, f_q^{-1}\right)$ acts as a private key, and the public key is calculated as

$$h = f_q^{-1} * g \,(\mathrm{mod}\, q) = 38x^6 + 3x^4 + 32x^3 + 25x^2 + 16x + 9$$

We have chosen the random polynomial r besides the message m as follows:

$$m(x) = x^4 + x^3 - x^2 + 1$$

$$r(x) = x^5 - x^4 + x^3 + 1$$

The message m is encrypted using public key as

$$c(x) = pr * h + m \,(\mathrm{mod}\, q) = 19x^6 + 13x^5 + 22x^4 + 10x^3 + 33x^2 + 12x + 16$$

Now, on receiving the ciphertext (c), the message is decrypted using the private key and centering modulo to q as

$$a = f * c \,(\mathrm{mod}\, q) = 5x^6 + 39x^4 + 6x^3 + 30x^2 + 10x + 35$$

After centering the coefficients, it becomes

$$b = 5x^6 - 2x^4 + 6x^3 - 11x^2 + 10x - 6$$

Finally, compute $z = f_p^{-1} * b \,(\mathrm{mod}\, p)$ and centering the coefficients using mod p

$$z = f_p^{-1} * b \,(\mathrm{mod}\, p) = x^4 + x^3 + 2x^2 + 1$$

The next step is to lessen the coefficients of z to mod p, and it becomes

$$z = x^4 + x^3 - x^2 + 1$$

which is equal to the original message m.

NTRU cryptosystem is secure and efficient on comparing with the best-known public-key cryptosystems such as RSA and ECC. There exist several attacks against NTRU scheme: brute force, multiple transmission attacks, and alternate private keys are used to decrypt the message m and lattice-based attacks. (Mersin, 2007) compared the NTRU cryptosystem with RSA and ECC cryptosystems. It has been shown that the performance of NTRU cryptosystem is far better than other algorithms. Thus, the key generation, encryption, and decryption processes of NTRU take less time. But, on comparing with the McEliece public-key cryptosystem, it has been investigated that the encryption and decryption processes of NTRU cryptosystem take more time.

9.4.2 Goldreich–Goldwasser–Halevi (GGH)

It is the most instinctive encryption scheme based on lattices. There exist many components of GGH encryption scheme in other lattice-based cryptosystems, but original GGH

TABLE 9.5

GGH Cryptosystem

Construction of Keys

- For private key: Choose a full-column rank integer matrix V such that columns are orthogonal to each other

$$V = [v_1, v_2, v_3, \ldots, v_n], v_j \in Z^n, 1 \le j \le n$$

- Select a random $n \times n$ unimodular matrix (U), such that $\det(U) = \pm 1$
- For public key, compute a matrix $W = VU$

Encryption

- Choose a perturbation vector that acts as an ephemeral key
- Select a plaintext message m of the same dimension that belongs to lattice
- Encrypt the message m using public key W and ephemeral key r, and compute the ciphertext

$$c = Wm + r$$

Decryption

- Decrypt the closest vector v that belongs to the lattice of ciphertext (c) using private key V and any decoding algorithm
- Compute the inverse of V to decrypt the ciphertext such that $x = c \cdot V^{-1}$
- Then, retrieve the original message m by multiplying with the inverse of the unimodular matrix such as $m = x \cdot U^{-1}$. In the end, its correctness can be checked by differentiating m with the hashed value

---

is practical which offers simplicity. It is introduced as a lattice equivalent of McEliece cryptosystem in which short orthogonal vectors are used as a private key by Goldreich et al. (1997). It acts as a good lattice basis. Till now, asymptotically good attack to GGH public encryption scheme is not identified because the security and correctness depend upon the selection of private basis and error vector. It is based on solving the problem "close vector problem (CVP)" in lattice, i.e., finds the lattice point closest to a given vector. As compared to RSA, Diffie–Hellman, and ElGamal, it offers high performance due to simple matrix operations (Table 9.5).

The security of GGH cryptosystem depends upon selecting a suitable perturbation vector r. If it is selected very small, then the closest vector v can be easily retrieved without any difficulty. If it is selected very large, then it may be not possible to decrypt using the private key. Hence, the perturbation vector r must be chosen balanced, i.e., relatively small as compared to the vectors in public key W. The advantages of lattice-based cryptography are as follows: till now, any quantum attacks do not exist to break lattice-based cryptosystems, and it is one of chief substitutes for post-quantum cryptosystems. The main disadvantages of lattice-based cryptography are as follows: lattice-based cryptosystems are not applied much yet due to security reasons (Nguyen and Regev, 2009). NTRU is efficient in implementation but lacks in security.

The first step is to create keys, for instance, consider a lattice $L \subset R^2$ with the basis (V), i.e., full-column rank integer matrix in which all columns are orthogonal to each other. Furthermore, we have selected a random unimodular matrix U:

$$V = \begin{bmatrix} 9 & 0 \\ 0 & 7 \end{bmatrix}, \quad U = \begin{bmatrix} -1 & 1 \\ -2 & 1 \end{bmatrix} \tag{9.10}$$

In the next step, we have computed public key as $W = U \cdot V$:

$$W = \begin{bmatrix} -9 & 7 \\ -18 & 7 \end{bmatrix} \qquad (9.11)$$

Suppose we wish to send a message $m = \begin{bmatrix} -11 & 5 \end{bmatrix}$ to other party and the key is selected as $r = \begin{bmatrix} -1 & 1 \end{bmatrix}$. Then, we encrypt it using W and ciphertext c is formed as

$$c = m \cdot W + r = [8 - 41]$$

On receiving the ciphertext, receiver decrypts it using the inverse of private key V as

$$x = c \cdot V^{-1} = [0.8889 - 5.8571]$$

On rounding, it becomes $x = \begin{bmatrix} 1 & -6 \end{bmatrix}$. At last, we retrieved the original message m by multiplying decrypted message x with the inverse of U, it becomes

$$m = x \cdot U^{-1} = \begin{bmatrix} -11 & 5 \end{bmatrix}$$

which is equal to the original message m.

Nguyen and Regev (2009) have shown the imperfection in the design of GGH cryptosystem. They solved GGH challenges and came with partial information about the lattice. No doubt, GGH cryptosystem offers security with a suitable choice of parameters. But, on choosing the high dimension of the lattice, it shows great improvement. Although the key size increases quadratically, it has been investigated that GGH cryptosystem can guarantee the security if the dimension of the lattice is larger than 350 (Micciancio, 2001).

9.5 Multivariate Cryptography

Multivariate cryptography schemes are based on nonlinear equations that are difficult to solve over finite fields. Therefore, it is said to be very light-weight cryptography that is efficient and effective to be used in embedded devices (microcontrollers). The security of multivariate cryptosystems is based on NP-hardness of the problem to solve nonlinear equations over a finite field. In the last two decades, the rigorous development occurs in multivariate cryptography. It is defined as the study of a public-key cryptosystem (Diffie and Hellman) in which trapdoor one-way function is based on the mapping of multivariate quadratic equations over a finite field. There exist several multivariate public-key cryptography schemes such as oil and vinegar (OV) signature scheme (Patarin, 1997), Rainbow signature scheme $(2^8, 18, 12, 12)$ (Ding and Schmidt, 2005), and Quartz signature scheme $(2, 129, 103, 3, 4)$ (Patarin et al., 2001). Such multivariate public-key encryption schemes are built in many ways, and all have their own advantages and disadvantages. It can also be called as trapdoor multivariate quadratic because most of the schemes are constructed using higher order quadratic polynomial equations.

Generally, the public key is a set of quadratic polynomials over a finite field:

$$p_1(x_1,\ldots,x_n) = \sum_{1 \le i \le j \le n} a_{ij}^{\ 1} x_i x_j + \sum_{1 \le i \le n} b_i^{\ 1} x_i + c^1$$

$$p_2(x_1,\ldots,x_n) = \sum_{1 \le i \le j \le n} a_{ij}^{\ 2} x_i x_j + \sum_{1 \le i \le n} b_i^{\ 2} x_i + c^2$$

$$\vdots$$

$$p_m(x_1,\ldots,x_n) = \sum_{1 \le i \le j \le n} a_{ij}^{\ m} x_i x_j + \sum_{1 \le i \le n} b_i^{\ m} x_i + c^m$$

(9.12)

where $p_1, p_2, \ldots, p_m$ are m quadratic polynomials with n variable such as $x_1, x_2, \ldots, x_n$ The multivariate public-key cryptosystem encrypts the message faster as compared to RSA and ECC (Chen et al., 2009). Its security depends upon the multivariate quadratic polynomial problem (MQP) where we need to find a vector $x' = (x'_1, x'_2, \ldots, x'_n)$ such that $p_1(x') = \cdots = p_m(x') = 0$. Such MQP problem is NP-hard over any field (Fraenkel and Yesha, 1979). So far, various multivariate public-key cryptosystems have been proposed and their descriptions are as follows.

9.5.1 OV Signature Scheme

Initially, the concept of multivariate quadratic systems namely OV signature scheme is introduced by Patarin (1997). The reason behind the name OV is that they cannot be mixed in terms of quadratic variables.

Let k be a finite field, oil (o) and vinegar (v) are two integers such that $n = o + v$. We set $O = \{v + 1, \ldots, n\}$ and $V = \{1, \ldots, v\}$, where $x_v + 1, \ldots, x_n$ and $x_1, \ldots, x_v$ are OV variables, respectively. Patarin (1997) suggested to select $o = v$. It is called a balanced OV scheme, if $v > 0$, then it is named as unbalanced OV scheme. The key generation consists a mapping F: $k^{0+v} \to k^0$ such that o quadratic polynomials is in form

$$f^k(x) = \sum_{i \in V} \sum_{j \in V} a_{ij}^{\ k} x_i x_j + \sum_{i \in V} \sum_{j \in O} b_{ij}^{\ k} x_i x_j + \sum_{i \in O \cup V} c_i^{\ k} x_i + d^k$$

(9.13)

where $x_j, j = 1, \ldots, o$ and $x_i, i = 1, \ldots, v$ are OV variables and $a_{ij}^{\ k}, b_{ij}^{\ k}, c_i^{\ k} x_i$ and d^k are randomly selected coefficients. It can be noted that mapping $F = (f_{v+1}(x), \ldots, f_n(x))$ is simply invertible. First, we need to select vinegar variables randomly. Furthermore, we solve the linear equations with o variables by using the Gaussian elimination method. If we do not get any solution, then we need to select vinegar variables again.

9.5.2 Rainbow Signature Scheme

Ding and Schmidt (2005) proposed a new effective rainbow signature scheme formed on the concept of OV scheme. The main purpose is to refine the performance of unbalanced OV scheme. The key idea is to replace the single layer with multiple layers in the OV scheme, which form the rainbow.

Let k be a finite field, S be the set $\{1, 2, \ldots, n\}$, and $0 < v_1 < v_2 < \cdots < v_{u+1}$ be a sequence of integers. We define $V_j = \{1, \ldots, v_j\}$, $O_j = \{v_j + 1, \ldots, v_{j+1}\}$ to be a linear space spanned by polynomials in the form

$$f^k(x) = \sum_{i \in V_l} \sum_{j \in V_l} a_{ij}{}^k x_i x_j + \sum_{i \in V_l} \sum_{j \in O_l} b_{ij}{}^k x_i x_j + \sum_{i \in O_l \cup V_l} c_i{}^k x_i + d^k \qquad (9.14)$$

where $l \in \{1, 2, \ldots, u\}$ such that $k \in O_l$. The main steps of rainbow signature scheme's construction are explained as follows:

- *Key creation*: Public key contains k and map of the form $P(x) = S_1 \cdot F \cdot S_2(x)k^n \to k^m$, where S_1, S_2 are two invertible or linear maps $S_1 : k^m \to k^m$ and $S_2 : k^n \to k^n$. Private key contains (F, S_1, S_2), where $F = \left(f^{v_1+1}, \ldots, f^n \right)$ is a rainbow central map.
- *Signature a document*: To sign a document d, we require to observe a solution of the following equation:

$$S_1 \cdot F \cdot S_2(x_1, x_2, \ldots, x_n) = F'(x_1, x_2, \ldots, x_n) = T'$$

On applying the inverse of S_1, we have

$$F \cdot S_2(x_1, x_2, \ldots, x_n) = S_1^{-1}T' = T''$$

We have computed recursively an equation to the inverse of F such that

$$F(x_1, x_2, \ldots, x_n) = T'' = \left(y_1'', \ldots, y_{n-v_1}'' \right)$$

In the end, we apply the inverse of S_2 such that $z = S_2^{-1}(y)$, which gives us the signature T' of document d, i.e., $T' \in k^n$.

- *Verification*: To demonstrate the authenticity of signature d, we have to check $F'(x_1', x_2', \ldots, x_n') = T'$.

If there is a need to sign a large-size document, then we have to apply the hash function and compute the hash value to verify its authenticity. The rainbow signature scheme offers simplicity due to simple matrix operations (multiplication and inversion) over a finite field. It is more efficient than OV scheme due to small key and concise signatures in size. Although there exist several attacks, Rainbow-Band-Separation attack (Fraenkel and Yesha, 1979) and MinRank attack (Kipnis and Shamir, 1999) find the linear mapping to change the polynomials into quadratic mapping.

9.5.3 Quartz Signature Scheme (2, 129, 103, 3, 4)

Quartz signature scheme was proposed by Patarin et al. (2001). It is based on the HFEv-trapdoor function. It was designed to produce very short signatures, i.e., of only 128 bits. It has been designed for specific applications. However, there are already several classical algorithms (RSA, ECC, DSA, etc.) that can generate signatures of length greater than or equal to 320 bits. The public and private keys are HFEv that maps with the following parameters: $(F, d, n, a, v) = (GF(2), 129, 103, 3, 4)$, where F is a finite field, d signifies the degree of the polynomial, n is the size of extension field, a represents the number of equations removed, and v denotes the number of vinegar variables.

In quartz algorithm, we will use a public key (Pu_k) that maps quadratic equations from F^{107} to F^{100} and give input $n - a = 100$ bits. First, we have to compute four signatures for the messages $m_0 = \text{SHA} - 1(m)$, $\text{SHA} - 1(m_0 \| 0x00)$, $\text{SHA} - 1(m_0 \| 0x01)$

and SHA $- 1(m_0 \parallel 0x02)$, where SHA stands for secure hash algorithm. Then, combine all of them into one 128-bit long signature. During its verification, we have to apply public key (Pu_k) four times. Generally, there are two known attacks: MinRank proposed by Kipnis and Shamir (1999) and direct algebraic attacks. It has not been used much in practice due to production of very short signatures and slow signature generation process.

9.6 Hash-Based Cryptography

The security of currently used DSAs is based on the hardness of factorization of sizeable composite numbers. Such algorithms are not quantum resistant. Therefore, hash-based cryptography is based on a cryptographic hash function as an alternative solution and its security depends upon the collision resistance of hash function. Hash-based cryptosystems are the prominent candidate of post-quantum cryptography due to their minimal security requirements. There exist various hash-based cryptography schemes that are beneficial for the era of quantum such as the Lamport–Diffie one-time signature scheme (LD-OTS) (Lamport, 1979) and the Winternitz one-time signature scheme (W-OTS) (MErkle, 1989). These one-time digital signature schemes are not suitable for practical states because every pair of the key can be utilized per signature only. Therefore, MErkle (1989) introduced MErkle's tree authentication scheme based on a complete binary hash tree to reduce the validity of one-time verification keys.

9.6.1 Lamport–Diffie One-Time Signature Scheme (LD-OTS)

Lamport (1979) developed a one-time signature scheme (LD-OTS). For security reasons, most of the signature schemes are based on hash functions, whereas security of the Lamport signature scheme is based on one-way functions $f : \{0,1\}^n \rightarrow \{0,1\}^n$ and cryptographic hash function $h : \{0,1\}^n \rightarrow \{0,1\}^n$. The construction of Lamport–Diffie one-time signature scheme is explained as follows:

- *Key creation*: The signature key (S_k) and verification key (V_k) are selected randomly which consists of $2n$ bit strings of length n such that

$$S_k(x) = \left\{ x_{n-1}[0]x_{n-1}[1], \dots, x_1[0]x_1[1], x_0[0]x_0[1] \right\} \in \{0,1\}^{(n,2n)} \qquad (9.15)$$

$$V_k(y) = \left\{ y_{n-1}[0]y_{n-1}[1], \dots, y_1[0]y_1[1], y_0[0]y_0[1] \right\} \in \{0,1\}^{(n,2n)} \qquad (9.16)$$

  where $y_j[k] = f\left(x_j[k]\right)$, for $0 \le j \le n-1$, $k = 0,1,\dots$. Thus, the key generation needs $2n$ assessments of f.
- *Signature generation*: Consider a document $d \in \{0,1\}^*$ which is signed by exploiting signature key (S_k). Suppose a message digest be a $g(D) = M\{m_{n-1}, \dots, m_0\}$. Then, its signature becomes

$$\sigma = \left\{ x_{n-1}[d_{n-1}], \dots, x_1[d_1], x_0[d_0] \right\} \in \{0,1\}^{(n,n)} \qquad (9.17)$$

  where σ is a group of n bit strings of length n, which are selected as an $f(D)$. So, the length of the signature becomes n^2.

- *Verification*: In order to verify the signature (σ), the verifier computes the message digest $M = \{m_{n-1}, \ldots, m_0\}$. It needs n evaluations of f to check equality such that

$$(f(\sigma_{n-1}), \ldots, f(\sigma_0)) = (y_{n-1}[d_{n-1}], \ldots, y_0[d_0]) \tag{9.18}$$

9.6.2 Winternitz One-Time Signature Scheme (W-OTS)

After the introduction of LD-OTS, MErkle has written that Winternitz suggested him the method and named it as Winternitz one-time signature scheme. The main notion is to use a string in (S_k) to sign various bits in message digest at the same time. It uses the same one-way functions and cryptographic functions like LD-OTS, and produces shorter signatures efficiently. The construction of the original W-OTS is explained as follows:

- *Key creation*: First, set the number of bits to be signed at the same time, i.e., $w \geq 2$. Then, compute

$$t_1 = \left\lceil \frac{n}{w} \right\rceil, t_2 = \left\lceil \frac{\lfloor \log_2 t_1 \rfloor + 1 + w}{w} \right\rceil, t = t_1 + t_2 \tag{9.19}$$

The signature key (S_k) is selected at random such that

$$S_k(x) = (x_{t-1}, \ldots, x_1, x_0) \in \{0,1\}^{(n,t)} \tag{9.20}$$

and the verification key (V_k) is computed by applying one-way function $f : \{0,1\}^n \to \{0,1\}^n, 2^w - 1$ times to each bit string in S_k.

$$V_k(y) = (y_{t-1}, \ldots, y_1, y_0) \in \{0,1\}^{(n,t)} \tag{9.21}$$

where $y_i = f^{2^w - 1}(x_i)$ for $0 \leq i \leq t - 1$.

- *Signature generation*: Consider a message digest $g(D) = M = (m_{n-1}, \ldots, m_0)$ to be signed. In order to divide the message digest d with w, prepend a minimum number of 0s to it. Now, the string d is spat into $t - 1$ bit strings such that $d = a_{t-1} \| \ldots \| a_{t-t_1}$, where $\|$ signifies concatenation of strings and a_i are integers belongs to $(0,1,\ldots,2^w - 1)$. Next, the checksum is computed

$$c = \sum_{i=t-t_1}^{t-1} (2^w - a_i) \tag{9.22}$$

Then, in order to divide a binary representation by w, we have to prepend a minimum number of 0s and the string is spat into t_2 groups such that

$$c = a_{t_2-1} \| \ldots \| a_0 \tag{9.23}$$

In the end, the signature is calculated as

$$\sigma = \left\{ f^{a_{t-1}}(x_{t-1}), \ldots, f^{a_1}(x_1), f^{a_0}(x_0) \right\} \tag{9.24}$$

- *Verification*: In order to verify the signature above, the bit strings $a_{t-1}, \ldots, a_1, a_0$ are computed and check the equality such that

$$\left(f^{2^w - 1 - a_{t-1}}\left(\sigma_{n-1}\right), \ldots, f^{2^w - 1 - a_0}\left(\sigma_0\right) \right) = \left(y_{n-1}, \ldots, y_0 \right) \tag{9.25}$$

If the computed signature is logically correct, then the following equation holds:

$$f^{2^w - 1 - a_i}\left(\sigma_i\right) = f^{2^w - 1}\left(x_i\right) = y_i \tag{9.26}$$

where $i = t - 1, \ldots 0$. It needs $t\left(2^w - 1\right)$ assessments of f in the worst case.

Till now, various generalizations of W-OTS have been occurred such as W-OTS (used in MAC) and W-OTS⁺ proposed by (Hülsing, 2013). The main purpose is to upgrade the W-OTS to increase the security and shorten the size of signatures.

9.7 Conclusion

Post-quantum cryptography offers secure alternatives, which consists many cryptographic systems that can be used for today's Internet communication. During the last decade, we have seen dynamic and secure developments in theory and its applications producing original ideas and algorithms. But there is a need of new hardware for their execution, which can be expensive of a large number of users. There is need of more effort to enhance the understanding and morale to use post-quantum cryptography widely. It is clear that in order to shift to the era of post-quantum cryptography, there are many interesting challenges and important questions that need to be resolved yet. Particularly, there is a need to shorten the public-key size and to improve the efficiency and execution of quantum safe algorithms in the current systems. At present, there are lots of things happening in the field of post-quantum cryptography. Thus, as the new enhancements and algorithms are introduced, they need to be investigated as well.

References

Bernstein, D. J., Introduction to post-quantum cryptography, In *Post-quantum Cryptography*, Springer, Berlin, Heidelberg, 2009, 1–14.

Bu, S., and Zhou, H., A secret sharing scheme based on NTRU algorithm, In *Wireless Communications, Networking and Mobile Computing, WiCom'09, 5th International Conference on*, IEEE, Beijing, China, 2009, 1–4.

Buchmann, J., *Introduction to Cryptography*, 2nd Edition, Springer, New York, 2013.

Buchmann, J., and Williams, H. C., A key-exchange system based on imaginary quadratic fields, *Journal of Cryptology*, 1988, 1(2), 107–118.

Cai, J. Y., and Cusick, T. W., A lattice-based public-key cryptosystem, In *International Workshop on Selected Areas in Cryptography*, Springer, Berlin, Heidelberg, 1998, 219–233.

Chen, A. I. T., Chen, M. S., Chen, T. R., Cheng, C. M., Ding, J., Kuo, E. L. H., and Yang, B. Y., SSE implementation of multivariate PKCs on modern x86 CPUs, In *Cryptographic Hardware and Embedded Systems-CHES*, Springer, Berlin, Heidelberg, 2009, 33–48.

Chen, L., Jordan, S., Liu, Y. K., Moody, D., Peralta, R., Perlner, R., and Smith-Tone, D., Report on post-quantum cryptography, National Institute of Standards and Technology Internal Report, NISTIR, 2016, 8105.

Diffie, W., and Hellman, M., New directions in cryptography, *IEEE Transactions on Information Theory*, 1976, 22(6), 644–654.

Ding, J., and Schmidt, D., Rainbow, a new multivariable polynomial signature scheme, In *Conference on Applied Cryptography and Network Security - ACNS 2005*, Lecture Notes in Computer Science, Springer, New York, 2005, 3531, 164–175.

Feynman, R. P., Simulating physics with computers, *International Journal of Theoretical Physics*, 1982, 21(6), 467–488.

Fraenkel, A. S., and Yesha, Y., Complexity of problems in games, graphs and algebraic equations, *Discrete Applied Mathematics*, 1979, 1(1–2), 15–30.

Goldreich, O., Goldwasser, S., and Halevi, S., Public-key cryptosystems from lattice reduction problems, In *Annual International Cryptology Conference*, Springer, Berlin, Heidelberg, 1997, 112–131.

Grover, L. K., A fast quantum mechanical algorithm for database search, In *Proceedings of the 28th annual ACM symposium on Theory of computing*, ACM, Philadelphia, PA, 1996, 212–219.

Hoffstein, J., Pipher, J., and Silverman, J. H., NTRU: a ring based public key cryptosystem, In *Proceedings of ANTS-III*, Springer, Portland, OR, 1998, 1423, 267–288.

Hülsing, A., W-OTS+–shorter signatures for hash-based signature schemes, In *International Conference on Cryptology in Africa*, Springer, Berlin, Heidelberg, 2013, 173–188.

Kipnis, A., and Shamir, A., Cryptanalysis of the HFE public key cryptosystem by relinearization, In *Annual International Cryptology Conference*, Springer, Berlin, Heidelberg, 1999, 19–30.

Koblitz, N., Elliptic curve cryptosystems, *Mathematics of Computation*, 1987, 48(177), 203–209.

Kwiat, P. G., Mitchell, J. R., Schwindt, P. D. D., and White, A. G., Grover's search algorithm: an optical approach, *Journal of Modern Optics*, 2000, 47(2–3), 257–266.

Lamport, L., Constructing digital signatures from a one way function, Technical Report SRI-CSL-98, SRI International Computer Science Laboratory, 1979.

Lenstra, A. K., and Lenstra, Jr., H. W., The development of the number field sieve, Lecture Notes in Mathematics, Springer, Berlin, 1993, vol. 1554.

Liu, B., Efficient architecture and implementation for NTRU based systems, Master's Thesis, University of Windsor, 2015.

Loidreau, P., Strengthening McEliece cryptosystem, In *International Conference on the Theory and Application of Cryptology and Information Security*, Springer, Berlin, Heidelberg, 2000, 585–598.

Loidreau, P., and Sendrier, N., Weak keys in the McEliece public-key cryptosystem, *IEEE Transactions on Information Theory*, 2001, 47(3), 1207–1211.

Löndahl, C., Some notes on code-based cryptography, 2015.

Malekian, E., and Zakerolhosseini, A., NTRU-like public key cryptosystems beyond dedekind domain up to alternative algebra, In *Transactions on Computational Science X*, Springer, Berlin, Heidelberg, 2010, 25–41.

McEliece, R. J., A public-key cryptosystem based on algebraic coding theory, *Deep Space Network Progress Report*, 1978, 44, 114–116.

Merkle, R. C., A certified digital signature, In *Conference on the Theory and Application of Cryptology*, Springer, Houthalen, Belgium, 1989, 218–238.

Mersin, A., The comparative performance analysis of lattice based NTRU cryptosystem with other asymmetrical cryptosystems, Master's thesis, İzmir Institute of Technology, 2007.

Micciancio, D., Improving lattice-based cryptosystems using the Hermite normal form, In *Cryptography and Lattices*, Springer, Berlin, Heidelberg, 2001, 126–145.

Nguyen, P. Q., and Regev, O., Learning a parallelepiped: Cryptanalysis of GGH and NTRU signatures, *Journal of Cryptology*, 2009, 22(2), 139–160.

Niederreiter, H., Knapsack-type cryptosystems and algebraic coding theory, *Problems of Control and Information Theory*, 1986, 15, 19–34.

Patarin, J., The oil and vinegar signature scheme, In *Presented at the Dagstuhl Workshop on Cryptography*, 1997.

Patarin, J., Courtois, N., and Goubin, L., Quartz, 128-bit long digital signatures, In *Cryptographers' Track at the RSA Conference*, Springer, Berlin, Heidelberg, 2001, 282–297.

Peikert, C., A decade of lattice cryptography, *Foundations and Trends® in Theoretical Computer Science*, 2016, 10(4), 283–424.

Rivest, R. L., Adleman, L., and Dertouzos, M. L., On data banks and privacy homomorphisms, *Foundations of Secure Computation*, 1978b, 4(11), 169–180.

Rivest, R. L., Shamir, A., and Adleman, L., A method for obtaining digital signatures and public-key cryptosystems. *Communications of the ACM*, 1978a, 21(2), 120–126.

Sendrier, N., On the concatenated structure of a linear code, *Applicable Algebra in Engineering, Communication and Computing*, 1998, 9(3), 221–242.

Shor, P. W., Algorithms for quantum computation: discrete logarithms and factoring, In *Proceedings of 35th Annual Symposium on Foundations of Computer Science*, Sante Fe, NM, IEEE, 1994, 124–134.

Sidelnikov, V. M., Vladimir, M., and Shestakov, S. O., On insecurity of cryptosystems based on generalized Reed-Solomon codes, *Discrete Mathematics and Applications*, 1992, 2(4), 439–444.

10

Identity-Based Encryption

Tanvi Gautam, Aditya Thakkar, and Nitish Pathak
Guru Gobind Singh Indraprastha University

CONTENTS

10.1 Introduction to Identity-Based Encryption.. 160
    10.1.1 History of Identity-Based Encryption 160
        10.1.1.1 Cocks IBE Scheme.. 160
        10.1.1.2 Boneh–Franklin IBE Scheme... 161
    10.1.2 Brief Discussion on IBE... 162
        10.1.2.1 Pros and Cons of IBE 162
10.2 Key Terminology and Definitions.. 163
    10.2.1 Public-Key Encryption... 163
    10.2.2 Identity-Based Encryption 163
        10.2.2.1 Algorithms of IBE 164
    10.2.3 Boneh–Boyen IBE... 164
        10.2.3.1 Algorithm of BB_1-IBE 165
        10.2.3.2 BB-IBE Security 166
    10.2.4 Hierarchical IBE.. 167
        10.2.4.1 Algorithms of HIBE 167
    10.2.5 Boneh–Boyen HIBE (BB-HIBE) 168
        10.2.5.1 Algorithm of BB-HIBE (Basic)....................................... 168
10.3 Application, Extension and Related Primitives.. 170
    10.3.1 Identity-Based Signature 170
        10.3.1.1 Algorithms of IBS 170
    10.3.2 Boneh–Lynn–Schasham Scheme... 170
        10.3.2.1 Algorithm of BLS Scheme 171
    10.3.3 Hierarchical Identity-Based Signature 171
        10.3.3.1 Algorithm of HIBS Scheme 172
    10.3.4 Identity-Based Key Agreement... 173
        10.3.4.1 Algorithm of IBE with Key Agreement....................... 173
    10.3.5 Public-Key Encryption with Keyword Search (PEKS)............... 175
    10.3.6 Identity-Based Conditional Proxy Re-Encryption 176
        10.3.6.1 Algorithm of IB-CPRE 176
10.4 Other IBE Schemes 176
    10.4.1 IBE with Wildcards (WIBE)... 176
        10.4.1.1 Algorithm of WIBE 177
    10.4.2 Wicked Identity-Based Encryption 177
        10.4.2.1 Limitations of WKD-IBE and WIBE.............................. 177
        10.4.2.2 Algorithm Steps of WKD-IBE 178

10.4.3   Fuzzy Identity-Based Encryption 178
        10.4.3.1   Applications of FIBE.. 178
        10.4.3.2   Algorithm of FIBE Scheme.. 179
10.5   Conclusion 179
References.. 180

10.1   Introduction to Identity-Based Encryption

10.1.1   History of Identity-Based Encryption

Adi Shamir, the Israeli cryptographer and the co-inventor of RSA algorithm, introduced the concept of identity-based cryptography in 1984 (Shamir, 1979). The idea behind it is to lower the intricacy of cryptography system by using the unique identity of the user, such as name, email id, and mobile number, rather than digital certificates for encrypting the message (Shamir, 1985).

For identity-based signature (IBS) scheme, Shamir had already proposed the RSA function, but the problem of identity-based encryption (IBE) is not solved. In 2001, two types of research, one on elliptic curves and pairings of Boneh and Franklin and the other based on quadratic residues by Clifford Cocks, gave the solution to the IBE problem (Chatterjee & Sarkar, 2006b).

10.1.1.1   Cocks IBE Scheme

Some of the main characteristics of the Clifford Cocks scheme are as follows:

- The ciphertext consists of two characters $(c, \hat{c})$ in $\mathbb{Z}_n$, where $n = pq$ ($p \equiv q \equiv 3 \bmod 4$). These two characters $(c, \hat{c})$ are encrypted by the public key R and $-R$, respectively. According to Jacobi, R is 1. The message information is contained exactly by one c and $\hat{c}$, so we can assume $R = r^2$ (Cocks, 2001).
- Let m be the encrypted message, $m \in \{\pm 1\}$,

$$m = t/n, \quad \text{where } t \xleftarrow{\$} \mathbb{Z}_n,$$

$$\text{let } c = t + R/t.$$

- Decrypt with $2r + c$, if $r^2 = R$.
  We know, $p \equiv q \equiv 3 \bmod 4$, i.e., $(-1/p) = (-1/q) = -1$ and,
  $R/n \Rightarrow (R/p) = (R/q)$, where quadratic residue is either R or $-R$.
  So, the square root of r is R or $-R$,

$$r^2 = \left(R^{(n+5-p-q)/8} \right)^2$$

$$= \left(R^{(n+5-p-q\,-\Phi(n))/8} \right)^2$$

$$= \left(R^{(n+5-p-q-(p-1)(q-1))/8} \right)^2$$

$$= \left(R^{(n+5-p-q-n+p+q-1)/8} \right)^2$$

$$= \left(R^{4/8} \right)^2$$

$$= \pm R.$$

For the case R or $-R$ is the quadratic residue,

$$(2r + c)/n = \left(t + Rt^{-1} + 2r\right)\big/n$$

$$= \left(t\left(1 + Rt^{-2} + 2rt^{-1}\right)\right)\big/n$$

$$= \left(t\left(1 + rt^{-1}\right)^2\right)\big/n$$

$$= (t/n)(\pm 1)^2$$

$$= t/n = \boldsymbol{m}.$$

10.1.1.2 Boneh–Franklin IBE Scheme

On the other hand, the Boneh–Franklin IBE scheme includes (Cocks & Pinch, 2001; Boneh & Franklin, 2003):

Pairings include two groups G_1 and G_2,

Bilinear map, $e: G_1 \times G_1 \to G_2$. {$G_1$ and G_2 groups have a similar order of prime p}.
Elliptic curve, $E: y^2 = x^3 + 1$ over

G_1 and G_2 are the subgroups of elliptic curve and multiplicative groups, respectively.

- Group G_1 (having P as generator) and G_2 with size q,
  $K_{\text{public}} = sP$,
  Public Hash Function:

$$h_1 = \{0,1\}^* \to G_1$$

$$h_2 = G_2 \to \{0,1\}^n, \quad \text{for some fixed } n.$$

For the encrypted message,

$$\mathcal{M} = \{0,1\}^n,$$

for ciphertext,

$$C = G_1^* \times \{0,1\}n,$$

Private-key generator (PKG) computes for getting the public key,

$$ID \in \{0,1\}^*,$$

$$Q_{ID} = h_1(ID),$$

sQ_{ID} is the private key to the user.

- At the time of encryption,

$$m \in \mathcal{M},$$

$$c = \left(rP, m \oplus h_2\left(g^r{}_{ID}\right)\right),$$

where $g_{ID} = e(Q_{ID}, K_{public}) \in G_2$ and r is random $\in \mathbb{Z}_q^*$.

- For decryption,

$$c = (x, y) \in C, \quad \text{using private key to get plaintext.}$$

$$m = y \oplus h_2\big(e(sQ_{ID}, x)\big).$$

- In the end, Boneh and Franklin (2003) verify that both sender and recipient have same value at the end,

$$h_2\big(e(sQ_{ID}, x)\big) = h_2\big(e(sQ_{ID}, rP)\big)$$

$$= h_2\big(e(Q_{ID}, P)^{rs}\big)$$

$$= h_2\big(e(Q_{ID}, sP)^r\big)$$

$$= h_2\big(e(sQ_{ID}, K_{public})^r\big)$$

$$= h_2\big(g^r{}_{ID}\big).$$

10.1.2 Brief Discussion on IBE

IBE is a tool of public-key encryption which is used to overcome the major problem that is getting the receiver's public-key authentication before the information/message is being assigned to him/her by the sender (Chatterjee & Sarkar, 2006). In IBE scheme the authentic public key can be any arbitrary string such as email address, name or dates. Let us take an example, Alice email id is alice@a.com, which is her public key (Shamir, 1985). The Alice obtains the private key corresponding to this email-id from PKG. Bob uses alice@a.com and some public specification of the PKG to encrypt a message to Alice. Alice decrypts the message using her private key. In the future, for decrypting all the messages received by Alice, she will use this private key only (see Figure 10.1).

10.1.2.1 Pros and Cons of IBE

- **PROS**:
  - The encrypted message received by the recipient requires no certification as his identity help in deriving the public key.
  - There is no need to revoke the key as they expire automatically.
  - Automatically terminates rendering information after a particular time and for future decryption enables postdating of messages.
- **CONS**:
  - The sender and the receiver should have a secure path for exchanging messages and for imparting the private-key use from IBE server.
  - As some of the keys are initialized by IBE centralized server, the risk of disclosure increases.

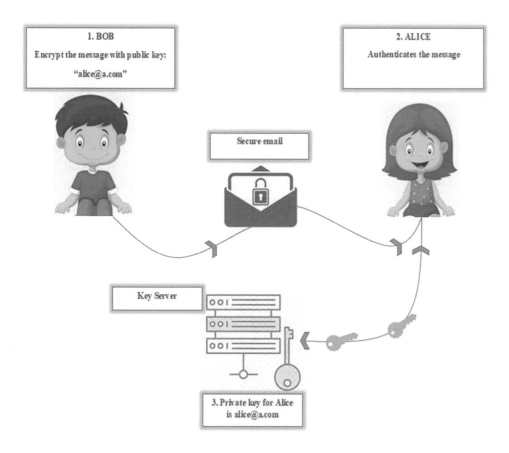

FIGURE 10.1
How IBE works.

10.2 Key Terminology and Definitions

10.2.1 Public-Key Encryption

Public-key encryption or asymmetrical encryption includes two keys: one is the public key which is known by all, and the other is the private or secret key which is known by the receiver. Both keys are related to each other as the encryption of the message is done by a public key and the decryption of the message is done by the corresponding private key. It is developed by Whitfield Diffie and Martin Hellman in 1976, and also known as the Diffie–Hellman encryption.

Example: *The secure message sent by Alice to Bob*. Alice will use Bob's public key for the encryption of the message, and Bob uses his own private key to decrypt it (see Figure 10.2).

10.2.2 Identity-Based Encryption

As discussed in Section 10.1.2, IBE is used for getting the authentication of the public key of the receiver before sending the message by the sender (Shamir, 1985). The public key is an arbitrary string which may be an email address, physical IP address, etc.

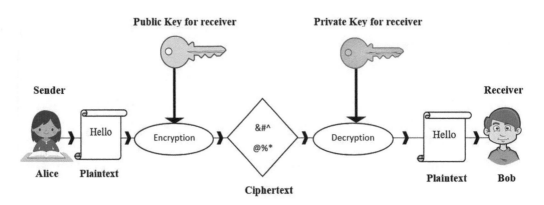

FIGURE 10.2
Public-key encryption.

IBE allows any third party to generate a public key. PKG is a trusted third party which gives the corresponding private key.

10.2.2.1 Algorithms of IBE

The following are the four algorithms that make the IBE scheme (Al-Riyami & Paterson, 2005) (Chatterjee & Sarkar, 2006):

a. **Setup**: Sets the parameters and master key for all system.
b. **Extraction**: The IBE private key initializes by using the PKG master key identical to the random chosen public key string $ID \in \{0, 1\}^*$.
c. **Encryption**: Using the IBE public key from system parameters the message is encrypted.
d. **Decryption**: Using the IBE private key from PKG master secret key the message is decrypted.

Correction: The correctness is very important in order to make the IBE scheme work

$$Decrypt\ (Extract, Parameter, Encrypt) = message$$

10.2.3 Boneh–Boyen IBE

The Boneh–Boyen IBE scheme is "commutative Blinding" scheme. This name is arisen due to the commuting of coefficients that we get, when we compute the ratio of the two pairings (Boneh & Boyen, 2004a):

$$\frac{e(aP,bQ)}{e(bP,aQ)}$$

Using the public parameters of the Boneh–Boyen IBE scheme, the value is calculated by the sender and this value is used to encrypt the plaintext. On the other hand, the receiver calculates the ciphertext and his private key by calculating such a ratio of pairings to get the same value as the sender have (Boneh & Boyen, 2004b).

In the Boneh–Boyen scheme, to perform encryption and decryption operations, we use an identity *ID* which is hashed to an integer value which makes it faster as it avoids modular exponentiation. This scheme is defined a hashing scheme and is resistant to chosen-ciphertext attacks and adaptive chosen-identity attacks (Boneh & Boyen, 2004c).

Note: Boneh and Boyen described the two schemes: one is BB_1, and the other is BB_2. We will only discuss BB_1 IBE scheme.

10.2.3.1 Algorithm of BB₁-IBE

1. **Setup**

    Parameters of BB_1 IBE are as follows (Boneh & Boyen, 2007):

    a. q is the prime power

    b. G_1 and G_T are the cyclic groups

    c. $\hat{e}$ is the pairing, i.e., $\hat{e}$: $G_1 \times G_1 \rightarrow G_T$

    d. n is length of plaintext (in bits).

    e. g, g_1, g_2, and g_3 are the elliptic curve such as

    $$g \in G_1,$$

    $$g_1 = g^\alpha$$

    $$g_2 = g^\beta$$

    $$g_3 = g^\gamma$$

    f. H_1, H_2, and H_3 are the cryptographic hash function such as

    $$H_1 : \{0,1\}^* \rightarrow \mathbb{Z}_p$$

    $$H_2 : G_T \rightarrow \{0,\ 1\}^n$$

    $$H_3 : G_T \times \{0,\ 1\}^n \times G_1 \times G_1 \rightarrow \mathbb{Z}_p$$

    g. $v = \hat{e}(P_1, P_2)$

    $$= \hat{e}(\alpha P, \beta P)$$

    $$= \hat{e}(P, P)^{\alpha\beta}$$

    Master secret for BB_1 IBE,

    $$\alpha, \beta, \gamma \in \mathbb{Z}$$

2. **Extraction**

    Extraction of private key

    a. Private key associated with identity *ID*,

    Calculate: $q_{ID} = H_1 (ID)$, where $q_{ID} \in \mathbb{Z}_p$

    b. Random per-user value,

$$r \in \mathbb{Z}_p$$

    c. The two components of the private key are calculated with the use of r,

$$d_{ID} = \left(g_1 q_{ID}^{\ r} g_2^{\ \alpha} g_3^{\ r}, g^r \right)$$

$$= (d_0, d_1)$$

Note: Random value is used for the private key.

3. **Encrypt**
    The following are the steps that the sender of the message will perform for the encryption of the message M.
    a. Pick random $s \in \mathbb{Z}_p$
    b. Calculate $k = v^s$
    c. Set the ciphertext, $C = (c, c_0, c_1, t)$ where

$$c = M \oplus H_2(k)$$

$$c_0 = g^s$$

$$c_1 = g_1 q_{ID}^{\ s} g_3^{\ s}$$

$$t = s + H_3(k, c, c_0, c_1)$$

4. **Decrypt**
    The following are the steps perform by the receiver to decrypt $C = (c, c_0, c_1, t)$.
    a. Calculate $k = \hat{e}(c_0, d) \div \hat{e}(c_1, d_1)$.
    b. Calculate $s = t - H_3(k, c, c_0, c_1)$.
    c. Check whether

$$k = v^s \text{ and}$$

$$c_0 = g^s.$$

    The error will raise if either of the condition fails and it will get an exit.
    d. Calculate $M = c \oplus H_2(k)$.

10.2.3.2 BB-IBE Security

For attacking the encrypted message with the Boneh–Boyen IBE scheme, the attacker has to access (Boneh, Joux, & Nguyen, 2000).

$$g, g_1 = g^{\alpha}, g_3 = g^{\gamma} \text{ and}$$

$$v = \hat{e}(g, g)^{\alpha\beta}, \text{ from the public parameters.}$$

Also the attacker inspects the ciphertext

$$g^s \text{ and } g_1{}^q{}_{ID}{}^s g_3{}^s = g^{s(\alpha q_{ID} + \gamma)}$$

After performing the above steps, he will retrieve

$$v^s = \hat{e}(g,g)^{\alpha\beta s}$$

This can be done in the following two ways:

Way 1: Calculate s from g^s, where g^s can be solved from discrete logarithm in G_1. Then calculate v^s with the result.

Way 2: Calculate

$$v^s = \hat{e}(g,g)^{\alpha\beta s}$$

where the value of β is defined as the discrete logarithm of

$$v = \left(\hat{e}(g,g)^\alpha\right)^\beta \quad \text{in } G_T.$$

The message can be decrypted with the Boneh–Boyen scheme if the attacker is able to calculate discrete logarithm in either G_1 or G_T.

This is very close in solving BDHP, and for the elimination of the random oracle model, the Boneh–Boyen system is most convenient for it, and in the standard model, they can manage all the usual security properties.

10.2.4 Hierarchical IBE

Hierarchical identity-based encryption (HIBE) was described first by Harwitz and Lynn. The concept of HIBE is to create hierarchies of PKGs. In this hierarchy, at a particular level, the working of a PKG depends on its above level PKG working (see Figure 10.3; Boneh, Boyen, & Goh, 2005).

Note: For each level of the hierarchy, user can have different identities. Therefore, identity *ID* with a maximum of l levels is $ID = (ID_1, ID_2, ..., ID_l)$.

It allows different security protocols that can be enhanced by the organization, easy to recover from a compromise as the recreation is done only for that affected part and not for the whole system (Chatterjee & Sarkar, 2006a).

10.2.4.1 Algorithms of HIBE

Unlike single IBE scheme, HIBE has five algorithms (Chatterjee & Sarkar, 2006b):

   a. **Root setup**: For the working of the top level of the hierarchy, parameters are created.
   b. **Lower-level setup**: For the working of each of the other level and for the one-time execution of the lower level, additional parameters are created here.

Extract, encryption, and **decryption** have the same functionality as they have in single IBE scheme.

10.2.5 Boneh–Boyen HIBE (BB-HIBE)

In this section, we will discuss the single-level Boneh–Boyen IBE (BB-IBE) scheme which is extended to an *l*-level hierarchical IBE (HIBE) scheme using the technique developed by Boneh, Boyen, and Goh (BBG) with parameters = $(G_1, G_2, \hat{e}, n, P, P_1, P_3, H_1, H_2, v)$ and master secret α assuming that $|G_1| = p$ (Boneh, Boyen, & Halevi, 2006).

The resulting HIBE scheme security depends on the difficulty of the BDHP (bilinear Diffie–Hellman problem). The attacker who is able to decrypt such a system can also be able to crack down BDHP. If we take BDHP as hard to crack down or solve, the basic BBG scheme is resistant to chosen-ciphertext attacks and adaptive (Chatterjee & Sarkar, 2006c) chosen-identity attacks. On the other hand, the full BBG is resistant to chosen-ciphertext and adaptive chosen-identity attacks.

Thus, we will discuss BBG-HIBE (Basic).

Note:

1. For limiting the HIBE construction to a single-level, PKG initiates the BB-IBE scheme discussed in Section 10.2.3.

2. There is a property of HIBE which defines that the size of the ciphertext increases with the increase in the number of levels in the HIBE construction.

   But in the BB-HIBE scheme, the size of ciphertext is constant, i.e., as the number of level changes in HIBE construction does not affect the size of ciphertext as it is constant (Boyen & Waters, 2006).

10.2.5.1 Algorithm of BB-HIBE (Basic)

1. **Setup**

   BB-IBE scheme parameters = $(G_1, G_2, \hat{e}, n, P, P_1, P_3, H_1, H_2, v)$.

   HIBE scheme parameters = $(Q_1, Q_2, ..., Q_l) \in G_1$, where parameter Q is generated randomly (Chatterjee & Sarkar, 2007).

   So, BBG-HIBE parameters = $(G_1, G_2, \hat{e}, n, P, P_1, P_3, Q_1, Q_2, ..., Q_l, H_1, H_2, v)$.

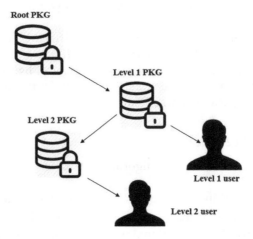

FIGURE 10.3
HIBE scheme.

2. **Extraction**

For the calculation of private key, we obtain $D_{ID} = (D_1, D_2)$, where identity ID is

$$ID = (ID_1, ID_2, \ldots, ID_l)$$

a. Randomly generated $r \in \mathbb{Z}_p$ by root PKG.
b. Calculate $q_{ID} = H_1(ID_1)$ {$1 \leq i \leq k$}
c. Then calculate

$$D_1 = rP \text{ and}$$

$$D_2 = \alpha P_2 + r \sum_{i=l}^{k} q_{IDi} \cdot Q_k$$

3. **Encryption**

For the encryption of message M, where $ID = (ID_1, ID_2, \ldots, ID_k)$ for any $k \leq l$.

a. Random $s \in \mathbb{Z}_p$.
b. Ciphertext $c = (c_0, C_1, C_2)$ where

$$c_0 = M \oplus H_2(v^s)$$

$$C_1 = sP \text{ and}$$

$$C_2 = sr \sum_{i=l}^{k} q_{IDk} \cdot Q_k$$

4. **Decryption**

For the decryption of ciphertext, calculate

$$c_0 \oplus H_2\left(\left(\hat{e}(C_1, D_2)\right) \div \left(\hat{e}(D_1, C_2)\right)\right)$$

$$= c_0 \oplus H_2\left(\hat{e}\left(sP, \alpha P_2 + r\left(\sum_{i=l}^{k} q_{IDi} \cdot Q_i + P_3\right)\right) \div \left(\hat{e}\left(rP, s\left(\sum_{i=l}^{k} q_{IDi} \cdot Q_i + P_3\right)\right)\right)\right)$$

$$= c_0 \oplus H_2\left(\hat{e}(sP, \alpha P_2)\right)$$

$$= c_0 \oplus H_2\left(\hat{e}(P, P_2)^{\alpha s}\right)$$

$$= c_0 \oplus H_2\left(\hat{e}(\alpha P, P_2)^s\right)$$

$$= c_0 \oplus H_2(v^s)$$

$$= M \oplus H_2(v^s) \oplus H_2(v^s) = M.$$

10.3 Application, Extension and Related Primitives

10.3.1 Identity-Based Signature

The most basic primitive in the world of cryptography is the signature. As mentioned earlier in the chapter, every user has his/her own pair of keys: one is a public key, and the other is private or secret key (Boneh & Katz, 2005; Chatterjee & Sarkar, 2005).

Public-key infrastructure (PKI) is a setup which is responsible for mapping the public keys to real-world identities like the name or email address of the user. These keys are certified by the Certification Authority (CA) (Boneh, Lynn, & Shacham, 2004).

PKI requirements get simplified by IBE scheme as KGC (Key Generation Centre) distributes the corresponding private key that eliminates the PKI and certification costs.

IBS uses public information let say the user's identifier to verify the digital signature.

10.3.1.1 Algorithms of IBS

1. **Setup**: The master entity runs this algorithm on input 1^k to get master and a secret key pair (mpk, msk). (Here, k is a security parameter, which has its unary notation as 1^k) (Boyen, Mei, & Waters, 2005).

2. **Extract**: For the user identity ID,

$$ID \in \{0,1\}^*,$$

This algorithm runs on the input of msk and ID to generate private usk.

3. **Sign**: This algorithm returns signature σ of message M, on the input of usk and M.

4. **Verify**: Given mpk, ID, M and signature σ. The algorithm verifies as it accepts if σ is valid for ID and M or it rejects it otherwise.

Correction: Generating mpk, M, usk and,

$$\textbf{Verify}\,(mpk, ID, M, \textbf{Sign}\,(usk, M)) = 1,$$

for $k \in \mathbb{N}$ and $ID, M \in \{0, 1\}^*$.

For security, the IBS scheme is resistant to chosen-message attacks and adaptive chosen-identity attacks.

10.3.2 Boneh–Lynn–Schasham Scheme

Authentication is a very important characteristic in cryptography. The signature scheme of Boneh–Lynn–Schasham (BLS) in cryptography provides us with a service in which it verifies whether the signer is authentic or not. The BLS signature scheme is also known by the name of BLS short-signature scheme (Boneh, Boyen, & Shacham, 2004).

In digital signature, short-signature schemes are easy to perform by a human as compared with long-signature schemes such as RSA and DSA.

The long-signature schemes, RSA and DSA, have 1,024 bits long key, whereas BLS short-signature scheme has 170 bits long key. Their security level is the same as that of long signature schemes. This BLS scheme is secure against existential forgery under a chosen-message attack (random oracle model), with taking assumption that the CDHP (computational Diffie–Hellman problem) is hard to solve.

The verification is done by using bilinear pairing and the elliptic curves group have the elements as signatures. The elliptic curves group gives a protection against the *index calculus* attacks.

10.3.2.1 Algorithm of BLS Scheme

The algorithm of BLS short-signature scheme is defined as follows:

1. **Setup**: A group in which CDHP is difficult to solve but DDH problem is comparatively easy to solve is called a **Gap** group.
   Through Bilinear pairing,
   a. $e: G \times G \to G_T$ where
      $G, G_T \in r$ (r is prime) and e is non-degenerate and efficiently computable.
   b. g is the generator of G.
   c. Illustrate the CDH problem g, g^x, g^y. e does not compute g^{sy}, i.e., the solution to CDH problem. Check, $g^z = g^{xy}$, where $e(g^x, g^y) = e(g, g^z)$.
   d. Using bilinear property, we check

   $$e\left(g^x, g^y\right) = e\left(g, g\right)^{sy} = e\left(g, g\right)^z = e\left(g, g^z\right)$$

   If it is true, and also we know that G_T is prime order group so,

   $$xy = z$$

2. **Extract**:
   Random chosen, $x \in [0, r-1]$. Private key x holder will publish g^x, i.e., the public key.
3. **Signing**:
   For message M,
   a. Hash function, $h = H(M)$
   b. Output, $\sigma = h^x$ where, σ is signature.
4. **Verification**:
   Given σ and g^x check,

   $$e(\sigma, g) = e\left(H(M), g^x\right).$$

10.3.3 Hierarchical Identity-Based Signature

Hierarchical identity-based signature (HIBS) scheme is similar to IBS scheme which has a hierarchy in it. Delegation can be done using this hierarchical approach scheme in a natural way. It has one more form of itself that is multi-keys HIBS (MK-HIBS) (Chatterjee & Sarkar, 2007).
   In multi-keys HIBS (Boneh, Gentry, Lynn, & Shacham, 2003):

a. The user has more than one identifier and so it carries a set of corresponding private key.
b. On selected message, the combo of private keys is used to get a single signature.
c. In the hierarchy, the location of identifiers may be situated at an arbitrary position.

Example: Consider an employee, **X** in an IT company **Y** who have given roles, $\mathbf{r_1}$ = **Manager**, $\mathbf{r_2}$ = **Project Head** and $\mathbf{r_3}$ = **Employee**. So, the role identifier in a hierarchical space can be in the form "**Y, $\mathbf{r_i}$**" where $1 \leq i \leq 3$.

To accept the principle, **X** can choose the role that he/she wants to perform from the given options. Now, assume that **X** wants to see some restricted digital documents using roles $\mathbf{r_1}$ and $\mathbf{r_3}$. This can be done by signing the request using the private key $\mathbf{S_1}$ and $\mathbf{S_3}$ corresponding to role identifiers "**Y, $\mathbf{r_1}$**" and "**Y, $\mathbf{r_3}$**", respectively. MK-HIBS uses a pair of private keys $\mathbf{S_1}$ and $\mathbf{S_3}$ to sign the access request.

10.3.3.1 Algorithm of HIBS Scheme

1. **Root setup**: PKG generates,
   a. Cyclic groups G, G_T, where G and $G_T \in q$ (q is prime).
      And on the input of λ (λ is security parameter),

$$e : G \times G \rightarrow G_T \; \text{(Bilinear Map)}.$$

   b. Choose a generator $P_0 \in G$
   c. Randomly chosen, $s_0 \in \mathbb{Z}_q^*$ and set $Q_0 = s_0 \, P_0$.
   d. Hash function,

$$H_1 : \{0,1\}^* \rightarrow G$$

$$\text{and } H_2 : \{0,1\}^* \rightarrow G.$$

   e. Master secret is set to be r.

2. **Lower-level setup**:
   Randomly chosen, secret $s_t \in \mathbb{Z}_q^*$ at level $t \geq 1$ for a lower-level PKG (Boneh & Waters, 2007).

3. **Extract**:
   For identity ID with level t,

$$ID_t = id_1, \ldots, id_t.$$

   a. Compute $P_t = H_1 (ID_t) \in G$
   b. $S_t = \displaystyle\sum_{i=l}^{t} s_{i-1} \cdot P_i$
   c. $Q_i = s_0 \, P_0$ for $1 \leq i \leq t - 1$
   d. Secret key $(S_t, Q_1, \ldots, Q_{t-1})$

4. **Sign**:
   Given any $n \geq 1$. Set

$$SK = \left\{ \left(S^j{}_{tj}, Q^j{}_1, \ldots, Q^j{}_{tj-1} \right) : 1 \leq i \leq n \right\}$$

   Given message M,

  a. Secret value, $s_\varphi \in \mathbb{Z}_q^*$
  b. Solve $P_M = H_2\left(ID^1{}_{t1}, \ldots, ID^n{}_{tn}, M\right);$
  c. Calculate

$$\varphi = \sum_{j=l}^{n} S^j{}_{tj} + s_\varphi P_M \text{ and}$$

$$Q_\varphi = s_\varphi P_0$$

  d. The signature is as follows:

$$\sigma = \left(\varphi, Q, Q_\varphi\right) \quad \text{where,}$$

$$Q = \left\{Q^j{}_i : 1 \le i \le t_j - 1; 1 \le j \le n\right\}$$

5. **Verification**:
  a. $P^j = H_1\left(ID^j\right)$ for $1 \le i \le t_j$ and $1 \le j \le n$.
  b. $P_M = H_2\left(ID^1{}_{t1}, \ldots, ID^n{}_{tn}, M\right)$
  c. Valid if,

$$e\left(P_0, \varphi\right) = \prod_{j=1}^{n} \prod_{i=1}^{tj} \left(eQ^j{}_{i-1}, P^j{}_i\right) \cdot e\left(Q_\varphi, P_M\right)$$

10.3.4 Identity-Based Key Agreement

Key agreement is defined as an agreement done between the two or more users to exchange messages and compose a common session key. It was first described in 1976 by Diffie and Hellman. They proposed that the two users in the communicating channel can exchange the public information and then use their corresponding private key to compute a common key and received information. This Diffie–Hellman key agreement scheme lacks in authentication and the attack like man-in-the-middle can be done (Diffie & Hellman, 1976).

Since the Diffie–Hellman key agreement faces problem in authentication, many types of research are done, and therefore, the scheme now is mainly known by Authentication Key Agreement (AKA). In the late 1990s, cryptography was well studied and researchers initialize a tool which plays a very important role in cryptography. This tool is pairing which results in the discrete logarithm problem in elliptic curves (Abdalla, Miner, & Namprempre, 2001).

Neil P. Smart (2002) proposed the key agreement in IBE with help of pairing using the concept of Weil pairing.

10.3.4.1 Algorithm of IBE with Key Agreement

1. **Setup**: The private key of the user is generated and distributed by PKG.

$$e : G \times G \rightarrow G_T, \quad \text{where } G, G_T \in q \text{ (q is prime).}$$

  a. P is the generator of G.
  b. Secret master $s \in \mathbb{Z}_q^*$.

   c. Evaluates master public key $P_{\text{public}} = sP$.

   d. Hash function,

$$H_1 = \{0,1\}^* \times G \rightarrow \mathbb{Z}_q^*.$$

So, parameters $= (G, G_T, e, q, P, sP, H_1)$

2. **Extract**: For identity ID_i, where $ID \in \{0, 1\}^*$

   a. Randomly chosen r such that

$$r_i \in \mathbb{Z}_q.$$

   b. Compute $R_i = r_i P$

   c. Solving,

$$h = H_1(ID_i \| R_i) \text{ and}$$

$$S_i = 1/(r_i + hs)$$

3. **Key agreement**: Suppose two users **X** and **Y** with private keys (S_X, R_X) and (S_Y, R_Y), respectively (Smart, 2003).

   To generate session key of X and Y,

   a. X and Y generate temporary private key $x, y \in \mathbb{Z}_q^*$.

   b. X sends (ID_X, R_X) to Y, Y send backs (ID_Y, R_Y, T_{YX}) to X where,

$$T_{YX} = y\left[R_x + H_1(ID_X \| R_X)P_{\text{public}}\right]$$

And for X,

$$T_{XY} = x\left[R_y + H_1(ID_Y \| R_Y)P_{\text{public}}\right]$$

   c. X computes:

$$K_{XY} = e\left(S_X, T_{YX}, P_{\text{public}}\right)^x$$

Similarly for Y computes,

$$K_{YX} = e\left(S_Y, T_{XY}, P_{\text{public}}\right)^y$$

Correctness: Using Bilinear Pairing to verify (Yi, 2003),

$$\boldsymbol{K_{XY} = K_{YX}}$$

So,

$$K_{XY} = e\left(S_X, T_{YX}, P_{\text{public}}\right)^x$$

$$= e\left[\left(r_x + hs\right)^{-1} y\left(r_x P + hsP\right), sP\right]^x$$

$$= e\left[\left(r_x + hs\right)^{-1} y\left(r_x + hs\right)P, sP\right]^x$$

$$= e\left(yP, sP\right)^x$$

$$= e\left(P, P\right)^{sxy}$$

$$K_{YX} = e\left(S_Y, T_{XY}, P_{\text{public}}\right)^y$$

$$= e\left[\left(r_y + hs\right)^{-1} x\left(r_y P + hsP\right), sP\right]^y$$

$$= e\left[\left(r_y + hs\right)^{-1} x\left(r_y + hs\right)P, sP\right]^y$$

$$= e\left(xP, sP\right)^y$$

$$= e\left(P, P\right)^{sxy}$$

Hence, $K_{XY} = K_{YX} = K$. Thus, X and Y shared a session key as

$$sk = H\left(ID_X \| ID_Y \| \text{Trans} \| K\right)$$

here $\text{Trans} = T_{XY} \| T_{YX}$.

10.3.5 Public-Key Encryption with Keyword Search (PEKS)

Boneh, Di Crescenze, Ostrovsky, and Persiano describe a keyword-search scheme with public-key encryption (PEKS) which helps the user in searching the encrypted keywords in the messages. This scheme does not affect the security of the original message (Boneh, Gentry, & Waters, 2005).

Example: User Alice, who is an employee in an *XYZ* company, is on a vacation for a few days and she has a smartphone with her in which she reads all her emails. During the vacation she only wants to read the important emails. Now with a smartphone, she also has Laptop and Desktop.

The mail gateway of Alice sends emails to the particular device according to the keywords in the email(s), such as when the sender, Bob, sends an email with the keyword "Hi" it goes to the desktop, "urgent" it goes to the smartphone and "Read" goes to the laptop.

To get the messages according to the keyword, Alice will send a "trapdoor" to the server to enable the selection of her important emails. For instance, when Bob sends a mail using PEKS schemes in which PEKS ciphertext contains keyword "urgent," the email will appear to the smartphone of Alice. Security is ensured to her as all the emails of her will be encrypted using her public key.

10.3.6 Identity-Based Conditional Proxy Re-Encryption

Proxy re-encryption scheme is an extended scheme of IBE. The scheme coverts the public key of the encrypted message of one user into an encryption intended to other user through a proxy. The encrypted message can be temporarily forward without giving the private key (Harrison, Page, & Smart, 2002). During the conversion, the proxy uses *proxy key* or *re-encryption keys* to perform the translation without knowing the secret key of any user and it also does not read the plaintext and then these proxy keys are apply to the ciphertext and automatically re-encrypt from one identity to another.

10.3.6.1 Algorithm of IB-CPRE

a. **Setup**: $e: G \times G \to G_T$ where, $G, G_T \in q$ (q is prime).
   i. g is the generator G.
   ii. Master secret *msk*.
b. **Extract**: On input of the identity $ID \in \{0, 1\}^*$ and the master key, the decryption SK_{ID} will be output.
c. **Encrypt**: On input of the identity $ID \in \{0, 1\}^*$ and the message M, the output ciphertext C_{id} with the specified identity.
d. **Re-Extract**: On input of SK_{ID} and $ID_1, ID_2 \in \{0, 1\}^*$ outputs the re-encryption key, i.e., $RK_{ID_1 \to ID_2}$.
e. **Re-Encrypt**: Generates an output if C_{ID_2} on the input of ciphertext under ID_1, i.e., C_{ID_1} and re-encryption key $R_{ID_1 \to ID_2}$.
f. **Decrypt**: For decryption, use secret key SK_{ID} decrypts the ciphertext C_{ID} will output the message M.

Note: We define security as to ensure that there should be no advantage against the non-colluding users from the key-holders of colluding users set (Hess, Smart, & Vercauteren, 2006).

10.4 Other IBE Schemes

10.4.1 IBE with Wildcards (WIBE)

IBE plays a vital role in the segment of exchanging mails. For instance, using IBE scheme in which the public is the identity of the user, such Alice sends an email to Bob by encrypting the message under his identity bob@ibe.univ.edu. Bob is the only one who can decrypt this message using his secret key that he gets from a trusted key distributer (Abdalla et al., 2006).

Now suppose Alice wants to send an important email to the entire IBE research department in the university, she can do this by encrypting the email under identity #@ibe.univ.edu and all the recipients having related identities can decrypt this message. This concept is known as identity-based encryption with *wildcards* (WIBE), which is firstly proposed by Michel Abdalla. In the above example "#" is the wildcard (Abdalla, An, Bellare, & Namprempre, 2002).

WIBE scheme can only encrypt the messages that are about 160 bits long. It is generally proposed to overcome the limitation of multi-key HIBE (MK-HIBE) (as discussed in Section 10.3.3) as it encrypts the message individually to each level.

10.4.1.1 Algorithm of WIBE

1. **Setup**:
   a. Pattern P $(P_1, ..., P_l) \in \{0, 1\}^*$ for $l \leq L$, where L is the maximum number of levels. Identity ID $(ID_1, ..., ID_{l'})$ matches the pattern P, if and only if $l' \leq l$ for $1 \leq i \leq l'$, $ID_i = P_i$.
   b. PKG generates master key pair (msk, mpk).

2. **Extract**:
   a. On input of secret key D_{ID} for identity $ID = (ID_1, ..., ID_l)$ it generates $D_{ID'}$ for identity $ID' = (ID_1, ..., ID_{l+1})$.
   b. User uses $D_\varepsilon = msk$ as private key for deriving key for single-level identities, where identity $\varepsilon = (ID_1, ..., ID_l)$.

3. **Encrypt**:
   a. On input of msk, pattern P and the message M, where

$$M \in \{0,1\}^* \text{ i.e. } M \in P.$$

   b. Generates the ciphertext C.

4. **Decrypt**: Using secret key D_{ID} for $ID_i = P_i$, decrypts the ciphertext C and returns the corresponding message M.
   The decrypt algorithm rejects if the encryption is invalid (Cao, 2006).

Correctness:

$$\textbf{\textit{Decrypt}}\left[\textbf{\textit{Extract}}(msk, ID), \textbf{\textit{Encrypt}}(mpk, P, M) \right] = M = 1.$$

10.4.2 Wicked Identity-Based Encryption

As discussed in Section 10.4.1, WIBE is proposed to overcome the HIBE scheme's limitations. Similarly, the hierarchical key derivation of WIBE scheme has also some limitations.

To generalize the mechanism of key delegation in HIBE scheme and to overcome WIBE limitations, Abdalla, Klitz, and Neven proposed a new concept of identity-based encryption with wildcards key derivation (WKD-IBE).

WKD-IBE scheme is seen as a dual scheme of WIBE. *The pattern* consists of an identity string that is associated with secret keys. The *wildcard* is also associated with secret keys, and the derivation of keys is done by key owner for any identity that engages the same *pattern* affiliated with this secret key.

10.4.2.1 Limitations of WKD-IBE and WIBE

i. Suppose a situation that has the maximum hierarchy length L and is constant only holds the security proof of WKD-IBE. Therefore, only those cases that accept such restriction can use this scheme. For instance, building an identity-based broadcast encryption schemes by using WKD-IBE scheme, the maximum length of the target group will get a direct impact due to the limitation on the maximum hierarchy length.

ii. The application in which the recipient anonymity of an encrypted message needs to be getting saved is not possible using this scheme, as the *pattern* with the ciphertext cannot be hidden by the WKD-IBE scheme solution.

These shortcomings or limitations can be solved by the pairing of identity-based encryption with wildcards and with wildcard key derivation, i.e. WW-IBE.

WW-IBE scheme is a universal primitive that is used in both key derivations and in encryption as it allows more general *patterns*.

10.4.2.2 Algorithm Steps of WKD-IBE

- Wildcards are initiated with the ciphertext to make ciphertext decrypt by a whole range of the user; they are initiated with secret keys to allow more general *pattern*.
- Secret keys are combined with *patterns* rather than combing with identity vectors.
- Pattern $P = (P_1, ..., P_l) \in \{0, 1\}^*$ where

  a. $l \leq L$

  b. $*$ is wildcard

  c. L is the maximum length of the WKD-IBE scheme.

So, in *pattern P*, each element is either a wildcard or a specific identity string.

For a given *pattern P*, the user in the hold of secret keys can produce secret keys for *pattern P'* which matches the *pattern P*.

$$P' = (P'_1, ..., P'_{l'}) \text{ matches } P \text{ if and only if,}$$

a. $l' \leq l$,

b. $P_i = P'_l$ and

c. $P' \in_* P$.

10.4.3 Fuzzy Identity-Based Encryption

We know that in IBE scheme the identities are the arbitrary string of characters, but in the new type of IBE, i.e. **Fuzzy** identity-based encryption (FIBE) a set of descriptive attributes are the identities (Chatterjee & Sarkar, 2006a). FIBE works on a certain distance between the users which is measured by some metric system. Suppose Alice with ω as the secret key will only able to decrypt the ciphertext which is encrypted with ω' as public key if and only if ω and ω' are within a certain distance.

10.4.3.1 Applications of FIBE

a. Biometric can be used as identity in FIBE. *Example*: User's iris scan can use as an identity to describe several attributes and the user can encrypt the message using their biometric identity.

b. Attribute-based encryption is defined as the application in FIBE in which the user can encrypt the information to all other users having the certain set of attributes. The FIBE provides the advantage by storing the document on an untrusted server.

10.4.3.2 Algorithm of FIBE Scheme

1. **Setup**:

$$e: G \times G \to G_T \quad \text{where, } G, G_T \in q \ (q \text{ is prime}).$$

a. g is the generator G.

b. Universe $\mu \in \mathbb{Z}_q$ of size $|\mu|$.

c. Randomly chosen $t, y \in \mathbb{Z}_q^*$. Public parameter,

$$T_1 = g^t {}_1, \dots, T_{1\mu1} = g^{1\mu}, Y = e(g, g)^y$$

Master key: $t_1, \dots, t_{1\mu1}, y$

2. **Extract**: Generate private key for identity ω.
   Randomly chosen $d - 1$ degree polynomial p,

$$D_i = g^{p(i) \cdot t} {}_i, \quad \text{for } i \in \omega \text{ and } p(0) = y.$$

3. **Encrypt**:
   Public key ω' encryption with message $M \in G_T$,

a. Random value $s \in \mathbb{Z}_q$

b. Ciphertext $E = \left(\omega', E' = MY^s, \left\{E_i = T^s {}_j\right\}\right), i \in \omega'$.

4. **Decrypt**: Computes ciphertext E,

$$E' \Big/ \prod_{i \in s} \left(e(D_i, E_i)\right)^{\Delta i, S(0)}$$

$$= Me(g, g)^{sy} \Big/ \prod_{i \in s} \left(e\left(g^{p(i)/t} {}_i, g^{st} {}_i\right)\right)^{\Delta i, S(0)}$$

$$= Me(g, g)^{sy} \Big/ \prod_{i \in s} \left(e(g, g)^{sp(i)}\right)^{\Delta i, S(0)}$$

$$= M$$

10.5 Conclusion

IBE scheme is a promising solution for resolving the problems emerging with the symmetric and asymmetric key management schemes. While there are issues, the comparative simplicity of its architecture makes IBE an attractive proposition for diverse computer systems including mobile computing. Moreover, the network world points how significantly lower the total cost of ownership of the IBE system is as compared to a total public-key system.

References

Abdalla, M., An, J. H., Bellare, M., & Namprempre, C. (2002). From identification to signatures via the Fiat-Shamir transform: Minimizing assumptions for security and forward-security. Lars R. Knudsen, editor, *Advances in Cryptology—EUROCRYPT 2002.* Lecture Notes in Computer Science, Germany, 2332, 418–433. doi:10.1007/3-540-46035-7_28.

Abdalla, M., Catalano, D., Dent, A. W., Malone-Lee, J., Neven, G., & Smart, N. P. (2006). Identity-based encryption gone wild. In Michele Bugliesi et al., editors, *Automata, Languages and Programming.* Lecture Notes in Computer Science, 4052, 300–311. doi:10.1007/11787006_26.

Abdalla, M., Miner, S., & Namprempre, C. (2001). Forward-secure threshold signature schemes. In David Naccache, editor, *Topics in Cryptology—CT-RSA 2001.* Lecture Notes in Computer Science, 441–456. doi:10.1007/3-540-45353-9_32.

Al-Riyami, S. S., & Paterson, K. G. (2005). CBE from CL-PKE: A generic construction and efficient schemes. In Serge Vaudenay, editor, *Public Key Cryptography—PKC 2005.* Lecture Notes in Computer Science, 398–415. doi:10.1007/978-3-540-30580-4_27.

Boneh, D., & Boyen, X. (2004a). Efficient selective-ID secure identity-based encryption without random oracles. In Christian Cachin and Jan Camenisch, editors, *Advances in Cryptology—EUROCRYPT 2004.* Lecture Notes in Computer Science, 3027, 223–238. doi:10.1007/978-3-540-24676-3_14.

Boneh, D., & Boyen, X. (2004b). Secure identity based encryption without random oracles. In Matthew Franklin, editor, *Advances in Cryptology—CRYPTO 2004.* Lecture Notes in Computer Science, 443–459. doi:10.1007/978-3-540-28628-8_27.

Boneh, D., & Boyen, X. (2004c). Short signatures without random oracles. In Christian Cachin and Jan Camenisch, editors, *Advances in Cryptology—EUROCRYPT 2004.* Lecture Notes in Computer Science, 3027, 56–73. doi:10.1007/978-3-540-24676-3_4.

Boneh, D., & Boyen, X. (2007). Short signatures without random oracles and the SDH assumption in bilinear groups. *Journal of Cryptology*, 21(2), 149–177. doi:10.1007/s00145-007-9005-7.

Boneh, D., Boyen, X., & Goh, E. (2005). Hierarchical identity based encryption with constant size ciphertext. In Ronald Cramer, editor, *Advances in Cryptology—EUROCRYPT 2005.* Lecture Notes in Computer Science, 3494, 440–456. doi:10.1007/11426639_26.

Boneh, D., Boyen, X., & Halevi, S. (2006). Chosen ciphertext secure public key threshold encryption without random oracles. In David Pointcheval, editor, *Topics in Cryptology—CT-RSA 2006.* Lecture Notes in Computer Science, 226–243. doi:10.1007/11605805_15.

Boneh, D., Boyen, X., & Shacham, H. (2004). Short group signatures. In Matthew Franklin, editor, *Advances in Cryptology—CRYPTO 2004.* Lecture Notes in Computer Science, 3152, 41–55. doi:10.1007/978-3-540-28628-8_3.

Boneh, D., & Franklin, M. (2003). Identity-based encryption from the Weil pairing. *SIAM Journal on Computing*, 32(3), 586–615. doi:10.1137/s0097539701398521.

Boneh, D., Gentry, C., Lynn, B., & Shacham, H. (2003). Aggregate and verifiably encrypted signatures from bilinear maps. In Eli Biham, editor, *Advances in Cryptology—EUROCRYPT 2003.* Lecture Notes in Computer Science, 2656, 416–432. doi:10.1007/3-540-39200-9_26.

Boneh, D., Gentry, C., & Waters, B. (2005). Collusion resistant broadcast encryption with short ciphertexts and private keys. In Victor Shoup, editor, *Advances in Cryptology—CRYPTO 2005.* Lecture Notes in Computer Science, 3621, 258–275. doi:10.1007/11535218_16.

Boneh, D., Joux, A., & Nguyen, P. Q. (2000). Why textbook ElGamal and RSA encryption are insecure. In Tatsuaki Okamoto, editor, *Advances in Cryptology—ASIACRYPT 2000.* Lecture Notes in Computer Science, 30–43. doi:10.1007/3-540-44448-3_3.

Boneh, D., & Katz, J. (2005). Improved efficiency for CCA-secure cryptosystems built using identity-based encryption. In Alfred Menezes, editor, *Topics in Cryptology—CT-RSA 2005.* Lecture Notes in Computer Science, Verlag, 87–103. doi:10.1007/978-3-540-30574-3_8.

Boneh, D., Lynn, B., & Shacham, H. (2004). Short signatures from the Weil pairing. *Journal of Cryptology*, 17(4), 297–319. doi:10.1007/s00145-004-0314-9.

Boneh, D., & Waters, B. (2007). Conjunctive, subset, and range queries on encrypted data. In Cynthia Dwork, editor, *Theory of Cryptography*. Lecture Notes in Computer Science, 4392, 535–554. doi:10.1007/978-3-540-70936-7_29.

Boyen, X., Mei, Q., & Waters, B. (2005). Direct chosen ciphertext security from identity-based techniques. *Proceedings of the 12th ACM Conference on Computer and Communications Security—CCS 05*, Verlag. doi:10.1145/1102120.1102162.

Boyen, X., & Waters, B. (2006). Anonymous hierarchical identity-based encryption (without random oracles). *Advances in Cryptology—CRYPTO 2006*. Lecture Notes in Computer Science, 290–307. doi:10.1007/11818175_17.

Cao, Z. (2006). Analysis of one popular group signature scheme. In Bahram Honary, editor, *Advances in Cryptology—ASIACRYPT 2006*. Lecture Notes in Computer Science, 460–466. doi:10.1007/11935230_30.

Chatterjee, S., & Sarkar, P. (2005). Trading time for space: Towards an efficient IBE scheme with short(er) public parameters in the standard model. In Dongho Won and Seungjoo Kim, editors, *Information Security and Cryptology—ICISC 2005*. Lecture Notes in Computer Science, 3935, 424–440. doi:10.1007/11734727_33.

Chatterjee, S., & Sarkar, P. (2006a). Generalization of the selective-ID security model for HIBE protocols. In Moti Yung et al., editors, *Public Key Cryptography—PKC 2006*. Lecture Notes in Computer Science, 3958, 241–256. doi:10.1007/11745853_16.

Chatterjee, S., & Sarkar, P. (2006b). HIBE with short public parameters without random oracle. In Xuejia Lai and Kefei Chen, editors, *Advances in Cryptology—ASIACRYPT 2006*. Lecture Notes in Computer Science, 4284, 145–160. doi:10.1007/11935230_10.

Chatterjee, S., & Sarkar, P. (2006c). New constructions of constant size ciphertext HIBE without random oracle. In Min Surp Rhee and Byoungcheon Lee, editors, *Information Security and Cryptology—ICISC 2006*. Lecture Notes in Computer Science, 310–327. doi:10.1007/11927587_26.

Chatterjee, S., & Sarkar, P. (2007). Constant size ciphertext HIBE in the augmented selective-ID model and its extensions. *Journal of Universal Computer Science*, 13(10), 1367–1395.

Cocks, C. (2001). An identity based encryption scheme based on quadratic residues. *Cryptography and Coding*. Lecture Notes in Computer Science, 360–363. doi:10.1007/3-540-45325-3_32.

Cocks, C., & Pinch, R. G. (2001). ID-based cryptosystems based on the Weil pairing, Unpublished manuscript.

Diffie, W., & Hellman, M. (1976). New directions in cryptography. *IEEE Transactions on Information Theory*, 22(6), 644–654. doi:10.1109/tit.1976.1055638.

Harrison, K., Page, D., & Smart, N. P. (2002). Software implementation of finite fields of characteristic three, for use in pairing-based cryptosystems. *LMS Journal of Computation and Mathematics*, 5, 181–193. doi:10.1112/s1461157000000747.

Hess, F., Smart, N., & Vercauteren, F. (2006). The eta pairing revisited. *IEEE Transactions on Information Theory*, 52(10), 4595–4602. doi:10.1109/tit.2006.881709.

Shamir, A. (1979). How to share a secret. *Communications of the ACM*, 22(11), 612–613. doi:10.1145/359168.359176.

Shamir A. (1985). Identity-based cryptosystems and signature schemes. In: Blakley, G. R., Chaum, D. (eds) *Advances in Cryptology. CRYPTO 1984*. Lecture Notes in Computer Science, 196. Springer, Berlin, Heidelberg.

Smart, N. P. (2003). Access control using pairing based cryptography. In: Joye, M. (ed) *Topics in Cryptology—CT-RSA 2003. CT-RSA 2003*. Lecture Notes in Computer Science, 2612. Springer, Berlin, Heidelberg.

Yi, X. (2003). An identity-based signature scheme from the Weil pairing. *IEEE Communications Letters*, 7(2), 76–78.

11

Attribute-Based Encryption

Tanvi Gautam, Aditya Thakkar, and Nitish Pathak

Guru Gobind Singh Indraprastha University

CONTENTS

11.1 Introduction 183
    11.1.1 History of Attribute-Based Encryption 183
    11.1.2 Brief Discussion on ABE.. 184
11.2 Key Terminology, Definitions, and Related Terms 184
    11.2.1 Identity-Based Encryption 184
    11.2.2 Fuzzy Identity-Based Encryption 184
    11.2.3 Attribute-Based Encryption 185
        11.2.3.1 Principle of an Ideal Attribute-Based Encryption Scheme 185
        11.2.3.2 Features of ABE 185
        11.2.3.3 Architecture of Data Sharing 185
        11.2.3.4 Algorithm of ABE... 187
11.3 Key-Policy ABE... 188
    11.3.1 Introduction to KP-ABE Scheme 188
    11.3.2 Features of KP-ABE Scheme 189
    11.3.3 Algorithm of KP-ABE 189
11.4 Ciphertext-Policy ABE 190
    11.4.1 Introduction to CP-ABE.. 190
    11.4.2 Features of CP-ABE 191
    11.4.3 Algorithm of CP-ABE.. 191
    11.4.4 Comparison of KP-ABE and CP-ABE 192
11.5 Applications of Attribute-Based Encryption 193
    11.5.1 Maintaining Personal Health Record Security Using ABE Scheme......... 193
    11.5.2 Audit Log... 193
    11.5.3 Targeted Broadcast.. 193
11.6 Conclusion 194
References... 194

11.1 Introduction

11.1.1 History of Attribute-Based Encryption

First, the theory of attribute-based encryption (ABE) was proposed by two researchers: Brent Waters and Amit Sahai in the study of fuzzy identity-based encryption (FIBE; discussed in Chapter 10) in the year 2005, and later other researchers, Vipul Goyal and

Omkant Pandey, join their hands with Amit Sahai and Brent Waters in the developing of ABE. They have written an article on ABE, i.e., "Attribute-Based Encryption for Fined-Grained Control of Encrypted Data". Recently, many more researches have been performed on ABE with multiple authorities, where the user's private keys are jointly generated.

11.1.2 Brief Discussion on ABE

Almost one of the recent approaches in the study of cryptography is the concept of ABE which is also one of the genres of public key cryptography.

The traditional approach of public key cryptography includes the encryption of the messages for the particular receiver by using his/her public key. This traditional approach of public key cryptography was changed by the study of Adi Shamir on identity-based cryptography and specially getting more precise by identity-based encryption (IBE; discussed in Chapter 10) which includes the public key as an arbitrary string, e.g., *e-mail id*. Then, the concept of ABE came into existence, which takes the study further by defining the identity as a set of attributes rather than as an atomic value (Shamir, 1979). In ABE, data are decrypted using the private key where data are dependent on the set of the attributes, such as residence address of the receiver and type of subscription account (Sahai & Waters, 2005).

There are two main types of ABE schemes:

A. Ciphertext-policy ABE (CP-ABE)

B. Key-policy ABE (KP-ABE).

11.2 Key Terminology, Definitions, and Related Terms

11.2.1 Identity-Based Encryption

As discussed in Chapter 10, IBE is used for getting the authentication of the public key of the receiver before sending the message by the sender (Shamir, 1984). The public key is an arbitrary string which may be an e-mail address, physical IP address, etc.

IBE allows any third party to generate a public key. Private key generator (PKG) is a trusted third party which gives the corresponding private key.

11.2.2 Fuzzy Identity-Based Encryption

In 2005, Amit Sahai and Brent Waters introduced a new concept in IBE, i.e., FIBE—a set of descriptive attributes that are the identities (Chatterjee & Sarkar, 2006). FIBE works on certain distance (threshold distance) between the users. This distance is measured by some metric system. Suppose that Alice with ω as the secret key will only able to decrypt the ciphertext which is encrypted with ω' as public key if and only if ω and ω' are within a certain distance. ABE is defined as the application in FIBE in which the user can encrypt the information to all other users having the certain set of attributes. The FIBE provides the advantage by storing the document on an untrusted server like a cloud server.

11.2.3 Attribute-Based Encryption

Sahai and Waters introduced the concept of ABE in the year 2005. The keys are generated by the authority for encryption and decryption of the data of the sender and receiver, respectively. These generated keys, the public key and the master key, are according to the set of attributes and should be pre-defined (Sahai & Waters, 2005). The user cannot add any data to the system if attributes are not pre-defined, and in such a case, the attribute will be re-defined by the authority and will again generate the public key and the master key.

11.2.3.1 Principle of an Ideal Attribute-Based Encryption Scheme

i. **Data Confidentiality**:
   The data owner encrypts the data before uploading to the storage and that is why the unauthorized parties do not get any of the information of the encrypted data.

ii. **Fine-grained access control**:
   Every individual user has different access right even if they are in the same group.

iii. **Scalability**:
   The efficiency of the system does not affect by the increase in the number of the authorized user.

iv. **User Revocation**:
   The stored data cannot be accessed by the repudiated user as the ABE scheme revokes the access rights directly from the system.

v. **Collusion resistant**:
   Each attribute belongs to the polynomial or the random number; the user cannot collude the different attribute with each other to decode the encrypted data.

11.2.3.2 Features of ABE

1. ABE has the capacity to address complex access control policies.
2. There is no need to know the exact list of users as the knowledge of access policy is sufficient.
3. ABE scheme appeases in terms of collision resistance.

11.2.3.3 Architecture of Data Sharing

In this growing world of technologies, the Internet has become a part of each and everyone's life. Every user daily shares his/her sensitive data to other users by using services like e-mails, social websites, etc., in which framework of the network is used to grant access to the information of the shared storage. Accessing these services/applications, users acknowledge their personal information (like name, residence address, age, interests, gender, etc.) into the public domain which helps in automatically linking the people having the same attributes to each other. These systems possess weak security for the privacy of the data, and the undesirable parties can easily mine the data and misuse it. The ABE scheme gives user-controlled privacy according to the attributes, and it also provides

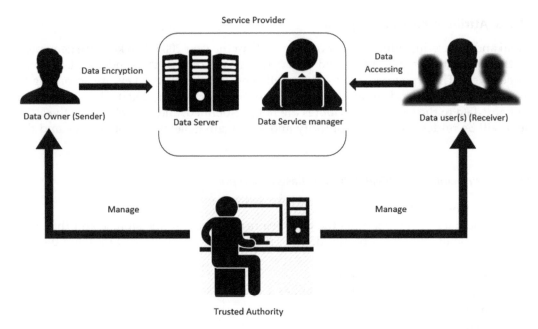

FIGURE 11.1
Data sharing system architecture.

scalability. Currently, a trusted central server is used to save all profile information of the user in the social networking websites, but in ABE-based systems, the profile information is used to save on an untrusted server which decreases the storage requirements and trafficking induced by the system (refer to Figure 11.1).

The data owner (sender) does the encryption before uploading the data to the server for keeping the data confidential. Then, the data user(s) (receivers) access the data by using the data decryption key(s). Cryptographic security can be achieved by many methods of network security from which ABE is one of the best to use (Gagné, Narayan, & Safavi-Naini, 2010).

Entities involved in Data Sharing System Architecture are as follows:

A. **Data Owner (Sender)**:

A person who owns data and upload the data in Storage Centre (like cloud storage), which makes sharing the data easy for them and also reduces the cost, is known as data owner. Encryption is done by the data owner before uploading the data. It helps in defining access policy and enforcement on its own data. The key for encrypting the file into a form called ciphertext is generated by the key generator.

B. **Data Storing Centre**:

It provides service for storing data and is responsible for controlling the accesses of the outside users to the stored data and provides the content according to their demand. The Data Storing Centre with Key Generator Centre (KGC) provides personalized user key to user per each attribute and repudiates the attribute group keys, and this invokes a fine-grained user access control. So, Data Storing Centre provides data storage, data eradication, and delivery services.

C. **Users (Receiver)**:

This entity denotes the person who accesses the stored data. If the set of attributes of the user appeases the access policy of the data encrypted by the data owner (sender) and also if it is not repudiated in any of the attribute group keys, the decryption of the ciphertext can be done by the user and he/she gets the data.

D. **Key Generation Centre**:

The generation of a public key and a secret key for ABE is done by Key Generation Centre. It is responsible for granting, repudiating, and updating attribute group keys for valid users. According to their attributes, the user gets the access rights. The encryption and decryption are done by the generated keys.

11.2.3.4 Algorithm of ABE

i. **Setup** (d):

ii. G_1 and G_2: Bilinear groups of prime order p.

iii. g is the generator of G_1.

iv. Bilinear map, e: $G_1 \times G_1 \rightarrow G_2$.

v. d is the threshold value.

vi. Randomly chosen $t_1, \ldots, t_n$.

vii. Public key, $PK = \left(T_1 = g^t_{1}, \ldots, T_n = g^t_{n}, Y = e(g,g)^y\right)$, where $y \in \mathbb{Z}_q$ and master key, $MK = (t_1, \ldots, t_n, y)$.

A. **KeyGen** (A_U, **PK, MK**):

  i. Private key generated by the authority for users U.

  ii. Randomly chosen, $d - 1$ degree polynomial q,

$$q(0) = y.$$

  iii. Private key D is

$$D_i = g^{q(i)/t}_{i}, \quad \text{where } i \in A_U.$$

B. **Encrypt** (A_{CT}, **PK, M**):

  i. Encrypt message $M \in G_2$ and set of attributes A_{CT}.

  ii. Randomly chosen, $s \in \mathbb{Z}_q$.

  iii. Encrypt data is published as

$$CT = \left(A_{CT}, E = MY^s = e(g,g)^{ys}, E_i = g^t_{i}^s\right) \quad \text{where } i \in A_{CT}.$$

C. **Decrypt** (**CT, PK, D**):

  i. Decryption of CT with private key D.

  ii. Randomly chosen d from $i \in A_U \cap A_{CT}$ to solve

$$e(E_i, D_i) = e(g, g)^{q(i)s}, \quad \text{if } |A_U \cap A_{CT}| \geq d,$$

and $Y^s = e(g, g)^{q(0)s} = e(g, g)^{ys}$ with the Lagrange coefficient.

$$M = E/Y^s.$$

For instance, for a set of descriptive attributes {*computer_dept, teacher, student, computer_sub*} of an encrypted data, the value of threshold is 3. The decryption of the encrypted data will only take place if the private key holds three or more number of attributes in the encrypted data. So, data user may have private key with attributes {*teacher, student, computer_sub*} to perform decryption and obtain the data (Delerablée & Pointcheval, 2008).

11.3 Key-Policy ABE

11.3.1 Introduction to KP-ABE Scheme

In 2006, Vipul Goyal introduces the KP-ABE scheme. In this scheme, the encrypted data are attached with the set of descriptive attributes, and it also forms end user's private key access policy. If this access design in a user's private key was appeased by the encrypted data's attributes, the message can be decrypted by the user to obtain the plaintext (refer to Figure 11.2).

Over the data attributes, the users are authorized with access tree-structure. The threshold gate is defined as the node of the access tree, and the leaf nodes are linked with the attributes. The user gets the private key from trusted attribute authority, including the policy in which it generates the key which can decrypt according to the type of the ciphertext

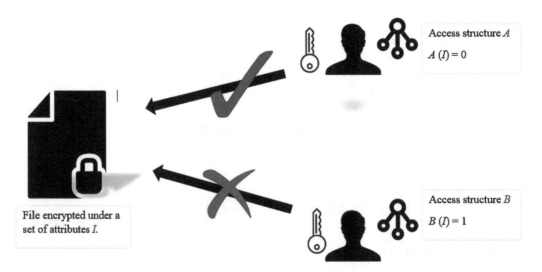

FIGURE 11.2
KP-ABE access control.

(Herranz, Laguillaumie, & Ràfols, 2010). KP-ABE access tool and re-encryption technique get together and used during the efficient revocation.

In the KP-ABE scheme, the problem that the data owner fails to decide is who is the user and who will decrypt the encrypted message. So, for some applications, the data owner has to trust blindly on the key generator.

11.3.2 Features of KP-ABE Scheme

   i. This scheme is designed for one-to-many communications.

   ii. This scheme is mainly suitable for the structured organization.

   iii. Applications like forensic analysis and targeted broadcast use KP-ABE scheme.

   iv. This scheme is collusion resistant.

   v. This scheme provides fine-grained access control.

   vi. KP-ABE scheme is more flexible than ABE scheme.

11.3.3 Algorithm of KP-ABE

There is a difference between KeyGen() algorithms of ABE scheme and KP-ABE scheme. The private key generation of the user is according to the access structure. The decrypt() algorithm is also different as the KeyGen() algorithm is different from ABE scheme. During the decryption, the attributes of encrypted data are used to execute decrypt node function (Goyal, Jain, Pandey, & Sahai, 2008).

For instance, the attributes of the encrypted data are {computer_dept ∧ student}, and the private keys of the user with access structure are {compter_dept ∧ (teacher ∨ student)}.

Decryption of the message/information can be done if and only if the access design of user's private key is appeased by the attribute, assign with the encrypted data.

A. **Setup (*d*):**

   i. G_1 and G_2: Bilinear groups of prime order p.

   ii. g is the generator of G_1.

   iii. Bilinear map, e: $G_1 \times G_1 \rightarrow G_2$.

   iv. d is the threshold value.

   v. Randomly chosen $t_1, \ldots, t_n$.

   vi. Public key, $PK = \left(T_1 = g^{t}{}_1, \ldots, T_n = g^{t}{}_n, Y = e(g,g)^y \right)$, where $y \in \mathbb{Z}_q$ and master key, $MK = (t_1, \ldots, t_n, y)$.

B. **KeyGen (*A_{U-KP}*, *PK*, *MK*):**

   i. Private key is generated by the authority for users U.

   ii. In the access design, the private key components were generated by the authority for each node x.

   iii. Randomly chosen, $d-1$ degree polynomial q_x such that

$$q_x(0) = q_{\text{parent}(x)}\big(\text{index}(x)\big),$$

where parent(x) is x's parent node and index(x) is associated with node x.

iv. Private key D is

$$D_x = g^{q_x^{(0)/t}}_i, \text{ where } i \text{ is equal to leaf node.}$$

C. **Encrypt (A_{CT}, PK, M)**:

i. Encrypt message $M \in G_2$ and set of attributes A_{CT}.
ii. Randomly chosen, $s \in \mathbb{Z}_q$.
iii. Encrypt data are published as

$$CT = \left(A_{CT}, E = MY^s = e(g,g)^{ys}, E_i = g_i^{t^s} \right) \quad \text{where } i \in A_{CT}.$$

D. **Decrypt (CT, PK, D)**:

The recursive method is used to obtain decryption.

i. Decryption of CT with private key D.
ii. If i is equal to the leaf node and is in the access design of the user's private key.

$$e(E_i, D_x) = e(g,g)^{s \cdot q_x^{(0)}}$$

The decrypt node function will also be called even if i is not equal to leaf node and computes $e(g,g)^{s \cdot q_x^{(0)}}$ using Lagrange coefficient.

$Y^s = e(g, g)^{q(0)s} = e(g, g)^{ys}$ will only compute, if the access design in user's private key appeases by the encrypted data's attributes.

The message $M = E/Y^s$ is obtained.

11.4 Ciphertext-Policy ABE

11.4.1 Introduction to CP-ABE

Ciphertext-policy was proposed in the year 2007 by J. Bethencourt with Amit Sahai and Brent Waters (Bethencourt, Sahai, & Waters, 2007). The CP-ABE scheme access control approach is very much relatable to the KP-ABE scheme, as in KP-ABE scheme, access policy is in user's private key, but in CP-ABE scheme, the access control is in encrypted data (Zhou & Huang, 2010), as shown in Figure 11.3.

In this scheme, the selected descriptive attributes are attached with the user's private key and form the access policy in encrypted data. For instance, private keys of the user with access design are {compter_dept ∧ (teacher ∨ student)} and the set of attributes of the encrypted data are {computer_dept ∧ student}; if the access design of user's encrypted data gets appease by the selected set of attributes in user's private key, then

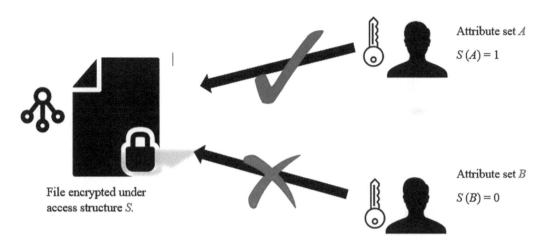

FIGURE 11.3
CP-ABE access control.

the message/information can be decrypt to obtain the plaintext (Emura, Miyaji, Nomura, Omote, & Soshi, 2009). Limitations that CP-ABE scheme faces are (i) in terms of organizing user attributes and defining policies and (ii) decryption keys only support logically specified a single set of user attributes.

11.4.2 Features of CP-ABE

i. It overcomes the disadvantage of KP-ABE scheme, i.e., data user cannot decide who can decrypt the message.
ii. In CP-ABE scheme, the access control is supported in a real environment.
iii. Increase the security in this study when applies with proxy re-encryption.
iv. It allows realizing implicit authorization.
v. In CP-ABE scheme, the user's private key is obtained back by them once data have been encrypted according to the policies.
vi. It provides fine-grained access control and flexibility.

11.4.3 Algorithm of CP-ABE

A. **Setup**:
   i. G_0: Bilinear group of prime order p.
   ii. g is the generator of G_0.
   iii. Randomly chosen exponents α, β from $\mathbb{Z}_q$.
   iv. Public key, $PK = (G_0, g, h = g^\beta, f = g^{1/\beta}, e(g, g)^\alpha)$, and master key, $MK = (\beta, g^\alpha)$.

B. **KeyGen (A_U, MK)**:
   i. Private key generated by the authority for users U.
   ii. Randomly chosen, $s \in \mathbb{Z}_q$ and s_j for each attribute j.

   iii. Randomly chosen, $d - 1$ degree polynomial q,

$$q(0) = y.$$

   iv. Private key D is

$$D = \left(DK = g^{(\alpha+s)/\beta}\right) \quad \text{where } j \in A_U$$

$$\text{and } D_j = g^s \cdot H(j)^s{}_j, D_j^* = g^s{}_j \text{ is output.}$$

C. **Encrypt** (A_{CT-CP}, *PK, M*):

   i. Randomly chosen, $y \in \mathbb{Z}_q$,

$$q_r(0) = y.$$

   ii. r is root node and I be set of leaf nodes in A_{CT-CP}
   iii. Encrypt data is published as

$$CT = \left(A_{CT-CP}, C' = Me(g,g)^{\alpha y}\right), C = h^y$$

$$\text{where } C_i = g^{q}{}_i^{(0)} \text{ and } C_I^* = H\left(att(i)^q\right)_i^{(0)}.$$

D. **Decrypt** (*CT, D*):

   i. Decryption of *CT* with private key *D*.
   ii. x is a leaf node and $k = att(x)$, $k \in A_U$.
   iii. Decryptnode (*CT, D, x*)

$$\left(e(D_k, C_x)\right) \div \left(e\left(D_k^*, C_x^*\right)\right) = e(g,g)^{sq}{}_x^{(0)}.$$

Using the Lagrange coefficient (*CT, D, s*)$= e(g, g)^{ys}$.
$C'/((e(C, DK)) \div (e(g, g)^{ys}) = M$ (Cheung & Newport, 2007).

11.4.4 Comparison of KP-ABE and CP-ABE (See Table 11.1)

TABLE 11.1

Comparison between KP-ABE and CP-ABE

Parameters	KP-ABE	CP-ABE
Efficiency	Average, high for the broadcast application	Average
Fine-grained access control	Low, high during re-encryption technique	Average
Collusion resistant	Good	Good
Computational overhead	Most of the computational overheads	Average of computational overheads

11.5 Applications of Attribute-Based Encryption

11.5.1 Maintaining Personal Health Record Security Using ABE Scheme

Personal Health Record (PHR) applications allow patients to manage, initialize, and share their personal information on health. Nowadays, PHR has become a trending topic in the area of healthcare technology. It is mainly stored in the cloud storage which reduces its cost and makes its mechanism easy to store and access. Over the data in PHR, there must be a fine-grained access control approach. The main challenges faced by the users in cloud services of PHR are the privacy, security, and data confidentiality on health. Losing physical control over the PHR data uploaded by the owner on cloud storage, the cloud server gets the access to the plaintext and challenges the security of PHR.

PHR recently chose the ABE scheme as their primitive encryption scheme which makes it more secure and effective, but it faces the challenge for the on-demand user revocation. So for the security of PHR, CP-ABE and KP-ABE are used: multi-authority ABE is used for reducing the key-management problem and the ABE scheme, and break glass ABE scheme is used for the urgent/emergency access to PHR.

11.5.2 Audit Log

Applications of forensic analysis are secured by the KP-ABE scheme. Electronic forensic analysis, "audit log", includes all account details like all activity on the system; therefore, their security concerns are high. To capture the enemy, the thorough audit log is a prize target. Encrypting an audit log is not enough to secure the systems specially the network system like Global Information Grid, because to access the audit log information only the secret key is needed and getting such a single key means accessing all the secret data (Waters, 2011). So, to counter such security issues, audit log uses KP-ABE scheme which makes the entries as a set of attributes such as data type, the name of the user, and time and date.

11.5.3 Targeted Broadcast

Targeted Broadcast is a new broadcast scenario which may have settings such as:

i. A broadcaster broadcast different items in a sequence where each item is described by the set of attributes.

For example, the broadcaster is broadcasting an episode of XYZ show, so the attributes describing this item are such as name, date, producer name, director name, season number, episode number, and genre.

ii. There are different packages, and each user signs up with different ones. Here access policy is the user's package which decides whether the user is able to access the item or not. Suppose that the user wants to see the episodes of XYZ show from season 1 and not from the current season 4, the encoded policy will be ("XYZ" **AND** ("Season 4" **OR** "Season 1")).

Targeted Broadcast allows the user to enjoy their shows, sports, news, etc., on their television according to their choices. The targeted Broadcast system is obtained from KP-ABE scheme, where each broadcast item is encrypted using the symmetric key and then KP-ABE scheme encrypt this symmetric key with the broadcast item's attributes.

11.6 Conclusion

As we discussed in this chapter, ABE is one of the genres of public key cryptography in which the identity is defined as the set of the attribute rather than an atomic value. In ABE, the public key and the master key are according to the set of attributes and are pre-defined, and if the user-predefined attributes don't match, the authority re-generates the master key and the public key. ABE is collision resistant, and the decryption operation of ABE requires a more convoluted approach to control the access, and hence, it makes ABE more secure. We conclude that the two schemes of ABE, (i) KP-ABE and (ii) CP-ABE, include some properties: (1) the attributes are encrypted in these schemes; (2) a public key, a private key, and a random polynomial are allotted with each attribute, so that the data cannot be recovered by any other user by using their own combination of attributes and therefore, no attack occurs; and (3) the access protocol accommodates a Boolean formula that includes flexibility in controlling user's access.

References

Bethencourt, J., Sahai, A., & Waters, B. (2007). Ciphertext-policy attribute-based encryption. *IEEE Symposium on Security and Privacy (SP 07)*, Berkeley, CA, USA, 321–334. doi:10.1109/sp.2007.11.

Chatterjee, S., & Sarkar, P. (2006). Generalization of the selective-ID security model for HIBE protocols. In Moti Yung et al., editors, *Public Key Cryptography—PKC 2006*. Lecture Notes in Computer Science, 3958, 241–256. doi:10.1007/11745853_16.

Cheung, L., & Newport, C. (2007). Provably secure ciphertext policy ABE. *Proceedings of the 14th ACM Conference on Computer and Communications Security—CCS 07*, 456–465. doi:10.1145/1315245.1315302.

Delerablée, C., & Pointcheval, D. (2008). Dynamic threshold public-key encryption. In David Wagner, editor, *Advances in Cryptology—CRYPTO 2008*. Lecture Notes in Computer Science, 317–334. doi:10.1007/978-3-540-85174-5_18.

Emura, K., Miyaji, A., Nomura, A., Omote, K., & Soshi, M. (2009). A ciphertext-policy attribute-based encryption scheme with constant ciphertext length. In *Proceedings of the 5th International Conference on Information Security Practice and Experience, ISPEC '09*, Springer-Verlag, Berlin, Heidelberg. Lecture Notes in Computer Science, 13–23. doi:10.1007/978-3-642-00843-6_2.

Gagné, M., Narayan, S., & Safavi-Naini, R. (2010). Threshold attribute-based signcryption. In J.A. Garay and R. De Prisco, editors, *Security and Cryptography for Networks*. Lecture Notes in Computer Science, 6280, 154–171. doi:10.1007/978-3-642-15317-4_11.

Goyal, V., Jain, A., Pandey, O., & Sahai, A. (2008). Bounded ciphertext policy attribute based encryption. In *Proceedings of the 35th International Colloquium on Automata, Languages and Programming, Part II, ICALP '08*, Springer-Verlag, Berlin, Heidelberg. Lecture Notes in Computer Science, 579–591. doi:10.1007/978-3-540-70583-3_47.

Herranz, J., Laguillaumie, F., & Ràfols, C. (2010). Constant size ciphertexts in threshold attribute-based encryption. In Phong Q. Nguyen and David Pointcheval, editors, *Public Key Cryptography—PKC 2010*. Lecture Notes in Computer Science, 19–34. doi:10.1007/978-3-642-13013-7_2.

Sahai, A., & Waters, B. (2005). Fuzzy identity-based encryption. *Advances in Cryptology—EUROCRYPT 2005*. Lecture Notes in Computer Science, 3494, 457–473. doi:10.1007/11426639_27.

Shamir, A. (1979). How to share a secret. *Communications of the ACM*, 22(11), 612–613. doi:10.1145/359168.359176.

Shamir, A. (1984). Identity-based cryptosystems and signature schemes. *Advances in Cryptology*. Lecture Notes in Computer Science, 47–53. doi:10.1007/3-540-39568-7_5.

Waters, B. (2011). Ciphertext-policy attribute-based encryption: An expressive, efficient, and provably secure realization. *Public Key Cryptography—PKC 2011*. Lecture Notes in Computer Science, 53–70. doi:10.1007/978-3-642-19379-8_4.

Zhou, Z., & Huang, D. (2010). On efficient ciphertext-policy attribute based encryption and broadcast encryption. *Proceedings of the 17th ACM Conference on Computer and Communications Security—CCS 10*. doi:10.1145/1866307.1866420.

12

Key Management

Jyotsna Verma

Central University of Rajasthan

CONTENTS

12.1 Introduction 197
    12.1.1 Background of Cryptographic Key System 198
    12.1.2 Importance of Key Management 199
    12.1.3 Challenges in Managing Cryptographic Key System 199
12.2 Key Exchange Cryptographic Algorithms 200
    12.2.1 Diffie–Hellman Key Exchange 200
    12.2.2 Elliptic Curve Diffie–Hellman Key Exchange Protocol 201
        12.2.2.1 Prerequisite 201
    12.2.3 Buchmann–Williams Key Exchange 202
    12.2.4 LWE Key Exchange 202
    12.2.5 Ring LWE Key Exchange 203
12.3 Key Distribution in Cryptosystem 204
    12.3.1 Symmetric Key Distribution 204
    12.3.2 Public-Key Infrastructure 205
    12.3.3 Public-Key Distribution 206
        12.3.3.1 Public Announcement of Public Keys 206
        12.3.3.2 Public Available Directory 206
        12.3.3.3 Public-Key Authority 207
        12.3.3.4 Public-Key Certificates 207
    12.3.4 X.509 Certificates 208
12.4 Conclusion 210
References 210

12.1 Introduction

Cryptographic system or cryptosystem is a suite of cryptographic algorithms, keys, and key management protocols for securing the message traveling over the unsecure networks between the entities. The cryptographic algorithms ensure the secrecy, privacy, and authenticity of the communicated message between the entities. It involves ciphers, keys, plaintext, and ciphertext for the encryption of the message. There are two types of cryptography algorithms in the cryptographic system: symmetric (secret key) and asymmetric (public key) cryptography algorithms. Symmetric key cryptography uses the same key commonly known as the secret key for the encryption and decryption of the message, whereas asymmetric key cryptography algorithm uses public and private keys for the encryption and decryption of the message. Hence, there are three types of keys involved in the cryptographic algorithm: secret key, private key, and public key for ensuring the secrecy of the message over the unsecured networks. For ensuring the secrecy of the message, cryptographic keys must be managed efficiently and is done by key management system.

Key management system or cryptographic key management system deals with the generation, distribution, exchange, storage, replacement, destruction, and management of keys in the cryptographic systems. Key exchange and key distribution are the most important and critical aspects of the cryptographic key management system. Key exchange or key establishment is the method to exchange the keys between the entities over the unsecure channel. There are various problems associated with the key exchange methods, like how to exchange the keys through the channel so that the keys can reach securely to the other end entities. Several key exchange algorithms are also present in the literature which will be discussed further in this chapter.

Key distribution is another aspect of a key management system which delivers the keys to the entities who wish to exchange messages between them, i.e., who want to communicate with each other. The key distribution method is used to reduce the risk of learning the keys by the intruder while exchanging the keys between the entities over the network. Literature witnesses various key management and distribution cryptographic algorithms for managing and distributing keys for the secure communications, but there is no generalized key management and key distribution technique which can be used for all the application scenarios.

12.1.1 Background of Cryptographic Key System

The concept of the cryptographic key system is not new; it was started way back thousands of years ago. Basically, cryptography is an art which is considered to be born along with the art of writing. The term "cryptography" was formed by combining two Greek words "Krypto" and "Graphene" which means secret writing. The roots of cryptography can be very much found in Egyptian and Roman civilizations. In Egyptian civilization, the Egyptians used to communicate with each other through messages written in "hieroglyph" writing system which is an old cryptographic technique. The messages written in hieroglyph were only understood by the scribes who used to transmit messages to the other end on behalf of the king. Similarly, the Romans used a popular Caesar cipher method to transmit the messages, and the first recorded use of Caesar cipher was by Julius Caesar, in which the plaintext characters are transformed monoalphabetically. Each character in the plaintext was transformed or shifts down by three characters alphabetically. The history of cryptography has witnessed many changes and can be found in the literature (Kahn, 1996; Singh, 2000; d'Agapeyeff, 2016). Till 1970s, cryptography was mostly used by the government agencies, but with the two major developments—public-key cryptography and the designing of Data Encryption Standards (FIPS PUB, 1999)—the nature of cryptography was completely changed and has brought them in the public domain (Davies, 1997).

Cryptography is both an art and science which transform the information into a form which is secure and immune to attacks. The components of cryptography include a plaintextwhich is the original text, the encryption algorithm that encrypts the plaintext, ciphertext which is formed after the encryption of the plaintext, and the decryption algorithm which decrypts the ciphertext into the plaintext. Encryption and decryption algorithms are sometimes referred to as ciphers. In the cryptographic technique, ciphers use keys to operate on the cryptographic algorithms to secure the messages. Keys are nothing but the numbers which are kept secret, and ciphers use them to encrypt and decrypt the messages by using encryption and decryption keys, respectively.

There are two categories of cryptographic algorithms: (i) symmetric key cryptographic algorithms and (ii) asymmetric cryptographic algorithms. In symmetric key algorithms, the same key is used by both the communicating entities and in asymmetric key cryptography; there are two types of keys used: private key and public key. The sender entity uses the public key of receiver entity to encrypt the plaintext, and the receiving entity uses its own private key to decrypt the ciphertext. The keys must be kept secret in the cryptographic algorithms to make the communication immune to attacks from the intruders, and hence, here comes the concept of key management which tries to manage and distribute the keys to the participating entities securely.

12.1.2 Importance of Key Management

For efficient and secure transmissions of messages between the entities in the cryptosystem, there is always a need of encryption and decryption keys. Hence, keys play a significant role in the cryptosystem, and so managing keys is a paramount area of research. Key management system, policies and rules help the cryptosystem to assure security (i.e., confidentiality, authenticity of the participating entities and integrity) to the communicating entities. Key management allows an efficient and systematic key generation, key distribution, key exchange, and storage of keys, so that the participating entities can be able to communicate with the secure environment and remain immune to attacks. In the absence of an efficient key management system, the communicating entities are more prone to come into the contact of intruders which results in an insecure cryptosystem.

12.1.3 Challenges in Managing Cryptographic Key System

Key management has various challenges that need to be addressed for an efficient implementation of cryptographic algorithms. The following are the major challenges associated with the key management system:

1. **Scalability**: There are billions of entities who want to communicate with each other, and hence they require the cryptographic algorithms that will help them to communicate over an unsecure network. For a large number of entities, a large number of keys are required, and managing such a large number of private and public keys in the cryptosystem is a tedious task. So, scalability is one of the issues of key management that needs to be addressed by the researchers.

2. **Availability**: Key management system should ensure data accessibilities for the authorized participating entities.

3. **Security**: Security is one of the biggest challenges of key management. Designing an efficient key management system is very necessary, as with the inefficient key management system, participating entities in the cryptosystem are more prone to the intruders and the keys are vulnerable from the intruders.

4. **Heterogeneity**: Development of key management techniques should be such that it can support multiple databases, different platforms, diverse range of applications, and standards.

12.2 Key Exchange Cryptographic Algorithms

Key exchange or key establishment in the cryptosystem is the method of exchanging keys between the communicating entities for secure communication, by using cryptographic key exchange algorithms. Broadly, key establishment protocols are divided into two types: key transport protocol, in which key created by one entity is transported to other communicating entity, and the second is key exchange protocols, in which keys are created by the joint effort of the communicating entities and are exchanged between them (Law et al., 2003). The keys can be exchanged in two ways—either in-band or out-of-band. In in-band key exchange, key is exchanged between the entities through the same communication channel in which the entities are communicating and encryption is taking place, whereas in out-of-band key exchange method, keys are exchanged through the channel other than the channel by which they are communicating, i.e., the channel that they use for encryption (Dulaney and Easttom, 2017). Key exchange is an important part of the key management system and is constantly explored by the researchers. This section discusses various popular key exchange methods, and more discussions can be found in subsequent subsections.

12.2.1 Diffie–Hellman Key Exchange

Diffie–Hellman Key Exchange (DHE) protocol was developed by Diffie and Hellman in (1976) for securely exchanging keys between the communicating entities even if the intruder is monitoring the communication channel. This is the first published public-key exchange protocol in which the communicating entities do not need to meet to agree on the keys rather they use session keys for the communication. The two entities before establishing the symmetric key must choose two numbers P and Q known to public. P is a large prime number with restriction that $((p - 1)/2)$ must also be the prime number and Q is also a prime number. The protocol works as follows:

1. Entity A chooses a large random number x and calculates $K_1 = Q^x$ mod P
2. Entity A now sends the K_1 to entity B over the channel.
3. Similarly, entity B chooses another large number y and calculates $K_2 = Q^y$ mod P
4. Entity B now sends the K_2 to entity A over the channel.
5. Both entities A and B after getting the K_2 and K_1, respectively, calculate the value of K which is a symmetric key. Entity A calculates $K = K_2{}^x$ mod P, and entity B calculates the $K = K_1{}^y$ mod P. In this way, they both share the symmetric key.

Predicting the value of the symmetric key is very difficult in DHE protocol because the values of x and y are very large and difficult to guess. Perhaps it will take a long time to interpret the key value by the intruder and hence it's a secure way to exchange the key, but still, it suffers from various attacks (Adrian et al., 2015). First, it suffers from Man-in-Middle attack in which the intruder can create two keys and send it to entity A, for creating the communication link between A and himself, similarly it does it with B. Suppose that entities A and B want to exchange the key and the intruder is trying to get the message by fooling entity A to be as B and similarly, entity B to be as entity A. The attack will work as follows:

1. Entity A chooses the random number x and calculates $K_1 = Q^x$ mod P and sends it to entities A and B.

2. The intruder will intercept the K_1 and choose his own number, say z, to calculate the $K_1 = Q^z$ mod P and send it to entity B over the channel.

3. Entity B receives the K_2 and thinks it was sent by entity A and calculates $K_3 = Q^y$ mod P by choosing the random number y.

4. Entity B now sends the K_3 to entity A over the channel which again intercepted by the intruder. Now, the intruder has both the values K_1 and K_3, and by using this, it will get the symmetric key K.

5. Entities A and B both calculate the key; entity A calculates the value of $E1 = K_2{}^x$ mod P and makes the communication between A and the intruder assuming that the key is shared with entity B. Similarly, B will calculate $E2 = K_2{}^y$ mod P and make the communication with the intruder assuming that the key is shared with A. In this way, both A and B make the communication with the intruder.

Various other exploitable issues in the DHE key exchange protocol configurations can be possible (Adrian et al., 2015). For instance, Adrian et al. found 23,631 servers to have browser-trusted certificates, and if any browser negotiates with any one of the listed servers about the DHE ciphersuites, then there may be a possibility of passive attack, which can compute the discrete log and can obtain the TLS session keys.

DHE protocol is a non-authenticated key protocol based on discrete exponential and logarithmic functions, but it forms the ground for public-key cryptography and various authenticated protocols (Hellman, 1978). Researchers have developed various exchange protocols to remove the drawbacks of DHE protocol which will be discussed in later subsections.

12.2.2 Elliptic Curve Diffie–Hellman Key Exchange Protocol

Elliptic Curve Diffie–Hellman (Miller, 1985) is another key exchange protocol, which is based on Elliptic Curve Cryptography (ECC). ECC is an approach based on the algebraic structure of elliptic curves over the finite fields of public-key cryptosystem and is more secure because it uses discrete elliptic curve logarithm functions (Koblitz, 1987). In this protocol, each communicating entity has a public and private key pair to create a symmetric key over an insecure channel.

12.2.2.1 Prerequisite (Miller, 1985)

1. Domain parameters of elliptic curve $T = (p, a, b, G, n, h)$ or $(m, f(x), a, b, G, n, h)$.

2. An elliptic curve private key d in the interval $[1, n - 1]$.

3. An elliptic curve public key Q, where $Q = dG$.

4. Entity A's key pair will be (d_A, Q_A), and Entity B's key pair will be (d_B, Q_B). The public key of the participating entity is known to public.

The following is the working of Elliptic Curve Diffie–Hellman Key Exchange protocol:

1. Entity A computes elliptic curve point $(x_k, y_k) = (d_A, Q_B)$, and entity B computes the point $(x_k, y_k) = (d_B, Q_A)$.

2. The shared secret key will be x_k; the x-coordinate of the elliptic point.

The shared key and the entities' private key are secret with the corresponding participating entities, and if an intruder tries to interpret the keys, then he has to solve Elliptic Curve Diffie–Hellman (ECDH) problem. ECDH is a very fast key exchange protocol (Schroeppel et al., 1995) and has several advantages over the other key exchange methods by providing security to the wireless devices with features like web browsing, email, and virtual private networking for the corporate networks (Lauter, 2004).

12.2.3 Buchmann–Williams Key Exchange

Buchmann William is an authenticated key exchange protocol, which is based on imaginary quadratic fields. Buchmann–Williams key exchange protocol was proposed by Buchmann and William (1988) and was later modified in 1989. In this protocol, the participating entities for exchanging the keys need to select a value D, which is very large ($\approx 10^{200}$), and an ideal value a, which can be made public. If two entities A and B want to exchange keys, then Buchmann–William key exchange protocol will work as follows:

1. Entity A will select a random integer x and compute an ideal reduced value b such that $b \sim a^x$ and send the b to entity B.
2. Entity B will also select a random integer y and compute an ideal reduced value c such that $c \sim a^y$ and send the c to entity A.
3. Now, entity A computes a reduced ideal value 1 (c^x), and similarly entity B computes a reduced ideal value $t2$ ($\sim b^y$), where $t1 = t2$ which can be seen in the theorem discussed in Buchmann and Williams (1988).

The idea of Buchmann-William key exchange protocol can also be used for designing public-key cryptography, but the complexity of the protocol is greater than the DHE protocol.

12.2.4 LWE Key Exchange

Learning With Errors (LWE) was introduced by Oded Regev as a problem in machine learning which is hard to be solved. LWE is used in creating various cryptosystems (Gentry et al., 2008; Regev, 2009; Peikert, 2009; Peikert and Waters, 2011). As LWE has been proven to provide the greater security and has a good asymptotical efficiency, Ding (2012) proposed a secure key exchange protocol based on LWE. For a communication between entities A and B over an unsecure channel, the LWE key exchange protocol will work as follows:

1. The public parameters are q, n, α where q is prime and $q > 2$ with a uniform random matrix $M \leftarrow Z_q^{n \times n}$.
2. Entity A chooses a secret vector $S_A \leftarrow D_{Z^n}, \alpha q$ and computes $P_A = MS_A + 2e_A \bmod q$, where $e_A \leftarrow D_{Z^n, \alpha q}$; then send P_A to entity B.
3. When entity B receives P_A, it will also choose a secret vector $S_B \leftarrow D_{Z^n}, \alpha q$ and an error $e'_B \leftarrow D_{Z, \alpha q}$.
4. It will compute $K_B = P^T_A \cdot S_B + 2e'_B \bmod q$ and $P_B = M^T \cdot S_B + 2e_B \bmod q$ while sampling $e_B \leftarrow D_{Z^n, \alpha q}$.
5. Now entity B computes and gets the shared secret key $SK_B = E(K_B, \sigma)$. Finally, entity B sends (P_B, σ) to entity A.

6. When entity A gets the (P_B, σ), then it will sample $e'_A \leftarrow D_{Z,\alpha q}$, compute $K_A = S^T_{\ A} P_B \cdot S_B + 2e'_A \bmod q$, and get the shared key $SK_A = E(K_A, \sigma)$.

By using a matrix secret form of LWE assumptions, one can get multiple shared secret bits with the protocol. The security of LWE key exchange lies with the hardness of solving LWE problem.

12.2.5 Ring LWE Key Exchange

A vast majority of public key algorithms available in literature, such as Diffie–Hellman algorithm, RSA algorithm, and ECDH, are easily broken by the quantum computer and are insecure. The Ring Learning With Errors (RLWE) key exchange (Ding et al., 2012) is one of the post-quantum public-key exchange algorithms that are designed to provide security against quantum computers. RLWE is a lattice-based cryptographic algorithm, and it is a widespread notion that lattice-based cryptographic algorithms provide security against quantum attacks and provide forward secrecy like Diffie–Hellman algorithm and ECDH.

In this protocol (Ding et al., 2012), when the communicating entities exchange key between them, then one of the entities will be an initiator say, entity A for the key exchange and other is a respondent entity say, entity B. The values of q, n, $a(x)$ are known to both entities A and B.

They generate a small polynomials x_α according to the discrete Gaussian distribution with the parameter α on the ring $R_q = Z_{q[x]} / \Phi(x)$.

The protocol begins with the initiator (A) and work as follows:

Initiation:

1. The initiator of the key exchange say entity A generates two polynomials S_A and E_A from the distribution x_α.
2. Entity A now computes $P_A = aS_A + 2E_A$.
3. Entity A sends the polynomial P_A to the responder entity, say entity B.

Response:

1. The responder, say entity B, also generates two polynomials S_B and E_B by sampling from the distribution x_α.
2. Entity B now computes $P_B = aS_B + 2E_B$.
3. Entity B generates E'_B from the distribution x_α and compute $K_B = P_B S_B + 2E'_B$. Then $P_B S_B + 2E_A S_B + 2E'_B$
4. The signal function Sig is used to find $w = Sig(K_B)$.
5. Respondent sides, i.e., entity B calculates the key $SK_B = \bmod_2(K_B, w)$ based on the w and the polynomial K_B.
6. Entity B sends P_B and w to entity A.

Finish:

1. After receiving P_B and w from entity B, sample E'_A from the distribution x_α compute $K_A = P_B S_A + 2E'_A = aS_A S_B + 2E_B S_A + 2E'_A$.
2. Initiator sides, i.e., entity A key is calculated as $SK_A = \bmod_2(K_A, w)$ based on the w and the polynomial K_A.

The signal function used in the RLWE key exchange is defined as follows:

The subset E $\{-q/4, \ldots, q/4\}$ of $Z_q = \{-((q-1)/2), \ldots, ((q-1)/2)\}$ is defined where signal function *Sig* is used to extract the approximate value of secret keys and the mod 2 operation defined above is used to eliminate error.

In the above-discussed key exchange protocol, the sender and the receiver have the same secret key with a small probability of failure to generate the same key which depends upon the choice of parameters. With a good choice of parameters, the failure rate will be less and will generate the same key.

12.3 Key Distribution in Cryptosystem

Key distribution is one of the major concerns of the key management system. In the key distribution scheme, the Key Distribution Center (KDC) is used which is responsible for securely distributing keys to the communicating entities. Symmetric keys can be distributed using symmetric encryption to the communicating entities, whereas the public keys can be distributed via a public announcement by using any public-key cryptographic algorithm, but the public announcement of public key can be forged by anyone and sometimes not desirable to use the public announcement method for distributing the keys. Public keys can also be distributed by using the public available directory maintained by a trusted organization. Each entity has its registered public keys with their names in the directory. This method also has vulnerabilities; if an intruder managed to get the private key of an entity, then it can get its public key from the available directory very easily. Public key authority and public key certificates are two other methods of distributing the keys to the communicating entities, in which an entity can get the public key via public key authority and public key certificates from the trusted organization, respectively. Both the methods of distributing the public keys have the vulnerabilities associated with it, which need to be addressed by the researchers. The above-mentioned distribution of private and public key method is discussed below in subsequent subsections.

12.3.1 Symmetric Key Distribution

In symmetric key cryptography, the same key is exchanged between the communicating entities prior to encryption and decryption of messages. The sender entity uses the same key to encrypt the message, and similarly receiver entity uses the same key to decrypt the message. But the major concern in this method is the secure distribution of the keys to the communicating entities. A secret key can be distributed to the entities through any of the following ways:

1. Keys can be personally distributed to the communicating entities.
2. Keys can be distributed by a trusted third party in person to the communicating entities.
3. The previously used secret key for the communication between the entities can also be used to distribute the newly created keys between the entities. The newly created keys can be encrypted using an old secret key.
4. The trusted third party can also deliver the keys to the communicating entities on the established encryption links to sender and receiver entities.

The first two methods are impractical because of manual delivery of the secret keys to the communicating entities. The efficiency of the third method depends on the level of security used for the previously used secret key by the communicating entities. The fourth method is somewhat practical, as this method involves a trusted third party for the delivery of the keys to the participating entities; this trusted third party can be called as a Key Distribution Center (KDC). In this method, for secure exchange of keys, two types of keys are used: a session key, which is a temporary key, used for encryption of message between the communicating entities for one logical session and is discarded after usage, and a master key, which is a long term key used to encrypt and distribute the session keys shared by entities and the KDC. For instance, if entity A wants to communicate with entity B, then entity A approaches to the KDC with a connection request packet, which is encrypted by the master key shared only by entity A and the KDC. The KDC then approves the connection request and generates a one-time session key and sends it to entity A by encrypting the session key with the master key; similarly, it shares that session key with entity B by encrypting it with the master key. Now both entities A and B have the on-time session key by which they can set up a logical connection and can communicate with each other.

Symmetric key distribution is simple to implement but can also be forged easily, because for the communication that takes place between each pair of communicating entities, it requires a unique symmetric key; for n entity, it requires $n(n-1)/2$ keys, which is a very large number of keys. Hence, an asymmetric/public-key cryptography comes into the picture which is discussed in the subsequent sections.

12.3.2 Public-Key Infrastructure

Public-Key Infrastructure (PKI) is a broad concept which came into existence for the universal use of public keys. PKI enables the exchange of information over an insecure network through the private- and public-key pair obtained through the certificate authority which is a trusted authority for issuing and distributing the public keys to the participating entities. PKI consists of the following elements (Vacca, 2004; Al-Janabi and Abd-Alrazzaq, 2011):

1. **Certificate authority**: Certificate authority (CA) is a trusted authority for issuing, distributing, storing, revoking, and renewing the certificate for the corresponding public key of an entity.
2. **Registered authority**: Registered authority verifies the identity of an entity; whether the entity requesting for the certificate or requesting for storing the certificate to the authority is a valid entity or not.
3. **Certificate revocation list**: Certificate revocation list (CRL) lists the revoked certificates issued to the participating entity and other certificate authorities.
4. **Certificate management system**: The certificate management system stores, distributes, and issues the certificates to the requesting entities.
5. **Certificate directories**: Certificate directories are responsible for storing and distributing the digital certificates and CRLs issued to the participating entities.

PKI is a combination of policies, standard, rules, entities, public-key cryptography, hardware, and software. It maintains the hierarchy of certificate authorities for getting the verification of the entities identity through the certificate issued by the CA.

12.3.3 Public-Key Distribution

In public key cryptography, there are two types of keys: private and public keys—which are used for the encryption and decryption of the messages, respectively. The private key in the cryptosystem must be kept secret by the participating entities, and the public key is announced publicly. Public keys can be distributed in many ways or techniques to the communicating entities for the secure communication to take place. Following are the ways for the distribution of public keys.

12.3.3.1 Public Announcement of Public Keys

One way of distributing the public keys to the communicating entities is through public announcement. Public keys are announced publicly so that the entities who want to participate in the communication can use the public keys for their secure communication. In this, the public key of each entity is announced in public forums, such as Internet mailing lists and USENET newsgroups for the users to use for their communication needs.

This approach is very simple and convenient to use, but it has one major drawback. If the public keys of the communicating entities were announced in public, it can be easily forged by some intruders (see Figure 12.1). The intruder pretends to be one of the communicating entities, say entity A, and sends its public key to another entity, say B. After receiving the public key from the intruder, entity B starts communicating with him assuming it to be entity A; by the time entity A discovers the forgery, the intruder has already taken all the encrypted messages destined for him/her.

12.3.3.2 Public Available Directory

Public available directory is another public-key distribution scheme in which publicly available dynamic directory is maintained. The distribution and maintenance of the publicly available directory are done by some trusted organizations or by some trusted entities (see Figure 12.2). In this scheme, each entity registers a public key in the publicly available directory maintained by the trusted authority. The trusted authority registers the entity in the directory with its name and a public key, say for an entity A, the entry in the directory will be (PK_a) and at any point time the entity may replace the existing key with the new key if for any reason key appears to be compromised or used for a large amount of data. The authority updates or publishes the entire directory to the public through circulated newspapers in hardcopy version or can be accessed by the entity in the electronic form.

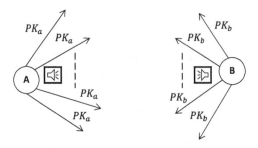

FIGURE 12.1
Public announcement of public keys.

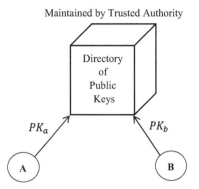

Maintained by Trusted Authority

Directory of Public Keys

PK_a

PK_b

A

B

FIGURE 12.2
Public available Directory.

The public available directories are more secure than a public announcement of public keys, but still suffer from a drawback. If an intruder manages to get the private key of the authority, then it can eavesdrop any messages sent by any communicating entities or can damage the records maintained by the trusted authorities.

12.3.3.3 Public-Key Authority

The distribution of public keys to the communicating entities from the publicly available directory must be done in a secure manner by having tighter control over the distribution of public keys. In this scheme, public-key authority has both public and private keys. The public key is known publicly to all the communicating entities, and the private key is known only to the authority. Now, if entity A wants to know the public key of entity B, then it will send a timestamped message to the public-key authority requesting the public key of entity B. Then, the authority responds to entity A with a private key-encrypted message containing entity B's public key, PK_b, the original message request of entity A and the original timestamp for the verification of the freshness of the message. Now, entity A stores the public key of entity B for the communication with entity B for future use through caching and the entity should ensure the freshness of the public key by requesting public key of the communicating entities from time to time. Entity A uses the public key of B (PK_b) for the encryption of the message destined for entity B containing a nonce (n_1) and an identifier of A (Id_a) to uniquely identify the transmission of the message. In the same manner, entity B also manages to get the public key of entity A (PK_a) and starts communicating with A by sending the encrypted message with A's private key; the message contains A's nonce (n_1) along with B's newly generated nonce (n_2). In order to ensure that the message is transmitted to entity A, A will return B's nonce (n_2) encrypted by B's public key. This is a seven-step process which needs seven message transmissions between two communicating entities and public-key authority for the communication between the communicating entities.

12.3.3.4 Public-Key Certificates

In the public-key authority scheme for the distribution of the public keys, there is a risk of tampering of records maintained by the trusted authority. An alternative approach is to use certificates created by a CA; the certificate contains the public keys of the participating entities and other related information. CA is nothing but a trusted third party which issues digital certificates to the participating entities. The issued certificate from the CA is then given to the communicating entities with their matching private keys, and this whole information is not made public, but rather they are kept secret. The CA has the only right to create and update the certificates, which contains the name and the public key of the communicating entity. Any communicating entity can read the certificates to determine the certificate's owner and can verify the certificate that it is issued by the CA or not. Later, Denning and Sacco (1981) added the requirement of verification of the certificate's freshness by any participating entities.

For ensuring the security of the certificates, so that the CA will not be able to issue any improper certificate, RFC 5280 has defined a revoked state in which if the CA issues an improper certificate or the private keys are compromised in any manner or any policies of CA are violated by the CA or by their users, then the certificate is revoked. There is also an authority revocation list, which is a kind of CRL that contains the list of revoked certificates from the issuing CA.

In public-key certificate scheme, each communicating entities request a certificate with a public key from the CA. Requests for the certificate made by the communicating entities to the authority are in person or through an authenticated secure communication. The CA sends the certificate to the entity which requests the certificate, say entity A. The certificate is encrypted by the CA's private key to prevent the certificate from getting forged, on receiving the encrypted certificate by the recipient, say entity B, the B will decrypt the message by the CA's public key to get the certificate element, i.e., identity of entity A and its public key. Its timestamp value indicates the freshness of the certificate. Certificates are assumed to be expired if it is sufficiently old where the timestamp serves as the date of expiry of the certificates. This scheme solves the problem of forgery, for example, if an entity A wants the public key of B, then it will download the corresponding certificate and the encrypted message. Then, the certificate is encrypted to make a message digest; the encrypted message then is decrypted through CA's public key and both the encrypted messages are compared; if both are equal, then no forgery happens and the certificate is valid.

12.3.4 X.509 Certificates

A public-key certificate scheme has solved the problem of the distribution of public keys to the communicating entities in a manner that it limits the forgery. But there is a problem associated with the certificate format issued from the CA. The issued certificates may have different formats, i.e., one certificate is in one format and another in another format, which creates the problem of compatibility; there must be some universal format followed universally.

To solve the above-said problem, a protocol called X.509 was developed by ITU which is accepted universally to structure the certificates by using Abstract Syntax Notation 1 (ASN.1). X.509 standard has a significant role in the certificate structure developed initially in 1988; later, it was issued in 2000 with the revised recommendation. The standard is based on the digital signature method and the public-key cryptography technique. The certificates are maintained by the CA. The CA has the responsibility to update the certificate and place them in their directory. The format of the X.509 standard for the certificates is provided in Table 12.1.

The certificates issued from CA have some period of validity, i.e., the certificate can be expired for various reasons, such as the private key of the communicating entity may be

TABLE 12.1

Format of the X.509 Standard Certificate

Fields	Description
Version	Version of X.509
Certificate Serial Number	A unique serial number of the certificate issued by CA
Signature Algorithm Identifier	The algorithm used to sign the certificates
Issuer Name	Name of the certificate authority defined by X.509
Period Validity	Expiry period of the certificate with start and end periods of the issued certificate
Subject Name	The name of the entity whose public key is being certified
Subject's Public-Key Information	Information of the public key and the algorithm identifier of an entity being certified
Issuer Identifier	A unique identifier of the issuer, i.e., CA
Subject Identifier	A unique identifier of the participating entities
Extensions	For later versions, extensions add one or more other fields
Signature	This field covers all other fields along with the hash code of the other fields

compromised, a certificate issued by a CA may be compromised, or the participating entities are no longer certified by CA. In this case, CA maintains the Certificate Revocation List (CRL) containing all the revoked certificates including certificates issued to both participating entity and other CA; the CA then posts the CRL to the directory with issuer name, issuer signature, the entry of each revoked certificate with the unique serial number, date of issue of CRL and the next scheduled date of the issue of the CRL.

Table 12.2 shows the format of CRL mentioned.

The communicating entities can get this information by checking the directory for the validity of the certificate when it receives the message certificate for its request. A local cache of certificates can also be maintained by the participating entities to improve their searches avoiding delays.

TABLE 12.2

Format of the Certificate Revocation List

Fields	Description
Signature Algorithm Identifier	The algorithm used to sign the certificates
Issuer Name	Name of the certificate authority defined by X.509
Update Date	The date of issue of CRL
Next Date	The next scheduled date of the issue of CRL
Certificate Serial Number	A unique serial number of the revoked certificate issued by CA along with date of revocation of the certificate
Signature	This field covers all other fields along with the hash code of the other fields

12.4 Conclusion

Cryptographic key management is a broad concept, and there are lots of challenges associated with it which need to be addressed by the research community. The chapter discussed various challenges involved in the cryptographic key management and problems associated with it. The NIST (National Institute of Standard and Technology) is looking forward and is making efforts to manage the keys and improving the cryptographic key system in order to secure the cryptographic keys. The exchange, storage, and distribution of keys in various networks like the delay-tolerant network open the door for further research. There is a serious requirement of key management protocols for the delay-tolerant networks (Menesidou et al., 2017). Furthermore, minimum standard for the secure handling in a heterogeneous and scalable environment is necessary and is a very significant challenge in view of future research.

References

Adrian, D., Bhargavan, K., Durumeric, Z., Gaudry, P., Green, M., Halderman, J. A., & VanderSloot, B. (October 2015). Imperfect forward secrecy: How Diffie–Hellman fails in practice. In *Proceedings of the 22nd ACM SIGSAC Conference on Computer and Communications Security* (pp. 5–17). ACM, Colorado, USA.

Al-Janabi, S. T. F., & Abd-Alrazzaq, H. K. (2011). Combining mediated and identity-based cryptography for securing e-mail. In *Digital Enterprise and Information Systems* (pp. 1–15). Springer, Berlin, Heidelberg.

Buchmann, J., & Williams, H. C. (1988). A key-exchange system based on imaginary quadratic fields. *Journal of Cryptology*, 1(2), 107–118.

Davies, D. (1997). A brief history of cryptography. *Information Security Technical Report*, 2(2), 14–17.

d'Agapeyeff, A. (2016). *Codes and Ciphers—A History of Cryptography*. Read Books Ltd., Uxbridge, United Kingdom.

Denning, D. E., & Sacco, G. M. (1981). Timestamps in key distribution protocols. *Communications of the ACM*, 24(8), 533–536.

Diffie, W., & Hellman, M. (1976). New directions in cryptography. *IEEE Transactions on Information Theory*, 22(6), 644–654.

Ding, J., Xie, X., & Lin, X. (2012). A simple provably secure key exchange scheme based on the learning with errors problem. In *IACR Cryptology ePrint Archive*, 688.

Dulaney, E., & Easttom, C. (2017). *CompTIA Security+ Study Guide: Exam SY0-501*. John Wiley & Sons, Hoboken, New Jersey, United States.

FIPS PUB (1999). Data encryption standard (DES). *FIPS PUB, 46-3*.

Gentry, C., Peikert, C., & Vaikuntanathan, V. (May 2008). Trapdoors for hard lattices and new cryptographic constructions. In *Proceedings of the Fortieth Annual ACM Symposium on Theory of Computing* (pp. 197–206). ACM, Victoria, BC, Canada.

Hellman, M. (1978). An overview of public key cryptography. *IEEE Communications Society Magazine*, 16(6), 24–32.

Kahn, D. (1996). *The Codebreakers: The Comprehensive History of Secret Communication from Ancient Times to the Internet*. Simon and Schuster.

Koblitz, N. (1987). Elliptic curve cryptosystems. *Mathematics of Computation*, 48(177), 203–209.

Lauter, K. (2004). The advantages of elliptic curve cryptography for wireless security. *IEEE Wireless Communications*, 11(1), 62–67.

Law, L., Menezes, A., Qu, M., Solinas, J., & Vanstone, S. (2003). An efficient protocol for authenticated key agreement. *Designs, Codes and Cryptography*, 28(2), 119–134.

Menesidou, S. A., Katos, V., & Kambourakis, G. (2017). Cryptographic key management in delay tolerant networks: A survey. *Future Internet*, 9(3), 26.

Miller, V. S. (August 1985). Use of elliptic curves in cryptography. In *Conference on the Theory and Application of Cryptographic Techniques* (pp. 417–426). Springer, Berlin, Heidelberg.

Peikert, C. (May 2009). Public-key cryptosystems from the worst-case shortest vector problem. In *Proceedings of the Forty-First Annual ACM Symposium on Theory of Computing* (pp. 333–342). ACM, Bethesda, MD, USA.

Peikert, C., & Waters, B. (2011). Lossy trapdoor functions and their applications. *SIAM Journal on Computing*, 40(6), 1803–1844.

Regev, O. (2009). On lattices, learning with errors, random linear codes, and cryptography. *Journal of the ACM*, 56(6), 34.

Schroeppel, R., Orman, H., O'Malley, S., & Spatscheck, O. (August 1995). Fast key exchange with elliptic curve systems. In *Annual International Cryptology Conference* (pp. 43–56). Springer, Berlin, Heidelberg.

Singh, S. (2000). *The Code Book: The Secret History of Codes and Code-Breaking*. Fourth Estate.

Vacca, J. R. (2004). *Public Key Infrastructure: Building Trusted Applications and Web Services*. CRC Press.

13

Entity Authentication

Hamza Mutaher and Pradeep Kumar

Maulana Azad National Urdu University

CONTENTS

13.1   Introduction...213
13.2   Password Authentication Protocol...214
13.3   Two-Factor Authentication 215
13.4   Biometric Authentication Method 216
     13.4.1   Physical Characters...216
     13.4.2   Behavioral Characters 217
13.5   Challenge-Handshake Authentication Protocol 218
13.6   Extensible Authentication Protocol 219
13.7   Three-Pass Protocol 220
     13.7.1   Shamir's Three-Pass Protocol...220
     13.7.2   Massey–Omura Cryptosystem 221
13.8   3D Secure 222
13.9   Conclusion 223
References...223

13.1 Introduction

Entity authentication is a mechanism that allows a connection between two parties authentically. The first party asks to establish the connection, whereas the second party accepts or denies the request after verifying the identity of the first party. The party which asks for connection is called the claimant; the party which checks the identity of the claimant is called the verifier. "An entity can be a person, a process, a client or a server" (Forouzan & Mukhopadhyay 2011).

A co-operative feature must be used to proceed and provide the authenticity called as credential and it can be formed as follows:

1. Something claimant knows: this credential is inexpensive, so it can be stored mentally like passwords and PINs (Personal Identification Numbers). This type of credentials is susceptible to lose due to humans' short memory or can be guessed by others.

2. Something claimant has: this type of credentials is hard to be revealed to unauthorized parties, for example, smart cards, magnetic stripe cards, and the software operating private and public keys.

3. Something inherited in claimant: this credential is a physical characteristic like fingerprints, voice, face recognition systems, and other biometrics (Kizza 2005; Van Tilborg & Jajodia 2016).

The purpose of entity authentication is to prevent unauthorized access to protect resources. Examples like outside party attempt to access school database to modify student's marks or attempts to withdraw money from somebody's bank account.

With the increment of security issues, a number of entity authentication protocols come up with different techniques and criteria applied to different applications. To consider any protocol as an authentication protocol, it must follow the rules of security definition which have been specified by its designers. Security definition is based on some security goals. Those goals guarantee the security. If a protocol meets those security goals, then it can be called as an authentication protocol (Ahmed & Jensen 2010). Some authentication protocols, such as password authentication protocol (PAP), challenge-handshake authentication protocol (CHAP), and extensible authentication protocol (EAP), and other authentication techniques will be discussed in this chapter.

13.2 Password Authentication Protocol

PAP is a simple authentication technique used to validate the identity of devices when point-to-point protocol (PPP) establishes a direct connection between two devices. PAP is a two-way handshake protocol that proceeds in the following two steps (Figure 13.1; Forouzan & Fegan 2007):

1. The first device user sends the authentication identity like username and password to the second device to get access permission.
2. The second device validates the identity of the user and then decides either to accept or to deny the connection.

Due to the improvement of network infrastructure, the world replaced access modem links, dial up-ISDN links, and T1 links by broadband and other new technologies. PAP became unsecured and replaced by new authentication protocols. On the other hand, PAP sends the password in plaintext; this could reduce the security goals.

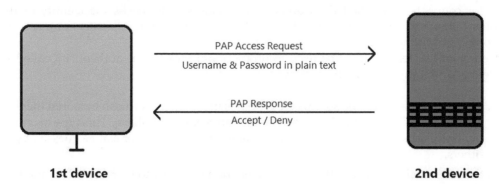

FIGURE 13.1
PPP connection session with PAP.

13.3 Two-Factor Authentication

Due to the improvements of hacking tools and malicious accesses to others accounts and degradation of security concept by fake users, thus allowing hackers to get access into the system. Users also prefer to use easily guessable passwords and keep the same password for multiple accounts, and some of them allow the browser to save their passwords by asking it to remember these passwords; thus, it becomes easy to a hacker to hack a user's accounts by using some attacks like dictionary attack and some tools like key-loggers or through social engineering (Sabzevar & Stavrou 2008; Vaithyasubramanian, Christy, & Saravanan 2015). Two-factor authentication (TFA) method came into the picture to solve these problems and ensure the authorized access of users on the Internet.

TFA is a cybersecurity method that ensures the authorized access of users to their own accounts. This method uses two phases of security which adds an extra level of security along with a static password during the login procedure. The extra level of security avoids hackers from comprising the account even if the hacker knows the username and the static password.

To apply TFA, users required to have threethings:

- Something user knows like [login IDs and passwords].
- Something the user has like [mobile, computer, email, and ATM card].
- Something user is like [fingerprint, iris recognition, voiceprint, and face].

These three requirements are helpful to apply the TFA method in which the first requirement is used by users to prove their identities while trying to access their accounts in social media or for some other proposes like ticket booking and money withdrawal using the second requirement. The third requirement, which is the most important step of TFA and also used by users, is to achieve the extra level of security. While the third requirement options are more expensive and rise the privacy concern, the One-Time Password (OTP) came into the picture as an alternative option for the TFA method (Vaithyasubramanian, Christy, & Saravanan 2015).

OTP is an authentication mechanism to secure user account from being accessed by unauthorized people. OTP adds more security to the login process by sending the user a password to be used once during logging in to his account. In OTP, the user will be sent a different password each time he attempts to log in to his account.

To implement OTP authentication, there are two main modes discussed in Xiao-rong, Qi-yuan, Chao, and Ming-quan (2005) as follows:

1. Challenge response mode: in this mode, a challenge will be generated to the user by the system during the login process. This challenge will be combined with the user password, and then the user will generate an OTP. User will send this OTP to the system and will be logged in successfully once.

2. Time synchronization mode: in this mode, the login time will be taken as a random number and will be combined with the user password to allow the user to access his account. In the mode, the user will be allowed to access his account only once and he has to repeat this process each time he attempts to log in to his account.

S/Sky mode is an OTP authentication which belongs to challenge-response mode (Haller 1995). The idea of S/Sky mode is to assign an account to each user. Each account will have three parameters [seed, sequence number, and secret passphrase]. By using these three parameters, the system can generate a series of OTPs. The user is allowed to use each OTP only once to log in to his account.

In S/Sky mode, the OTP-calculating program will be set up in both client and server sides. Then, in the time of login, the client will send his account to the server. The server will send back a client the seed and sequence number. Then, the OTP-calculating program in client side will generate an OTP and send it to the server along with the challenge and secret passphrase. The OTP-calculating program in server side will also generate an OTP. The both OTPs will be compared; if they are the same, then the user can log in to his account.

13.4 Biometric Authentication Method

Secure authentication methods are gaining more attention from research to secure access to various applications. Companies apply several types of biometric authentication method to ensure the authentic access to their applications. Every application requires a different type of biometrics depending on the level of security it needs and the type of biometric it suits: for example, to prove your attendance in the organization you work with, you have to use a fingerprint biometric; to enter to the military lab, you have to use a palm biometric; and to identify yourself, you may need to use a retina scan biometric.

The PIN of ATM cards and password can be easily observed by an intruder. Once the intruder observes the PIN and password, he can easily gain access to the user account, and the system cannot identify that he is the intruder, and this is one of the PIN and password limitations. In this case, the biometric authentication provides more reliable and accurate security to users and called "a something unique about the user" or a something user is, like a fingerprint, face, iris, voiceprint, and so on. The biometrics cannot be shared, repudiation unlikely, hard forging, and impossible to be disclosed, lost or stolen, and these are some features of biometric authentication (Ratha, Connell, & Bolle 2001).

Biometric authentication is an authentication process that depends on the biological characteristics of the user to identify whether he is the one who claims to be or not. Biometric system extracts the user biological characteristics and stores them in the database. Once the user inputs his biological characteristics into the system, the system compares both input and database-stored characteristics; if both are the same, then the authentication is confirmed and the user can access his account completely. Biometrics is used to manage the access procedure of physical and digital resources like computer devices, smartphones, cars, buildings, and apartments.

Biometric measurements have two types of characters: physical characters (such as fingerprint, palm, retina, iris, and face characteristics) and behavioral characters (such as signature and face) (Liu & Silverman 2001).

13.4.1 Physical Characters

1. Fingerprint: it uses the pattern of a fingertip. Many approaches are used to verify the fingerprint. Some use the traditional approach used by police to match the minutiae, while the others apply the straight pattern-matching device and use the

approach which is able to detect whether the finger is alive or not. Fingerprint authentication is mostly used for an in-house system where the system is executed in a fully controlled area.

2. Palm: it analyzes and measures the shape of a hand. It provides better performance and is easy to use. Palm authentication is usually used in places where the users use the system irregularly. The accuracy of palm authentication is very high as compared to fingerprint.

3. Retina: the retina biometric analyzes the blood vessels in the black area of the human eye. The users need to look into a given point in the scanner to be authenticated, and it is quite accurate. The people who wear glasses don't accept the retina biometric as they have to focus on the reading device which is a bit hard for them.

4. Iris: the iris biometric analyzes the colored ring of tissue around the eye pupil. This biometric has very high accuracy and requires no close distance between the user and scanning device, and the users who wear glasses are able to identify themselves. Iris biometric system is one of the most devices that perform well in authentication mode.

5. Face: face recognition scans and analyzes the human face characteristics. It needs a digital camera to support the authentication procedure to easily recognize the facial image. It is more expensive as compared to other biometric systems and is usually applied in the casino industry.

13.4.2 Behavioral Characters

1. Signature: signature authentication verifies and analyzes the movement of users while signing their names. This verification depends on two features: the first one is the speed in which the user signs, and the second one is the pressure he/she applies while signing. The digital signature in the device is compared with the static signature in the file to authenticate the user. Signature verification is usually used in the banks.

2. Voice: voice authentication does not mean voice recognition; rather, it means voice-to-print authentication. Voice-to-print authentication is very complicated as it requires to convert the voice into text. This technology requires no more new device because most computers have built-in microphones. This technology is more complicated as compared to other biometrics, and it is not user-friendly.

All biometric systems contain the same components listed below (Figure 13.2; Matyáš & Říha 2002):

1. User gateway: it is responsible to secure some assets. It is like the gate of a building; if users cross this gate and get authenticated, then he can get complete access to his account.

2. Central controlling unit: it is responsible to obtain the authentication requests from the users, manage and monitor the procedure of authentication, and send back the results of the authentication to the user.

3. Input device: the responsibility of the input device is biometric data acquisition by verifying the quality of the sample and users' liveness.

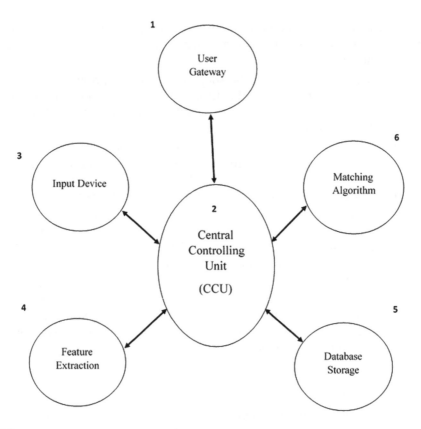

FIGURE 13.2
The model of a biometric system.

4. Feature extraction: it is responsible to process the biometric data. The feature extraction extracts the features of the users and tests the quality of biometric data gained by the input device and sends them to the matching algorithm.

5. Database storage: it is a database that stores the biometric templates. The biometric templates may also be stored in the users' medium like smartcards, and there should be a link between the biometric template and users.

6. Matching algorithm: it is reasonable to compare between the present user's biometric templates and biometric templates stored in the storage. After the comparison process, the matching result will be yes or no.

13.5 Challenge-Handshake Authentication Protocol

It is an authentication protocol that provides a secure PPP connection between two parties using a three-way handshake mechanism which shares a challenge message to check the authenticity between them. CHAP encrypts passwords that make it more secure than PAP. The password is encrypted by the hash function (Forouzan & Fegan 2007; Ghilen,

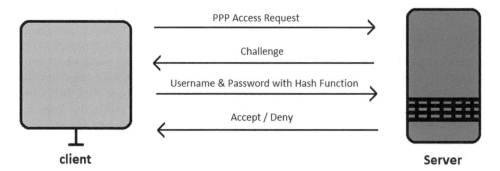

FIGURE 13.3
PPP connection session with CHAP.

Azizi, & Bouallegue 2015; Inamura 2015). The authentication process is explained in the following (Figure 13.3; Ghilen, Azizi, & Bouallegue 2015) (Inamura 2015):

1. After the PPP connection is established between server and client, the server generates a random challenge message and sends it to the client to check its authenticity. The client calculates the password with challenge message using a hash function and sends it as a response to the server.

2. The server calculates its own hash function and compares it with the response hash function. If both are the same, then accept the connection; if not, deny it and send a result acknowledgment to the client.

13.6 Extensible Authentication Protocol

EAP is an authentication protocol which is used to support login authentication process. It is considered as a framework that allows choosing one if there are multiple authentication methods. The authentication process in EAP occurs over the data link layer like PPP or IEEE 802 standard. As EAP operates under the data link layer, so it does not require an IP. EAP works in many environments such as switch circuits, wired and wireless networks. Elimination and retransmission of the authentication are supported by EAP where there is no fragmentation.

The implementation of EAP has been done with network devices like host and router which are linked up by switched circuits or dial-up lines using PPP. The implementation of access points and switches has been done using IEEE 802. EAP also supports the encapsulation of IEEE wired links, and it has been defined in [IEEE 802.1X] and also supports the encapsulation of IEEE wireless LAN which is defined in [IEEE 802.11i].

EAP is a flexible authentication protocol, and this is a feature of it. The other feature is that EAP allows choosing a certain authentication method. Consistently, the authenticator in EAP sends a request to get further information to decide which suitable authentication method to be used. In some cases, the authenticator cannot be updated to use all authentication methods; EAP allows the use of a backend authentication server. The backend authentication server can apply some or all authentication methods (Aboba, Blunk, Vollbrecht, Carlson, & Levkowetz 2004).

13.7 Three-Pass Protocol

Three-pass protocol (TPP) is a security mechanism which allows two parties to communicate with each other without exchanging their secret keys before establishing communication, and it depends on commutative encryption (Abdullah, Khalaf, & Riza 2015). In TPP, the message sender and receiver have encryption and decryption methods, respectively. These methods allow the message to be encrypted using two different secret keys and using the same keys to decrypt the message.

In TPP, sender A encrypts the message using his own unique secret key and sends it to receiver B. Receiver B receives the message and encrypts it using his own unique secret key and sends it back to receiver A. After that, sender A decrypts the message by his own secret key and sends it back again to receiver B. Receiver B decrypts the message by his own secret where there is only one level of security and can read the plaintext of the message. This kind of security mechanism can be performed if the commutative cipher or Last-In First-Out (LIFO) method is used (Siahaan 2016). The above security procedure is explained in Figure 13.4.

13.7.1 Shamir's Three-Pass Protocol

It is one of the various cryptography protocols that permit the secure communication with no need to exchange the secret keys in advance. This cryptography protocol ensures the secure communication where the hackers cannot modify or manipulate the message during transmission from the sender to the receiver. Both parties of communication (sender and receiver) must have a cipher system for secret keys, which has a commutative property of encryption function $E_{K_1}\left(E_{K_2}\left(X\right)\right) = E_{K_2}\left(E_{K_1}\left(X\right)\right)$, where X is a plaintext and K_1 and K_2 are secret keys. In this double encryption process, the result will be the same whether the encryption is started by the first party or second party. Shamir's TPP is explained in the following (Figure 13.5; Lang 2012):

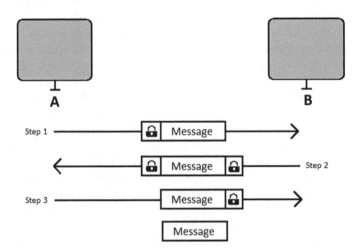

FIGURE 13.4
TPP authentication process.

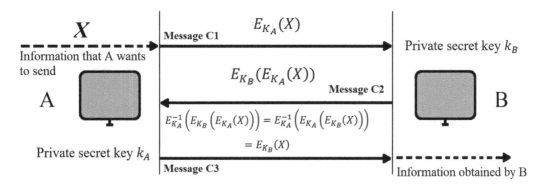

FIGURE 13.5
Illustration of Shamir's TPP.

1. There are two users, A and B. Both users will choose their own private secret keys K_A and K_B randomly.

2. A will encrypt the message by his/her own secret key K_A and send it to B. The cryptogram will be $C_1 = E_{K_A}(X)$.

3. When B receives the message, he/she will consider it as a plaintext and encrypt it by his/her own secret key K_B and send it back to A. The cryptogram will be $C_2 = E_{K_B}(C_1) = E_{K_B}(E_{K_A}(X))$.

4. When A receives the message, he/she will decrypt it using his/her own secret key K_A. According to the commutative property, $E_{K_1}(E_{K_2}(X)) = E_{K_2}(E_{K_1}(X))$ the cryptogram will be as:

$C_3 = E_{K_A}^{-1}(E_{K_B}(E_{K_A}(X))) = E_{K_A}^{-1}(E_{K_A}(E_{K_B}(X))) = E_{K_B}(X)$. Then, A will send C_3 to B.

5. B will receive the message and decrypt it using his/her own secret key K_B, and he/she will be able to read the message successfully.

13.7.2 Massey–Omura Cryptosystem

It is a cryptography mechanism that uses the private key cryptography to secure a message in the time of transmission. Massey–Omura cryptosystem has some features that make it vulnerable to cryptanalysts. This kind of cryptosystem deals with prime modulus and considered as an exponential system. Therefore, if a cryptanalyst could get the encryption key and modulus, then he/she is able to subtract the private key of correspondents and decrypts the message easily. Massy–Omura cryptosystem uses the commutative property function for encryption and decryption of messages, as discussed in Section 13.7.1. Massy–Omura cryptosystem with its correspondence protocol is explained in Winton (2007) as follows:

1. An alphabet should be selected, let's say A.

2. The maximum length N of the message must be determined. So, N is the maximum length of characters for each message.

3. The scheme S is selected to convert the message from alphabetic form to numerical form and vice versa.

4. The integer L is the largest integer and considered as a representor of the message. It is selected based on A, N, and S.

5. The prime $p > L$ is determined as the modulus of the network.

6. The integer w_i is determined as an encryption key $\gcd\{w_i, p-1\} = 1$ for every network member.

7. A decryption key $w_i, x_i = w_i^{-1}(\bmod\ p-1)$ is calculated for every integer. The availability of x_i is ensured since w_i is an organ of the group unit modulo $p - 1$, where the calculation of x_i is accomplished using the Euclidean algorithm.

8. Every member of the network will be provided with encryption and decryption keys w_i and x_i, respectively.

9. The A, N, S, L, and p parameters are published in the center directory. Furthermore, w_i and x_i are the private keys. These private keys are known to the key center and the members of the network to whom these keys are assigned.

In the correspondence protocol, there are two network members Bob and Sue. Let's assume that Bob has $w_1 = t$ and Sue has $x_j = u$ and $x_j = v$ keys. Bob will send a message to Sue in the following procedure:

1. Bob formulates his message by alphabet A where M should not exceed the maximum length N.

2. The message m will be converted from its original alphabetic form to numerical form $M \leq L$ by Bob using scheme S.

3. Bob encrypts M by calculating $M^r (\bmod\ p)$ and sends it to Sue.

4. Sue also encrypts M by calculating $\left(M^r (\bmod\ p)\right)^u (\bmod\ p) = M^{ru}(\bmod\ p)$ and sends it back to Bob.

5. Bob decrypts M partially by calculating $\left(M^{ru}(\bmod\ p)\right)^t (\bmod\ p) = \left(M^u\right)^{rt}(\bmod\ p)$ since $t = r^{-1}(\bmod\ p-1)$ then $rt \equiv 1(\bmod\ p-1)$, so that $rt = 1 + s(p-1)$ for some integer s. Therefore, $\left(M^u\right)^{rt}(\bmod\ p) = \left(M^u\right)^{1+s(p-1)}(\bmod\ p) = M^u \cdot \left(M^{p-1}\right)^{us}(\bmod\ p)$. However, since $M \leq L < p$ and p is a prime, then $\gcd\{M, p\} = 1$. Therefore, $M^{p-1} \equiv 1(\bmod\ p)$ by Fermat's theorem. Hence, $M^u \cdot \left(M^{p-1}\right)^{us}(\bmod\ p) \equiv M^u(\bmod\ p)$. Then, Bob sends it to Sue.

6. Sue continues the decryption procedure by calculating $\left(M^u (\bmod\ p)\right)^v (\bmod\ p) = M^{uv}(\bmod\ p) \equiv (\bmod\ p)$ the same as the fifth step above since $v = u^{-1}(\bmod\ p-1)$. Furthermore, $M(\bmod\ p) = M$ since $M \leq L < p$.

7. Sue converts M to its original alphabetic form m by scheme s, and she will be able to read the message.

13.8 3D Secure

It is also called Three-Domain Secure. It was developed by Visa International Service Association (2002a,b). The main idea of the 3D secure protocol is to allow the card issuer to authenticate the cardholder in the time of online payment and e-commerce transactions. 3D secure contains three domains and is explained in Assora and Shirvani (2006) as follows:

1. Issuer domain: this domain is responsible to maintain the relationship between the card issuer and the cardholder. This domain allows card issuer to authenticate cardholder using Access Control Server (ACS).

2. Acquirer domain: this domain is responsible to maintain the relationship between the merchant and the acquirer. The merchant must have a certain program called merchant plug-in (MPI) which is responsible to establish and maintain the authentication procedure of the messages.

3. Interoperability domain: this domain is responsible to maintain the relationship between the issuer domain and the acquirer domain. This domain has two elements. First is the visa directory server which is responsible to obtain a request from a certain merchant card number and transfer it to the ACS and then send the result back to the merchant from ACS. Second is the authentication history server which records each and every successful or unsuccessful authentication attempts.

13.9 Conclusion

User authentication is one of the most important aspects of network security. This aspect is the target of various researchers in order to ensure user privacy in all applications of information technology. In this chapter, we have discussed many of entity authentication protocols and techniques and their applications. The advantages of these protocols and techniques have been mentioned as well. This chapter is helpful for the researchers who target the authentication as a research area.

References

Abdullah, A. A., Khalaf, R., & Riza, M. 2015. "A realizable quantum three-pass protocol authentication based on Hill–Cipher algorithm." *Mathematical Problems in Engineering* 2015(481824): 6 pages. doi:10.1155/2015/481824.

Aboba, B., Blunk, L., Vollbrecht, J., Carlson, J., & Levkowetz, H. 2004. "Extensible authentication protocol (EAP).": 1–67.

Ahmed, N., & Jensen, C. D. 2010. "Definition of entity authentication." In *Security and Communication Networks (IWSCN), 2010 2nd International Workshop on IEEE*, Karlstad, Sweden: 7.

Assora, M., & Shirvani, A. 2006. "Enhancing the security and efficiency of 3-D secure." In *International Conference on Information Security*, Springer, Berlin, Heidelberg: 489–501.

Forouzan, B. A. & Fegan, S. C. 2007. *Data Communication and Networking*. 4th Edition. McGraw-Hill, New Delhi, India. http://books.google.com/books?id=E01eMAEACAAJ&pgis=1.

Forouzan, B. A., & Mukhopadhyay, D. 2011. *Cryptography and Network Security*. 2nd Edition. McGraw-Hill Education, New Delhi, India.

Ghilen, A., Azizi, M., & Bouallegue, & R. 2015. "Integration and formal security analysis of a quantum key distribution scheme within CHAP protocol." In *Computer Systems and Applications (AICCSA), 2015 IEEE/ACS 12th International Conference of IEEE*, Marrakech, Morocco.

Haller, N. 1995. "The S/KEY one-time password system." http://www.faqs.org/rfcs/rfc1760.html.

Inamura, M. 2015. "Expansions of CHAP: Modificationless on its structures of packet and data exchange." In *Information Systems Security and Privacy (ICISSP), 2015 International Conference on IEEE*, Angers, France.

Kizza, J. M. 2005. "Chapter 1: Network security." *Computer Network Security*. Springer Science & Business Media, New York. http://www.springerlink.com/index/10.1007/978-3-540-73986-9.

Lang, J. 2012. "A no-key-exchange secure image sharing scheme based on Shamir's three-pass cryptography protocol and the multiple-parameter fractional Fourier transform." *Optics Express* 20(3): 2386–98. www.osapublishing.org/oe/abstract.cfm?uri=oe-20-3-2386.

Liu, S., & Silverman, M. 2001. "Practical guide to biometric security technology." *IT Professional* 3(1): 27–32.

Matyáš, V., & Říha, Z. 2002. "Biometric authentication security and usability." In *Advanced Communications and Multimedia Security* (pp. 227–39), Springer, Boston, MA.

Ratha, N. K., Connell, J. H., & Bolle, R. M. 2001. "Enhancing security and privacy in biometrics-based authentication systems." *IBM Systems* 40(3): 614–34.

Sabzevar, A. P., & Stavrou, A. November 2008. "Universal multi-factor authentication using graphical passwords." In *2008 IEEE International Conference on Signal Image Technology and Internet Based Systems*, IEEE, Bali, Indonesia: 625–32.

Siahaan, A. P. U. 2016. "Three-pass protocol concept in Hill Cipher encryption technique." *Seminar Nasional Aplikasi Teknologi Informasi (SNATI)* 5(7) 31–35.

Vaithyasubramanian, A., Christy, A., & Saravanan, D. 2015. "Two factor authentications for secured login in support of effective information preservation." *ARPN Journal of Engineering and Applied Sciences* 10(5): 2053–56.

Van Tilborg, H. C., & Jajodia, S. (Eds.). 2016. *Encyclopedia of Cryptography and Security*. Springer Science & Business Media, New York, USA.

Visa International Service Association. 2002a. "3-D secure protocol specification: Core functions version 1.0.2."

Visa International Service Association. 2002b. "3-D secure protocol specification: System over-view version 1.0.2."

Winton, R. 2007. "Enhancing the Massey–Omura cryptosystem." *Journal of Mathematical Sciences and Mathematics Education* 7(1): 21–29.

Xiao-rong, C., Qi-yuan, F., Chao, D., & Ming-quan, Z. 2005. "Research and Realization of Authentication Technique Based on OTP and Kerberos." In *Proceedings—Eighth International Conference on High-Performance Computing in Asia-Pacific Region, HPC Asia 2005*: 409–13.

14

Message Authentication

Ajay Prasad and Jatin Sethi

University of Petroleum and Energy Studies

CONTENTS

14.1   Introduction...226
14.2   Message Authentication 227
       14.2.1   Message Integrity 227
       14.2.2   Data Origin Authentication 227
       14.2.3   Non-Repudiation 228
       14.2.4   Hash Functions 228
       14.2.5   Message Digest 228
14.3   Introduction to Hash Functions 229
       14.3.1   Introduction to Logic of Message Digest Algorithms............230
       14.3.2   MD2 230
       14.3.3   MD4 230
       14.3.4   MD5 231
       14.3.5   MD6 232
14.4   Secure Hash Algorithm 233
       14.4.1   SHA-1 233
       14.4.2   SHA-2 233
       14.4.3   SHA-256 234
14.5   Message Authentication Code 235
       14.5.1   Requirement for MAC 235
       14.5.2   Introduction to MAC Based on Block Ciphers 235
       14.5.3   Introduction to MAC Based on Iterated Hash Function..........236
                14.5.3.1   General Method..236
14.6   Hash Message Authentication Code (HMAC) 237
       14.6.1   MAC Based on Hash Functions 237
       14.6.2   Design Objectives of HMAC...237
       14.6.3   HMAC Algorithm 237
       14.6.4   HMAC Security 238
14.7   Cipher-Based Message Authentication (CMAC)................................239
       14.7.1   MAC Based on Block Ciphers...239
       14.7.2   Design Objectives of CMAC 240
       14.7.3   CMAC Algorithm 240
       14.7.4   CMAC Security 240
14.8   Whirlpool...241
       14.8.1   Design Objective of Whirlpool 241
       14.8.2   Whirlpool Algorithm 241
       14.8.3   Whirlpool Performance 242

14.9   RACE Integrity Primitives Evaluation Message Digest (RIPEMD) 242
      14.9.1   Introduction to RIPEMD 242
      14.9.2   Origination & Design Objective of RIPEMD............................... 242
      14.9.3   Algorithm for RIPEMD-160, 320 242
      14.9.4   Algorithm for RIPEMD-128, 256 243
14.10   BLAKE 244
      14.10.1 Introduction to BLAKE 244
      14.10.2 Origination & Design Objective of BLAKE............................... 244
      14.10.3 Algorithm for BLAKE-256, 512 244
              14.10.3.1   Initialization 244
              14.10.3.2   Round Function 244
              14.10.3.3   Finalization 245
14.11   Spectral Hash 245
      14.11.1 Introduction to Spectral Hash 245
      14.11.2 Origination and Design Objective of Spectral Hash................ 246
      14.11.3 Algorithm for Spectral Hash 246
14.12   Conclusion 247
References 247

14.1 Introduction

One of the most complex areas of cryptography is authentication of message. Message authentication works in three different modules. The first module is about key generation; the second module is the algorithm, which takes the key and message as input and then produces MAC; and the third module requires an algorithm, which will be used for validation, i.e., a verifying algorithm, which verifies the MAC at receiver's end. Cryptographic Hash Functions takes message as an input and produces a value that is called hash, which is used to validate message integrity. MAC and cryptographic hash functions have lot in common but they possess different security requirements. MAC at one end ensures authentication while hashing ensures integrity. But MAC and hashing could be combined to ensure both authentication and integrity.

In this chapter, we will focus on various methods used to calculate message authentication codes and various hashing algorithms. The technology/technical terms used in the chapter are explained wherever they appear or at the "Key Terminology & Definitions" section. Apart from regular References, additional References will be included in the "References for Advance/Further reading" for the benefit of advanced readers.

Hash functions play the core role in message authentication where hash functions are iterative algorithms that operate in an arbitrary length message and return a fixed-length output. There are several different algorithms that can implement the message digest such as MD2, MD4, MD5, MD6, Secure Hash Algorithms 1 (SHA 1), SHA 256, SHA 512, HMAC, and so on.

MAC functions do possess various security requirements. An MAC should be so secure such that the attacker should not be able to guess MAC without performing an infeasible amount of computation. MAC values are generated and verified by same secret key and hence it is different from digital signature which means that before the start of communication the sender and receiver agrees on one common secret private key which will be used for MAC. Hence, MAC doesn't provide non-repudiation feature while communication.

14.2 Message Authentication

Message Authentication is the mechanism which lets the sending party and the receiving party maintains the integrity of the messages being shared. It allows the sender to send the message to the receiver in such a way that if the message being sent is modified in the channel before reaching the receiver, the receiver will surely detect it. The following are the attacks identified in the context of communication through a network:

1. Disclosure: Data being disclosed to any unauthorized person.
2. Traffic analysis: Intercepting the traffic between the sending and the receiving party and analyzing the data collected to make deductions about it.
3. Masquerade: Sending fraudulent messages to the receiver and fraudulent acknowledgments to the sender
4. Content modification: Editing the data contained the message
5. Sequence modification: Changing the sequence of the messages shared.
6. Timing modification: Delaying or replaying of the messages which have been already sent. It may even involve recreating a whole session of conversation between the parties concerned.
7. Source repudiation: Source denies to transmit the message
8. Destination repudiation: Destination denies to have received the message.

Out of the attacks defined, Message Authentication can solve the attacks (3) to (6). Attack (7) can be dealt with digital signatures, which is also a type of message authentication mechanism.

14.2.1 Message Integrity

Integrity is one of the three major pillars of information security. Message Authentication helps in maintaining integrity as it ensures that the message which is received is exactly how the sender sent it—neither any data were added, or deleted, or modified. If there is not any mechanism in place to ensure the security of the network, the network is susceptible to man-in-the-middle attacks which severely affect the integrity of the messages being sent.

14.2.2 Data Origin Authentication

It is often very important for a person or party to verify the source of the message received. There should be a mechanism in place to verify that there is no man in the middle who is tampering with the messages and then forwarding the message to the receiver. For example, if a user sends the details of the deposit transactions to the bank's central computer and make it look like the bank's branch office has sent it, he can easily make a lot of money. An unauthorized user can escalate his privileges if he fools the machine into thinking that the messages it is receiving are being sent by the system administrator. So in absence of message authentication mechanisms, the receiver cannot verify the source of the message which leads to various attacks being performed by the hackers.

14.2.3 Non-Repudiation

Non-repudiation means to make sure that someone cannot deny their actions. If a legal document is being signed, witnesses are required so that person signing the document cannot deny signing it. Similarly, in regards to communication, if a party or a person has originated a message or sent some document, they cannot deny sending it and they cannot question the originality of their signature. Digital signatures are used to ensure non-repudiation.

Digital Signatures are a form of electronic signatures that basically refer to a signature that is electronically generated. It uses asymmetric cryptography to ensure integrity and non-repudiation.

A digital signature scheme usually contains three algorithms:

1. A key-generation algorithm which generates a random private key and a corresponding public key.
2. A signing algorithm takes the private key and the message as the input and generates a signature as the output.
3. A signature verification algorithm which takes the public key, message, and signature as input and tells if the message's authenticity is intact or not.

14.2.4 Hash Functions

A hash function is a function which takes variable-length input of data and returns a fix-sized hash value. The hash generated has even distribution and is random. The advantage of hash functions is that upon a change in the input data, hash value changes which is used for message authentication. A cryptographic hash function is used for security purposes. Since the hash is a one-way function, some amount of data are modified. It cannot generate the pre-specified hash value, and hence it becomes important to create hash functions which are hard to find out (Holden, 2013). There are no two data sets which generate the same hash value unless they are exactly the same. Due to these benefits, hash functions are used to check the authenticity of the messages that is if the contents of the messages have been modified or not. In cryptographic hash function, padding of data input is done to a fixed-length integer and the padding contains the length of the actual message in bits (Hsieh et al., 1999). Both sender and the receiver calculate hash value through the hash function, if the values matched before and after sending, the message's integrity is maintained otherwise, it has been tampered after sending (Figure 14.1).

14.2.5 Message Digest

A message digest is a cryptographic hash function that takes any arbitrary length input and converts into an output of predefined fixed length known as hash or digest. Digests

FIGURE 14.1
Hash functions.

are used to protect the integrity of the data and detect modifications or alteration in any part of the message. This output then replaces the original input. This has lots of benefits. First, the output will always have the same length which can help in better storing and processing of data. Secondly, the output will be much shorter, so the storing can be done much more efficiently. There are several different algorithms that can implement the message digest such as MD2, MD4, MD5, MD6, Secure Hash Algorithms 1 (SHA 1), SHA 256, SHA 512, HMAC, and many more.

14.3 Introduction to Hash Functions

Hash functions are extremely useful functions which have certain properties that make it extremely suitable for cryptography. These properties that rely on the mathematical algorithm make it unique. It is an algorithm that takes any arbitrary size input and produces an output of fixed size. Input here is either as message or media and the output is digest or message digest. The one-way hash function means it is infeasible to invert. This means that the input data can only be recreated through either brute forcing of possible inputs or expecting a collision or by using a rainbow table of matched hashes.

There are certain properties to a hash function. Some of them are described briefly below:

1. *Pre-Image Resistance*

   Suppose you are given a hash value h, no messages m should be easily discovered where **h = hash(m)**.

2. *Second pre-image resistance*

   Suppose we have a message $m1$, discovery of a different message $m2$ such that **hash(m1) = hash(m2)** should be difficult. In other words, decrypting a hash function should lead to a single message.

3. *Collision Resistance*

   Suppose we have an input message $m1$, its hash should not coincide with the hash of another message $m2$. In other words, **hash(m1) = hash(m2)** should be false.

   Informally, these properties about hash functions are meant to stop a malevolent adversary from replacing or modifying the input data without changing its digest. This establishes accuracy and correctness. Thus, if two strings have the same digest, they should be identical. Second pre-image resistance on the other hand stops an attacker from counterfeiting a data surface with the same hash as a document the attacker cannot control. Collision resistance, thus, prohibits a malevolent attacker from creating two different documents with the same digest.

14.3.1 Introduction to Logic of Message Digest Algorithms

The algorithm is given arbitrary length input as a message and thus creates an output—a fingerprint which is a 128-bit message hash of the input. It is the working of Message Digest that it is infeasible or difficult, computationally, to produce two inputs that have the same message digest or hash, or to produce any message having a predefined hash or digest.

The simple logic for the Message Digest algorithm is that an arbitrary input is passed through to the hash function on the sender's end. The hash function digests it and forms the message digest. On the receiver end, this message digest is decrypted by either a rainbow table of matched hashes or by know hashes. If the hash digest is the same as input digest, *t* is accepted as non-tampered data. If it does not match, it is discarded.

All Message digest algorithms create a 128-bit output hash.

14.3.2 MD2

The Algorithm works in the following five steps:

1. *Append Padding bytes*
   The message is padded by a multiple of 128 bit or 16 bytes which is the block length to make it a multiple of 16 bytes.

2. *Append Checksum*
   The message created in the previous step is appended by a checksum of 16 bytes. To do this, a 256-bit random permutation is used.

3. *Initialize MD buffer*
   A new buffer already initialized to zero is used. It is a 48-byte auxiliary buffer which is used to compute digest value.

4. *Process Message in 16-byte blocks*
   The 256-byte permutation similar to the one used in step 2 is used here. A loop is run which permutes each byte in the auxiliary block 18 times for every 16 input bytes processed.

5. *Output*
   When all blocks have been processed, the first partial block of the auxiliary block becomes the message digest.
   The MD2 message digest is simple to implement and provide a fingerprint of the message. The difficulty of two messages to come up with the exactly same hash is of the order 2^{64}, and coming up with a message from message digest is 2^{128}.

14.3.3 MD4

The MD4 algorithm was created to be quick on 32-bit machines. It does not require any large substitution tables as the MD2 does and the algorithm can be coded quite efficiently.

The Algorithm works in the following five steps:

1. *Append Padding bytes*

   The 448-bit hash value of any message is formed modulo 512. This padding is by adding a single "1" bit to the message, and appending zero bits, thus making the padded message congruent to 448, modulo 512.

2. *Append Length*

   A 64-bit original message representation (original message length before the padding bits were added) is appended to the result after padding. As a result, the message length is a multiple of 512.

3. *Initialize MD buffer*

   A 4-word buffer is used to calculate the message hash. All these words are 32-bit registers. These registers are initialized to some predefined values given in Table 14.1.

4. *Process Message in 16-byte blocks*

   This step defines three auxiliary functions that each process three input words to produce a 32-bit output word.

   $$F(B,C,D) = (B \wedge C) \vee (\neg B \wedge D)$$

   $$G(B,C,D) = (B \wedge C) \vee (B \wedge D) \vee (C \wedge D)$$

   $$H(B,C,D) = B \oplus C \oplus D$$

   $\oplus, \wedge, \vee, \neg$ denote XOR, AND, OR, and NOT operators, respectively.

5. *Output*

   A, B, C, D produced as the message digest is the output. Lower-order bytes of A are converted to a higher-order byte of D.

   The MD2 message digest is simple to implement, thus providing a fingerprint of the message. The difficulty of two messages coming up with the same message digest is of the order 2^{64} and coming up with a message from message digest is 2^{128}.

14.3.4 MD5

The MD5 algorithm is a further extension of the MD4 message-digest algorithm. In comparison of speed, it is slower but more conservative.

The Algorithm works in the following five steps:

1. *Append Padding bytes*

   The 448-bit hash value of any message is formed modulo 512. Padding is by adding a single "1" bit to the message, and appending zero bits, thus making it congruent to 448, modulo 512.

TABLE 14.1

Predefined Values to Initialize MD Buffer for MD 4 Algorithm

Word A	01	23	45	67
Word B	89	ab	cd	ef
Word C	fe	dc	ba	98
Word D	76	54	32	10

TABLE 14.2

Predefined Values to Initialize MD Buffer for MD 5 Algorithm

Word A	01	23	45	67
Word B	89	*ab*	*cd*	*ef*
Word C	*fe*	*dc*	*ba*	98
Word D	76	54	32	10

2. *Append Length*

A 64-bit original message representation (the original message length before the padding bits were added) is appended to the result after padding. As a result, the message has a length that is a multiple of 512.

3. *Initialize MD buffer*

A 4-word buffer is used to compute the message digest. All these words are 32-bit registers. These registers are initialized to some predefined values given in Table 14.2.

4. *Process Message in 16-byte blocks*

This step defines three auxiliary functions that each process three input words to produce a 32-bit output word.

$$F(B,C,D) = (B \wedge C) \vee (\neg B \wedge D)$$

$$G(B,C,D) = (B \wedge C) \vee (C \wedge D)$$

$$H(B,C,D) = B \oplus C \oplus D$$

$$I(B,C,D) = C \oplus (B \vee \neg D)$$

$\oplus, \wedge, \vee, \neg$ denote XOR, AND, OR, and NOT operators, respectively.

5. *Output*

A, B, C, D is a message digest produced as output. Lower-order bytes of A are converted to a higher-order byte of D.

The MD5 message digest is simple to implement and provide a fingerprint of the message. The difficulty of two messages coming up with the same message digest is of the order 2^{64} and coming up with a message from message digest is 2^{128}.

14.3.5 MD6

MD6 is a hash function in cryptography which uses a Merkel tree-like structure. This tree structure is used for immense parallel and simultaneous computation of hashes for long inputs. A hash tree or Merkle tree consists of every leaf node as the hash of one particular data block and every non-leaf node as the cryptographic hash of the labels of its child nodes. For use in large data structures, efficient and secure verification can be allowed through the hash trees.

The performance of MD6 for MD6-256 is as high as 28 cycles per byte on an Intel Core 2 Duo processor and offers some resistance against differential cryptanalysis.

The MD6 is still in development as a replacement for SHA-3 (Secure Hash Algorithm) but has since failed to replace it due to speed issues.

14.4 Secure Hash Algorithm

One-way hash functions are iterative algorithms that operate in an arbitrary length message and return a fixed-length output, h, which is called a message digest or hash value. It is theoretically impossible/near to impossible to get two identical hashes for the same input for the computational advancement that we have.

The following are some examples of SHA algorithms:

SHA 1

SHA 2

SHA 256 (from family of SHA 2)

SHA 192 (proposed)

14.4.1 SHA-1

SHA-1 is a cryptographic hash function published by the National Institute of Standards and Technology (NIST). The three SHA algorithms are SHA-0, SHA-1, and SHA-2. The SHA-0 algorithm was not used in many applications. On the other hand, SHA-2 differs from the SHA-1 hash function. SHA-1 is the most widely used hash function.

However, the most widely used are SHA 1 and SHA 256.

SHA-1 hash function performs 80 (0–79) iterations to produce the message digest and its transformation round includes simple additions, rotations, and Non-linear functions (NLFs) f [16].

$$f(t;B,C,D) = (B \text{ AND } C) \text{ OR } ((\text{NOT } B) \text{ AND } D) \qquad (0 \le t \le 19)$$
$$f(t;B,C,D) = B \text{ XOR } C \text{ XOR } D \qquad (20 \le t \le 39)$$
$$f(t;B,C,D) = (B \text{ AND } C) \text{ OR } (B \text{ AND } D) \text{ OR } (C \text{ AND } D) \qquad (40 \le t \le 59)$$
$$f(t;B,C,D) = B \text{ XOR } C \text{ XOR } D \qquad (60 \le t \le 79).$$

where B, C, and D are 32-bit words.

Concerning the required 80 W_t values, the first 16 are formed by decomposing the 512-bit incoming message block in 16 32-bit words that each one of those corresponds to one 32-bit W_t value.

The remaining 64 bits are produced through the following equation:

$$W_t = \text{ROTL1}(W_t - 3 \quad W_t - 8 \quad W_t - 14 \quad W_t - 16)$$

$16 \le t \le 79$ where ROTL_x stands for x times left circular bit rotation (Figure 14.2).

14.4.2 SHA-2

The SHA-2 hash standard specifies four secure hash algorithms, SHA-224, SHA-256, SHA-384, and SHA-512. All the four algorithms are iterative, one-way hash functions that can process a message to produce a hashed representation (see Table 14.3).

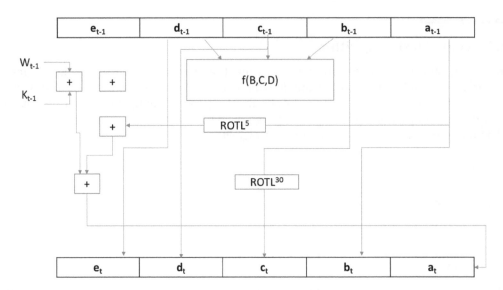

FIGURE 14.2
SHA 1.

TABLE 14.3

Comparison between Secure Hash Algorithms

Algorithm	Word	Message Size	Block	Digest
SHA 224	32	$<2^{64}$	512	224
SHA 256	32	$<2^{64}$	512	256
SHA 384	64	$<2^{128}$	1,024	384
SHA 512	64	$<2^{128}$	1,024	512

14.4.3 SHA-256

SHA-256 needs 64 iterations to produce its message digest. Its transformation round is similar to the SHA-1 round, including adders and logical functions arranged in order to produce the round's output values. The incorporated NLFs are as follows:

$$Ch(t;B,C,D) = (B \text{ AND } C) \text{ OR } ((NOT\ B) \text{ AND } D)$$

$$Maj(t;B,C,D) = (B \text{ AND } C) \text{ OR } (B \text{ AND } D) \text{ OR } (C \text{ AND } D)$$

$$Summation\ (x)(0\text{-}256) = ROTR2\,(x) \oplus ROTR13\,(x) \oplus ROTR22\,(x)$$

$$Summation\ (x)(1\text{-}256) = ROTR6\,(x) \oplus ROTR11\,(x) \oplus ROTR25\,(x)$$

where ROTRx stands for x times right circular bit rotation.

As in SHA-1, regarding the 64 W_t values needed, the first the SHA-1 and SHA-256 architectures consist of two computational paths, namely the data- and control-paths. The Control Unit blocks (a set of counters), which forms the control path, produce and send control signals to the transformation Rounds and W units that both form the data path (Figure 14.3).

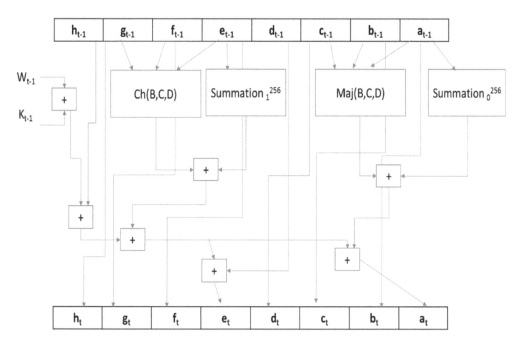

FIGURE 14.3
SHA 256.

14.5 Message Authentication Code

14.5.1 Requirement for MAC

The message Authentication code is a piece of authentication information used to verify the sending authority of the information and the integrity of the message.

It also protects the data's integrity and authentic characteristics by allowing and verifying any changes to the content of the message.

Message Authentication code: An essential requirement for sending sensitive and critical data from one system to another; it is considered to be the best practice for sending data over a network.

14.5.2 Introduction to MAC Based on Block Ciphers

MAC, if constructed from block ciphers, uses AES or similar algorithms to make the encryption more strong and reliable.

Now using AES in CBC (cipher block chaining mode) will make the message more secure.

CBC (Cipher Block chain mode): It is a kind of mode of operation for ciphers based on block, where the sequence of bits is encrypted as a single unit. Cipher block chain uses an initial vector of a fixed certain length; its key characteristic is that it uses a chaining mechanism that causes the decryption of a block of ciphertext which depend on all the previous ciphertext blocks.

MAC Generation using CBC

- Divide the message a into blocks a_i.
- Compute first iteration $b_1 = e_k(a_1 \oplus IV)$ where IV is the Initial Vector length
- Compute $b_i = e_k(a_i \oplus b_i - 1)$ for the next blocks
- Final block is the MAC value: $m = MAC_k(a) = b_n$

MAC Verification using CBC

- Repeat MAC computation (m')
- Compare results: If $m' = m$, the message is verified and is correct.
- If $m' \neq m$, the message and or the MAC value m have been altered during transmission

14.5.3 Introduction to MAC Based on Iterated Hash Function

The security of an MAC depends on the security of the underlying hash algorithm, and almost all hash functions are iterative.

Hash functions transform and map strings that are long or of variable length into strings of fixed length.

But using a hash function it requires the digest $h(m)$ to be protected, otherwise, the digest itself could tamper. This problem is solved by using MACs, as they are keyed hash functions. This is a symmetric cryptography schema: the sender and the receiver share a secret key (Figure 14.4).

14.5.3.1 General Method

The input a is padded to a multiple of the block size and is divided into t blocks denoted a_1 through a_t.

It is then processed block by block for each block. The intermediate result is stored in an n-bit ($n \geq m$) chaining variable denoted with H_i:

$$H_0 = IV \quad H_i = f(H_i - 1, a_i), 1 \leq i \leq t \quad h(a) = g(H_t), \text{ where } g \text{ is output transformation}$$

It is important to fix the value of IV (Initial Vector) and pre-code the message.

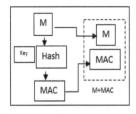
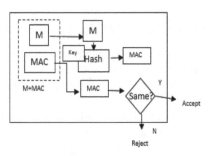

FIGURE 14.4
MAC.

14.6 Hash Message Authentication Code (HMAC)

14.6.1 MAC Based on Hash Functions

HMAC means applying a message authentication code algorithm using only cryptographic hash functions. Using Message authentication code with hash functions is more efficient than using Message authentication code with block ciphers because hash functions like MD5 and SHA-1 are much faster than block cipher cryptographic functions like DES. While designing hash functions like SHA, its usage as MAC was not kept in mind and hence it cannot be used for the same purpose directly as its usage does not include a secret key. Various proposals were presented so as hoe to incorporate secret key into hash functions. HMAC is compulsory to implement MAC for IP security and for many other Internet protocols.

14.6.2 Design Objectives of HMAC

The design objectives listed for HMAC are as follows:

1. To use those hash functions particularly whose code is available for free and they have efficient performance in software.
2. The embedded hash function should be easy to replace if a faster or more secure hash function is discovered.
3. The actual efficiency of the hash function should not be degraded.
4. Keys should be handled in a simple way.
5. Proper cryptanalysis of the strength of authentication procedure should be carried out based on the required assumptions related to the hash function.

The first two goals are very crucial because of software development purposes. The hash function is treated as a black-box by the HMAC. It has two advantages. Firstly, the hash function as a module can be used to implement HMAC. Due to this, most of the HMAC code is already compiled and ready to use with minor customization. Secondly, if there is a need to replace the existing hash function in the HMAC, the only thing required to be done is the replacement of one hash function's module with another one. Also, if the security of one hash function is compromised or the efficiency is reduced, the hash function can easily be replaced with the more secure or more efficient one.

14.6.3 HMAC Algorithm

The overall operation of HMAC can be explained as follows (Figure 14.5):

H = Embedded hash function.

I = Initial input value to the hash function

M = Input message to HMAC

Y_i = ith block of M, $0 \leq i \leq (N-1)$

N = Number of blocks in M

b = Number of bits in a block

n = Length of hash code produced by the hash function

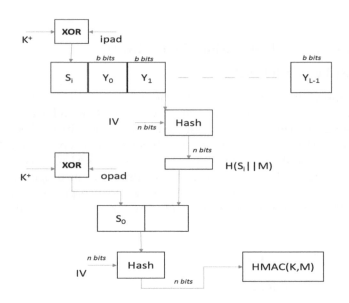

FIGURE 14.5
HMAC algorithm.

K = Secret key; length (K) should be greater than or equal to n

K^+ = K padded with zeros on the left to make the length in bits equal to b

iPad = 00110110 (36 in hexadecimal) repeated $b/8$ times

opad = 01011100 (5C in hexadecimal) repeated $b/8$ times

HMAC Algorithm

$$\text{HMAC}(K, M) = H\Big[\big(K^+ \oplus \text{opad}\big) \| H\big[\big(K^+ \oplus \text{ipad}\big) \| M\big]\Big]\text{P}$$

The algorithm can be explained as follows:

1. Zeros are appended to the left end of K to create a b-bit string K^+
2. XOR K^+ with iPad
3. Append M to S_i
4. Apply H to the stream generated in Step-3
5. XOR K^+ with opad to produce the b-bit block S_o
6. Append the hash result from Step 4 to S_o
7. Apply H to the stream generated in Step 6 and output the result.

The time taken for HMAC execution is approximately the same as the embedded hash for the long messages. HMAC deals with the addition of three executions of the hash compression function.

14.6.4 HMAC Security

The security of HMAC function is dependent on how secure the embedded hash function is. How secure an MAC function is shown in the terms of what is the effort, probability of a successful attack in a particular amount of time spent by the attacker? In a given amount of time, and a given amount of effort, the probability of an attack being successful is equal to the probability of an attack being successful on the hash function which is embedded. MAC can fail only if the hash function fails.

Hash function can be broken only in the following ways:

1. There are collisions in the hash function even after the key is secret and random.
2. The attacker computes an output of the compression function even though the key is secret, random, and unknown to the attacker.

Before designing the hash-based MAC, we made some assumptions based on the strength and efficiency of the hash functions. All these assumptions are defied if the attacker is successful in attacking the HMAC. The first attack is tough as looking for collisions in the hash function is easy but finding collisions in the hash function embedded in HMAC is much harder because the key is unknown in this case.

Hash functions' design principle is to ensure randomness. This is the major design methodology used for collision avoidance. If the second attack is successful, it means that the randomness of the hash function is poor.

In conclusion, HMAC fails only in case the assumptions on which the design of the hash functions is based, fails.

14.7 Cipher-Based Message Authentication (CMAC)

14.7.1 MAC Based on Block Ciphers

Basically, there are two types of MAC's which uses the block cipher mode of operation. One is Data Authentication Algorithm (DAA) which is not so in practice now, and another is CMAC which is designed to overcome the vulnerabilities of DAA.

Let's talk about DAA:

DAA has been one of the most used MAC's for a no. of years but now it is outdated. It is basically based on Data Encryption Standard. This algorithm uses CBC mode of operation taking initial vector Zero. The data (for example: Message, record, files, or programs) are first grouped into contiguous 64-bit blocks as $D1, D2, ..., Dn$.

If necessary, the last block is to be padded on the right with zeros to form a 64-bit block. Then the Data Authentication code (DAC) can be calculated as follows:

$$O_1 = E(K, D_1)$$

$$O_2 = E(K, (D_2 \oplus O_1))$$

$$O_3 = E(K, (D_3 \oplus O_2))$$

$$\vdots \qquad \vdots$$

$$O_N = E(K, (D_N \oplus O_{N-1}))$$

14.7.2 Design Objectives of CMAC

MAC has a limitation that only fixed-length messages are processed. (If n is the cipher block size and m is any fixed positive integer, then only fixed length of mn bits are processed.) As well as it is only secured under a reasonable set of security criteria.

This limitation could be overcome by using CMAC mode of operation with AES and triple DES.

14.7.3 CMAC Algorithm

CMAC algorithm is also known as OMAC1, i.e., One Key CBC-MAC version 1. This uses the very simple and efficient way of reusing underlying block cipher's machinery for its functioning, and applies a full encryption on each authenticated block. This design is very simple and a small variation of CBC-MAC and is alike CBC mode of operation for block ciphers.

The operation is described as follows:

The message to be authenticated is divided into no. of blocks say t; these blocks are iteratively encrypted and XORed with the following block to be processed; only the last block is treated in another manner. It is also XORed with one of two key-dependent constants (denoted $R1$ and $R2 = R1 * x$) before encryption and it yields the authentication tag.

These constants are calculated during CMACs initialization, and their purpose is to make sure that algorithm's security is maintained while processing variable-length messages. CBC-MAC itself is secure for only fixed-length messages (Figure 14.6).

14.7.4 CMAC Security

CMAC is totally based on the AES algorithm that is a strong cryptographic algorithm. We know that strength of any cryptographic algorithm lies in the secret key k and the accuracy of the implementation in all of the participating systems. So for the secure authentication and integrity of the message, the secret key shall not be compromised and be shared with care. It should be kept confidential as well as the secret key shall be created in such a way that it meets the pseudo-randomness requirements.

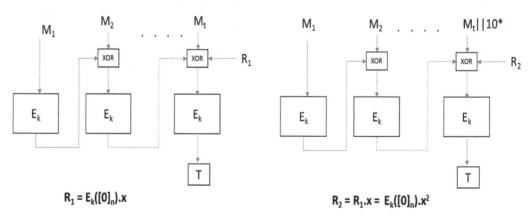

$$R_1 = E_k([0]_n).x \qquad\qquad R_2 = R_1.x = E_k([0]_n).x^2$$

FIGURE 14.6
CMAC algorithm.

14.8 Whirlpool

14.8.1 Design Objective of Whirlpool

For all digital signatures and message authentication processes, the most significant part is the hash function. A hash function accepts a variable-size message M as input and produces a fixed-size hash code $H(m)$, sometimes called a message digest, as output.

Although the use of block cipher was said to induce many security vulnerabilities, AES which is a block cipher is very secure. On the same lines for compression in whirlpool, block cipher is used.

Whirlpool has the following features:

1. The hash code length is 512 bits, equaling the longest hash code available with SHA.
2. The overall structure of the hash function is one that has been shown to be resistant to the usual attacks on block-cipher-based hash codes.
3. The underlying block cipher is based on AES and is designed to provide for implementation in both software and hardware that is both compact and exhibits good performance.

14.8.2 Whirlpool Algorithm

Given a message consisting of a sequence of blocks $m1, m2, ..., mt$ (Figures 14.7 and 14.8):

FIGURE 14.7
Whirlpool design.

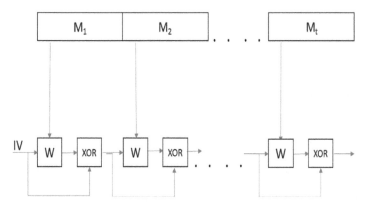

FIGURE 14.8
Whirlpool algorithm.

14.8.3 Whirlpool Performance

The NIST evaluation of Rijndael determined that it exhibited good performance (execution speed) in both hardware and software, and it is well suited to restricted-space (low memory) requirements. These criteria were important in the selection of Rijndael for AES. Because Whirlpool uses the same functional building blocks as AES and has the same structure, we can expect similar performance and space characteristics.

14.9 RACE Integrity Primitives Evaluation Message Digest (RIPEMD)

14.9.1 Introduction to RIPEMD

RIPEMD is a cryptographic hash function family designed in the late 1990s at COSIC research group, Belgium. It's tuned to 32-bit software implementations. It was funded by the European Union (EU) to find a long-term solution as RIPE Consortium independent research on MD4 and MD5 detected collisions based on their testing theory. Its design principles are of MD4 while performance factors are similar to SHA-1. It uses two independent and parallel chains simultaneously which were combined to form a result. RIPEMD-160, as suggested by the name, is a 160-bit hash function whereas RIPEMD-128 is the fast variant as it provides 128-bit hash only (Wang & Shen, 2014; Wang et al., 2005; Bosselaers et al., 1997).

14.9.2 Origination & Design Objective of RIPEMD

MD4 was famous in the 1990s but collision issues were encountered, so R. Rivest came up with a brand new version of MD4 called MD5. At the same time, RIPE Consortium independently worked on a project which proved in theories about the collision count can be found by a 10 million $ in 6 days and as per Moore's Law, computation power is doubled every year even reducing the cost of construction, so MD5 can't be a long-term security solution. As a result, they decided to build a secure version of MD4, called RIPEMD. RIPEMD is basically two parallel versions of MD4 running simultaneously with changes in shifts and message orders while the parallel instances differed only in round constants. Both the parallel instances were added in the last to form a 128-bit hash called RIPEMD-128. After a while, attacks were found possible on RIPEMD-128, so newer versions of RIPEMD were launched namely RIPEMD-160, RIPEMD-256, RIPEMD-320. RIPEMD found its applications in European bank projects, digital signatures, MAC, etc.

14.9.3 Algorithm for RIPEMD-160, 320

RIPEMD-160 breaks the input into blocks of 512 bit each which is further divided into 16 strings each of 4 byte. 4-byte strings are then converted using the little endian methodology to 32-bit numbers. A set of 5 32 bit entities form a 160-bit hash. To complete the hash a padding is added to the end to generate the hash. Next and most important part is compression function which consists of five parallel rounds performed in 16 rounds done for two hash, i.e., parallely. Therefore, 160 rounds as compared to 48 in MD4. Each step process a new value with four registers and a message word. At the end, the new state is found by adding to each word of the old state one register from the left half and one from the right half.

TABLE 14.4

Permutation for the Message Words

I	0	1	2	3	4	5	6	7	8	9	10	11	12	13	14	15
$p(i)$	7	4	13	1	10	6	15	3	12	0	9	5	2	14	11	8

TABLE 14.5

Order of the Message Words

Line	Round 1	Round 2	Round 3	Round 4	Round 5
Left	id	p	p^2	p^3	p^4
Right	π	πp	πp^2	πp^3	πp^4

1. Operations in one step. $A := (A + f(B, C, D) + X + K)^{\lll s} + E$ and $C := C^{\lll 10}$. Here$)^{\lll s}$ denotes cyclic shift (rotation) over s bit positions.
2. Ordering of the message words. Take the permutation ρ (Table 14.4) and further define the permutation π by setting $\pi(i) = 9i + 5$ (mod 16). The order of the message words is given in Table 14.5.

14.9.4 Algorithm for RIPEMD-128, 256

The important difference between RIPEMD-128 & RIPEMD-160 bit is we use the same hash result and the chaining variable of 128-bit, i.e., only four rounds are used.

1. *Operations in one step*

$$A := \left(A + f(B, C, D) + X + K \right)^{\lll s}$$

2. *Boolean functions*
   Define the Boolean functions in Table 14.6.

3. *Constants*
   Take the integer parts of the numbers as shown in Table 14.7.

TABLE 14.6

Boolean Functions in RIPEMD

Line	Round 1	Round 2	Round 3	Round 4
Left	F_1	F_2	F_3	F_4
Right	F_4	F_3	F_2	F_1

TABLE 14.7

Integer Parts of Constants in RIPEMD Algorithm

Line	Round 1	Round 2	Round 3	Round 4
Left	0	$2^{30} * \sqrt{2}$	$2^{30} * \sqrt{3}$	$2^{30} * \sqrt{5}$
Right	$2^{30} * \sqrt[3]{2}$	$2^{30} * \sqrt[3]{3}$	$2^{30} * \sqrt[3]{5}$	0

TABLE 14.8

Comparison between Various BLAKE Algorithms

Algorithm	Word	Message	Block	Digest	Salt
BLAKE-28	32-bit	$<2^{64}$ bit	512-bit	224-bit	128-bit
BLAKE-32	32-bit	$<2^{64}$ bit	512-bit	256-bit	128-bit
BLAKE-48	64-bit	$<2^{128}$ bit	1,024-bit	384-bit	256-bit
BLAKE-64	64-bit	$<2^{128}$ bit	1,024-bit	512-bit	256-bit

14.10 BLAKE

14.10.1 Introduction to BLAKE

BLAKE is a cryptographic hashing function available in 32-bit and 64-bit versions. It also has a 512- and 1024-bit block size. The details are given in Table 14.8.

14.10.2 Origination & Design Objective of BLAKE

BLAKE was proposed by Jean-Philippe Aumasson, Luca Henzen, Willi Meier, and Raphael C.-W. Phan. It is somewhat based on the ChaCha Stream Cipher put forward by Dan Bernstein. The algorithm was developed to participate in the NIST hash function competition for SHA-3 hashing algorithm organized by the US National Institute of Standards and Technology. It was among the 51 candidates competing in 2008 and advanced into the finals with four other functions but ultimately lost to Keccak which was chosen as SHA-3 in 2012.

In late December of 2012, BLAKE2, the successor of BLAKE, was announced with changes and improvements and was set to replace MD5 and SHA-1 hashing algorithms as they were widely used but not properly functional anymore.

14.10.3 Algorithm for BLAKE-256, 512

14.10.3.1 Initialization

16-word states are initialized such that different inputs give different initial states.
h-Chain value, C-constant, s-salt, t-counter

$$
\begin{vmatrix}
h_0 & h_1 & h_2 & h_3 \\
h_4 & h_5 & h_6 & h_7 \\
s_0 \oplus c_0 & s_1 \oplus c_1 & s_2 \oplus c_2 & s_3 \oplus c_3 \\
t_0 \oplus c_4 & t_0 \oplus c_5 & t_1 \oplus c_6 & t_1 \oplus c_7
\end{vmatrix}
\longrightarrow
$$

V0	V1	V2	V3
V4	V5	V6	V7
V8	V9	V10	V11
V12	V13	V14	V15

14.10.3.2 Round Function

The round function iterates 14 times (BLAKE-256) and 16 times (BLAKE-512) on the state v. Each round comprises of transformations based on single core function G:

$$G0(v0,v4,v8,v12) \quad G1(v1,v5,v9,v13) \quad G2(v2,v6,v10,v14) \quad G3(v3,v7,v11,v15)$$

$$G4(v0,v5,v10,v15) \quad G5(v1,v6,v11,v12) \quad G6(v2,v7,v8,v13) \quad G7(v3,v4,v9,v14)$$

where, at round *r*, *Gi*(*a*, *b*, *c*, *d*) sets

$a \leftarrow a + b + (m[j] \oplus n[k])$ // Step 1 (with input)

$d \leftarrow (d \oplus a) \ggg 16$

$c \leftarrow c + d$ // Step 2 (no input)

$b \leftarrow (b \oplus c) \ggg 12$

$a \leftarrow a + b + (m[k] \oplus n[j])$ // Step 3 (with input)

$d \leftarrow (d \oplus a) \ggg 8$

$c \leftarrow c + d$ // Step 4 (no input)

$b \leftarrow (b \oplus c) \ggg 7$

14.10.3.3 Finalization

New chain value $h'_0 - h'_{15}$ is extracted from the state $v0 - v15$

$h'_0 \leftarrow h0 \oplus s0 \oplus v0 \oplus v8$

$h'_1 \leftarrow h1 \oplus s1 \oplus v1 \oplus v9$

$h'_2 \leftarrow h2 \oplus s2 \oplus v2 \oplus v10$

$h'_3 \leftarrow h3 \oplus s3 \oplus v3 \oplus v11$

$h'_4 \leftarrow h4 \oplus s0 \oplus v4 \oplus v12$

$h'_5 \leftarrow h5 \oplus s1 \oplus v5 \oplus v13$

$h'_6 \leftarrow h6 \oplus s2 \oplus v6 \oplus v14$

$h'_7 \leftarrow h7 \oplus s3 \oplus v7 \oplus v15$

14.11 Spectral Hash

14.11.1 Introduction to Spectral Hash

Every other field in today's world, such as finance, media, forensics, education, medicine, etc., is using digital images extensively. Image acquisition and image storage have also become less complicated and less expensive nowadays and hence, used extensively.This leads to the retrieval problem of the images which led to the use of hash-based methods in image retrieval algorithms. These methods are time efficient than other ones because they work on the binary code of the images rather than using the image itself. But the problem with the digital images is not only their retrieval but also the trust factor. It can easily be tampered using any editing device. Image forging can be of many types such as image retouching, image splicing, and copy-move forgery.

Spectral hashing (SH) techniques are used to generate binary codes for the features extracted (Granty et al., 2016).

It involves copy-move forgery and image retrieval. Sparsification of the graph Laplacian and tampering detectionis performed using SH based PCT (Polar Cosine Transformation) and is used for image retrieval process.

14.11.2 Origination and Design Objective of Spectral Hash

The origination of Spectral Hash started with the booming of digital images being used by many organizations. As the retrieval of the images from the databases was a tedious task, "trust factor" was also a problem. Digital tampering Hash-based-algorithm doesn't store the image in the database but stores the binary code, which makes them easy to retrieve. There are several issues in image forgery such as image retouching, copy-move forgery, and image splicing. In the copy-move forgery, the parts of an image are copied and pasted on the same image with any wrong intention. There are several image tampering detection methods, including the active methods and passive methods.

Active methods are anything like watermarks or something, whereas passive methods do not require any of pre-embedded information and identify the changes in the image by tampering only.

This approach involves the copy-move forgery detection. The image retrieval process is being performed using sparsification of graph Laplacian, the tampering detection, and SH. The tampering detection uses a spectral-hashing-based PCT. The image is subdivided into covering patches, from which PCT features are extracted. SH is then used to identify the comparable patches.

The image retrieval process can only be examined and analyzed the large data sets containing a large number of training images and test images of almost 80 categories containing different kind and size of images. A large number of images (around 5,000 out of 6,000) are used for training sets, whereas around 1,000 images are used for data set.

Initialization → Feature Extraction & Graph Construction → Generation of Graph Lap Lacian Matrix → Sparsification of Graph lap Lacian → Spectral Hashing & Hamming Distance Calculation → Retrieval of similar images

14.11.3 Algorithm for Spectral Hash

It mainly retrieves similar images by using the features of the images. It extracts the features from the images and then analyses the binary equivalent of those features. For the training set of image, a feature vector is constructed for each image and then evaluated using a distance metric learning scheme. After that, using these matrices, a graph is formed. The pairwise similarities between the images and a graph are evaluated by the graph Laplacian and graph Laplacian matrix is created. Then, SH is used to obtain binary codes for the images. This process is done for all the images. Hamming distance between the binary code of image in training dataset and the query image is calculated and based on that values, images are retrieved which are most similar to the query.

The process includes the feature vector extraction followed by the process of graph generation in which the images are represented as the nodes along the graph perimeter. Then the nodes are connected to each other on the basis of least Euclidian distance. Then pairwise similarity is found and the graph Laplacian is formed. The average Hamming distance between similar images is calculated using the following equation (Zhang et al., 2010):

$$\sum w_{ij} \left| y_i - y_j \right|^2$$

To reduce the Hamming distance between similar images, the SH codes should satisfy the following conditions (Zhang et al., 2010):

$$\min \sum w_{ij} \left| y_i - y_j \right|^2$$

subject to $y_i \in \{-1, 1\}^2$, $\Sigma y_i = 0$, $(1/n)\Sigma y_i y_i^T = 1$
where w_{ij} is the affinity matrix that characterizes similarities between image x_i and x_j (Granty et al., 2016)

$$w_{ij} = \exp\left(-\left|x_i - x_j\right|^2 / \varepsilon^2\right)$$

where y_i is the binary code of x_i, respectively. Each bit has 50% of 0 or 1 satisfying the condition $(1/n)\Sigma y_i y_i^T = 1$ that requires bits to be uncorrelated.

14.12 Conclusion

In this chapter, we discussed almost all traditional and modern or upcoming Hash schemes and methodologies. With an increase in security threats, like in 2005 security defects were identified in hashing algorithms showing mathematical weakness. Later in 2007, NIST conducted a contest to design more secure hash algorithms. Therefore, the idea is still getting continued to do more research on developing algorithms, which will produce more secure rather only increasing hash size because that's not the proper solution. Message authentication and hashing techniques are of course used a lot and are more in use now because of the increase in online communication like digital payments, digital certificates, etc.

References

Bosselaers A., Dobbertin, H., Preneel, B. (January 1997) The RIPEMD-160 cryptographic hash function, *Dr. Dobb's Journal*, 22:1, 24–28.

Granty R. E. J., Kousalya, G. (2016) Spectral-hashing-based image retrieval and copy-move forgery detection, *Australian Journal of Forensic Sciences*, 48:6, 643–658. doi:10.1080/00450618.2015.112 8966.

Holden, J. (2013) A good hash function is hard to find, and viceVersa, *Cryptologia Journal*, 37:2, 107–119, Taylor & Francis.

Hsieh, T.-M., Yeh, Y.-S., Lin, C.-H., Tuan, S.-H. (1999) One-way hash functions with changeable parameters, *Information Sciences*, 118, 223–239.

Wang G., Shen Y. (2014) (Pseudo-) preimage attacks on step-reduced HAS-160 and RIPEMD-160. In: Chow S. S. M., Camenisch J., Hui L. C. K., Yiu S. M. (eds) Information Security. ISC 2014. Lecture Notes in Computer Science, vol. 8783, Springer, Cham.

Wang X., Lai X., Feng D., Chen H., Yu X. (2005) Cryptanalysis of the hash functions MD4 and RIPEMD. In: Cramer R. (eds) Advances in Cryptology – EUROCRYPT 2005. EUROCRYPT 2005. Lecture Notes in Computer Science, vol. 3494, Springer, Berlin, Heidelberg.

Zhang D., Wang J., Cai D., Lu J. (2010) Self-taught hashing for fast similarity search. *Proceedings of the 33rd Annual International ACM SIGIR Conference on Research and Development in Information Retrieval (SIGIR)*, Geneva, Switzerland.

Digital Signatures

Ajay Prasad and Keshav Kaushik

University of Petroleum and Energy Studies

CONTENTS

15.1 Introduction 250
15.2 ElGamal Signature Scheme 251
    15.2.1 Introduction to ElGamal Signature Scheme 251
    15.2.2 Signature Parameters of ElGamal Signature Scheme 251
    15.2.3 Key Generation and Signature Generation...............................251
        15.2.3.1 Key Generation 251
        15.2.3.2 Signing Process...252
    15.2.4 Verification, Correctness, and Forgery 252
        15.2.4.1 Verification...252
        15.2.4.2 Correctness and Forgery 254
15.3 Elliptic-Curve Digital Signature Algorithm 255
    15.3.1 Signature Generation and Verification Algorithm 256
        15.3.1.1 ECDSA Key Generation...256
        15.3.1.2 Signature Generation Algorithm for ECDSA 256
        15.3.1.3 Signature Verification Algorithm for ECDSA 257
    15.3.2 Security of ECSDA...258
    15.3.3 Comparison with RSA and DSA 258
        15.3.3.1 Comparison with RSA 258
        15.3.3.2 Comparison with DSA 258
15.4 Identity-Based Signature 259
    15.4.1 Introduction and Definition of ID-Based Signature................259
    15.4.2 Efficiency and Security Comparison 261
15.5 Bimodal Lattice Signature Schemes (BLISS) and Lamport Signature Scheme.......262
    15.5.1 Introduction to BLISS and Lamport...262
        15.5.1.1 BLISS Signature Scheme..262
        15.5.1.2 Key Generation, Signature, and Verification Algorithms...........263
        15.5.1.3 Signature Algorithm 264
        15.5.1.4 Verification Algorithm 264
        15.5.1.5 Lamport Signature Scheme..264
        15.5.1.6 Key Generation Signing and Verification....................264
    15.5.2 Security Comparison 266

15.6 Merkle Signature Scheme 267
    15.6.1 Key Generation (Merkle Tree) 267
    15.6.2 Signature Scheme 268
    15.6.3 Verification 268
15.7 Other Existing Signature Schemes 269
15.8 Conclusion 270
References 270

15.1 Introduction

As the world is moving away from paper documents and signing them manually with ink and stamps, digital signatures can provide with assurance to authenticity, integrity, and nonrepudiation. A digital signature mainly comprises three main algorithms: a key generation algorithm which selects a key randomly from a pool of private keys, a signing algorithm which is responsible for signing the document with the help of private key, and a signature verifying algorithm. The real concept behind digital signature is to avoid paper and keep the data secure in electronic form. Digital signature uses an accepted and standard format known as Public-Key Infrastructure (PKI). There are various certifying authorities that are responsible for issuing digital signature. Technically binary blocks of digital signatures are attached to a document or a message to prove signer's authenticity. This helps in reducing various cyber-attacks like Man-in-the-Middle attack, eavesdropping, etc. It also ensures that nonrepudiation is maintained for the author. Digital signatures have various other benefits, like message-wide authentication. This protects against the insertion of a skewed block in a message or document body stream by an attacker. Another benefit is the integrity of message, which confirms that a message has not been altered or modified during a communication. Choosing the method of digitally signing a digital document or a message is more like making a business decision. That is, the viability, vitality, and durability factors of a document have to be taken into consideration before choosing a scheme of signature. Digital signature users can range from a closed consortium or organizations to open communities and traders. Digital signatures exist for legal reasons. We have long been able to send documents electronically. However, many documents must be signed to be legal in court proceedings. Until recently, the legality of electronic signatures sent with documents was uncertain. This is the digital signature, in which the sender encrypts (signs) a message digest with his or her private key. If this digital signature can be verified with the true party's public key provided by a digital certificate from a reliable certificate authority, the only rational basis for nonrepudiation is that the true party's private key was stolen. Similarly, if someone falsely claims that a person signed a contract, a digital signature will prove that assertion false because verification with that person's public key will fail. Consequently, although many forms of electronic signature are permitted by law, only digital signatures provide strong legal protections.

Before going into further details, a brief understanding of digital signature is necessary. A digital signature scheme typically consists of the following three distinct procedures:

1. Key generation

2. Signature generation

3. Verification of signature.

The key generator mainly involves setting a public key and corresponding private key. The public key is supposed to be broadcast to all, and the private key is meant to be secured at the signer's side. The signing procedure involves steps to sign a digital document using the private key. The signature can be verified using the public key for its authenticity.

In this chapter, we will discuss about the various digital signature schemes, how they work, and what are the performance and limitations. The technology/technical terms used in the chapter are explained wherever they appear in the section.

15.2 ElGamal Signature Scheme

15.2.1 Introduction to ElGamal Signature Scheme

The ElGamal digital signature scheme was introduced by Elgamal (Elgamal, 1985). Since then, a lot of work has been done in advancing the proposed algorithm. For example, the US National Security Agency (NSA) introduced a variant known as digital signature algorithm (DSA) standard which was widely used digital signature standard (DSS) and adopted it as FIPS 186 in 1994 (FIPS PUB 186, 1994). Four revisions to the initial specification have been released: FIPS 186-1 in 1996 (FIPS PUB 186-1, 1998), FIPS 186-2 in 2000 (FIPS PUB 186-2, 2000), FIPS 186-3 in 2009 (FIPS PUB 186-3, 2009), and FIPS 186-4 in 2013 (FIPS PUB-186-4, 2013). ElGamal Signature Scheme is not used as its proposed form. However, its variants like DSS (FIPS PUB) (Schnorr, 1991; Boneh, 2011) were used extensively by various agencies.

The ElGamal signature scheme is based on an arithmetic modulo a prime. In fact, it was the first to use modular arithmetic. The main reasons for its low use/popularity are due to its longer signature size and hence longer computation time. As RSA (Rivest et al., 1978) thrives on the difficulty of factoring large integers, ElGamal signature scheme drives over the difficulty of computing discrete logarithms.

15.2.2 Signature Parameters of ElGamal Signature Scheme

The Elgamal was designed using discrete logarithms. One of the core parameters in the process of building signatures is a large prime number, which ensures that the discrete logarithms modulo p computation is difficult. Another parameter α, such that $\alpha < p$ as a congruence modulo p generator, is also to be known to the system at everyone. Also as in many variants, there will be a collision-resistant hash function H which will be used to hash the message before making the signature. More parameters will be discussed when few other variants are introduced after the next section.

15.2.3 Key Generation and Signature Generation

15.2.3.1 Key Generation

Elgamal (1985) described key generation by computing the following:

$$y \equiv \alpha^x \bmod p \tag{15.1}$$

For a user A signing a document m such that $0 \le m \le p - 1$, X_A being the private key and Y_A being the public key shared with all. The public key can be said to be a triplet known to all in the form (Y_A, α, p).

The signature for a message m will be a pair (r,s) such that $0 \le (r,s) < p-1$ and

$$\alpha^m \equiv y^r r^s \bmod p \qquad (15.2)$$

15.2.3.2 Signing Process

The signing procedure consists of the following three steps:

a. Choosing a uniform random number k, such that $\gcd(k, p-1) = 1$
b. Computing,

$$r \equiv \alpha^k \bmod p \qquad (15.3)$$

c. (15.2) substituting for (15.1) and (15.3)

$$\alpha^m \equiv \alpha^{xr} \alpha^{ks} \bmod p$$

d. s can be equated using

$$m \equiv xr + ks \bmod (p-1) \qquad (15.4)$$

Holding condition $\gcd(k, p-1) = 1$
   Thus,

$$s = k^{-1}(m - xr) \bmod p - 1 \qquad (15.5)$$

e. The signed message will be $(m, (r,s))$.

15.2.4 Verification, Correctness, and Forgery

15.2.4.1 Verification

Having $(m,(r,s))$ the process of verification first checks whether (i) $0 < r < p$ and (ii) $0 < s < p-1$, and then finally checks the equality of Equation 15.2

$$\alpha^m = y^r r^s$$

or

$$r = \alpha^{s^{-1}} y^{-rs^{-1}} \qquad (15.6)$$

Mostly m is replaced by $H(m)$, where H is a collision-resistant hash function. In this case, the signature generation equation will become

$$s = k^{-1}(H(m) - xr) \bmod p - 1 \qquad (15.7)$$

and the verification equation will become

$$\alpha^{H(m)} = y^r r^s \bmod p \qquad (15.8)$$

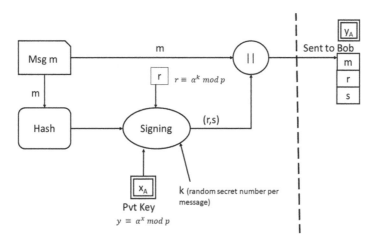

FIGURE 15.1
ElGamal signing at Alice.

In general, Elgamal can be described in the form of Bob and Alice communication as follows (Figure 15.1):

Alice wants to sign a digital document m, so she

1. Sets x_a and y_a where $y_a = \alpha^{x_a}\ 0 < x < p$, $\gcd(x, p-1) = 1$
2. Sends y_a to everyone
3. Picks k randomly, $0 < k < p$ with $\gcd(k, p-1) = 1$
4. Computes $r = \alpha^k$
5. Finds signature, $s = k^{-1}(m - xr) \bmod p - 1$
6. Sends $(m, (r, s))$ to Bob.

Bob on receiving $(m, (r, s))$ and already having y_a wants to verify the document. He

1. Verifies by computing r from m, y, s as $\alpha^{s^{-1}} y^{-rs^{-1}} = r$

Figure 15.2 illustrates the process (hashed variant) pictorially.

DSA (FIPS 186, 1994–2013) was proposed at federal information processing standard (FIPS) for digital signatures by National Institute of Standards and Technology (NIST) in August 1991. DSA is a variant of Elgamal digital signature using mark SHA1 (m) (Stevens, 2012) hashing and provides a digital signature of 320 bits on a message of 160 bits which is much smaller than original ElGamal signatures (Elgamal, 1985).

The Schnorr (Schnorr, 1991) is another variant of ElGamal which applies two prime numbers (as system parameters) p and q. The key generation involves choosing a such that $a^q \equiv 1 \bmod p$ where (a, q, p) also become a system parameter known to all. The public key y_A is extracted choosing a private key s_A such that

$$y_A \equiv a^{-s_A} \bmod q, \quad 0 < s_A < q$$

The signature is produced by computing r as in ElGamal and then computing a concatenation of message and the r.

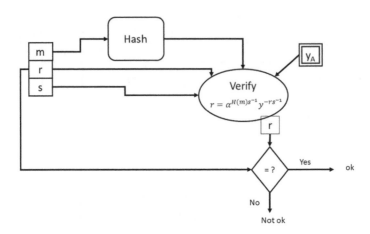

FIGURE 15.2
ElGamal verification at Bob.

$$e = H(M \| r)$$

Computing, $v = (k + se) \bmod q$

The signature pair is (e, y). Verification involves computing r as in ElGamal. Schnorr also generates much shorter signatures than ElGamal.

15.2.4.2 Correctness and Forgery

The ElGamal signature scheme is correct in the sense that a signature generated with the signing algorithm will always be verified as correct having the public key of the signer. The signature generation Equation 15.7 implies the following:

$$H(m) \equiv xr + sk (\bmod p - 1)$$

Using Fermat's little theorem, we have
$$\alpha^{H(m)} \equiv \alpha^{xr} \alpha^{ks}$$

$$\equiv (\alpha^x)^r (\alpha^k)^s, \text{ as in (15.2)}$$

$$\equiv y^r r^s \bmod p$$

Elgamal (1985) discussed various possible attacks on a digital signature system in the context of ElGamal signature scheme. He also formalized attacks in the following categories:

a. Recover private key
b. Forging signatures.

The possibility of an attack of recovering private key also questions the correctness of the signature scheme. The recovering of the key can be performed in several ways:

a. Recovering x by known message signature pairs: this would be difficult as it would require an exponential number of message signature pairs in order to recover $x \bmod q$ since $p - 1$ is chosen to have at least one large prime factor q. Also, if k (random number) generated is used twice in the signing, then the system of equations is uniquely determined and x can be recovered. So for the system to be secure, any value of k should never be used twice.

b. Solving $\alpha^m \equiv y^r r^s \bmod p$ or $y \equiv \alpha^x \bmod p$: this is equivalent to the complexity of computing discrete logarithms over $GF(p)$.

c. Extracting linear dependencies: this is also complex as computing discrete logarithms.

Forging ElGamal signatures can be tried in many ways as discussed in ElGamal (1985).

a. Find r, s to satisfy $\alpha^m \equiv y^r r^s \bmod p$: this again will fall into the category of computing discrete logarithms. For $y^r r^s \equiv A \bmod p$ by fixing s and computing r, the proof is not yet achieved so far.

b. Existential forgery (Pointcheval & Stern, 2000): An intruder knowing one legitimate signature and messages can generate other legitimate signature and messages in these attacks. Most of the state-of-the-art signature schemes allow existential forgery. However, existential forgery is a weak adversarial goal, and hence it can be avoided using hashing over the message before signing it.

15.3 Elliptic-Curve Digital Signature Algorithm

Elliptic-curve digital signature algorithm (ECDSA) is a significant variation in the DSA which uses elliptic-curve cryptography (ECC). ECDSA is also used by bitcoin as it ensures that funds can only be spent by their rightful owner. Actually, ECDSA is the elliptic curve which is analogous to DSA. It was proposed in the year 1992 by Scott Vanstone (Menezes and Vanstone, 1993). It was accepted by ISO (International Standards Organization) in 1998 and was accepted by ANSI (American National Standards Institute) in 1999. It was accepted as a standard algorithm by IEEE (Institute of Electrical and Electronics Engineers) in the year 2000 (Vanstone, 2003).

A few concepts related to ECDSA are as follows:

Private key: A private key is a secret number which is only known to a person who generates it. It is a randomly generated number which is used in bitcoin. A person with a private key in bitcoin can spend the funds in public ledger. In bitcoin, the size of the private key is single unsigned 256-bit integer, i.e., 32 bytes.

Public key: A public key is a number which corresponds to a private key, but it is not kept as a secret, rather exposed publically. It can be used to determine whether a signature is genuine or not without requiring the private key to be exposed.

Signature: A signature is a number which is responsible for the signing operation. It is generated mathematically from a hash of something which is to be signed plus a private key. The signature comprises two numbers: s and p. By using the public key in a mathematical algorithm, it can be determined that the signature was originally generated from a hash and a private key without knowing the private key.

In the ECC, generally the bit size for the public key needed for ECDSA is twice the size of security levels (in bits) (Hankerson et al., 2004). For example, for the security level of 90 bits (i.e., an attacker requires a maximum of about 2^{90} number of operations to get the private key), the size for a public key in ECDSA will be 180 bits, whereas in DSA, the public key size would be at least 20,124 bits. Signature sizes for both DSA and ECDSA are the same which is equal to $4t$ bits approximately, where t is the security level measured in bits, i.e., 360 bits for the security level of 90 bits.

15.3.1 Signature Generation and Verification Algorithm

15.3.1.1 ECDSA Key Generation

Let's assume that Ram wants to send a signed message to Sham. For successful execution, they must agree on the curve parameters (CURVE, B, n). We need field, equation of the curve and B (which is a base point of prime order on the curve), n (which is the multiplicative order of point B) in order to generate ECDSA key. Lets take an entity A, whose key pair is related to a particular set of EC domain parameters Dom = (q, FR, a, b, B, n, h). Given E as an elliptic curve which is defined over Zq, the point of prime order n in $E(Zq)$ is P. Every entity A does

1. Choose a randomized integer r in the interval $[1, n - 1]$
2. Calculate rP and assign it to Q, i.e., $Q = rP$
3. The public key of A is Q, and the private key of A is r (Figure 15.3).

15.3.1.2 Signature Generation Algorithm for ECDSA

Let m be a message, and we need to sign it. So, to do so, we need an entity A with domain parameters as Dom = (q, FR, a, b, B, n, h), and they are used as follows: k^{-1}

1. Choose a randomly generated integer k in the interval $[1, n - 1]$.
2. Calculate x_1, y_1, r and assign it to kP i.e., $kP = x_1, y_1, r = x_1 \bmod n$ (where x_1 is an integer between 0 and $q - 1$). If r equals to zero, then go back to step 1.
3. Calculate k raise to the power $-1 \bmod n$.
4. Calculate k raise to the power -1 $\{h(m) + dr\}\bmod n$ and assign it to $s = k^{-1}\{h(m) + dr\}\bmod n$, where h is the secure hash algorithm. Now, if s equals to 0, then go back to step 1.
5. Pair of integers (r, s) is the signature for the message m (Figure 15.4).

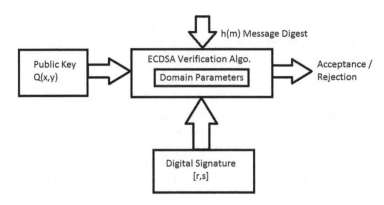

FIGURE 15.3
ECDSA overview.

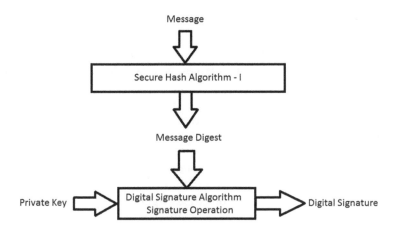

FIGURE 15.4
ECDSA signature generation process.

15.3.1.3 Signature Verification Algorithm for ECDSA

Integers (r, s) are the signatures of A on message m. Now, we need to verify those signatures. Another person (say W) somehow gets the authenticated copy of A's domain parameters Dom = (q, FR, a, b, B, n, h) and public key Q of A as well. In order to verify those signatures, here is an algorithm:

1. First make sure that integers (r, s) are in the interval $[1, n - 1]$
2. Calculate s raise to power -1 mod n and $h(m)$, and then assign it to v, i.e., $= s^{-1}$ mod n and $h(m)$ is assigned to v
3. Calculate $u_1 = h(m)v \bmod n$ and $u_2 = rv \bmod n$
4. Calculate $u_1 P + u_2 Q = (x_0, y_0)$ and $u = x_0 \bmod n$
5. The signature is accepted if and only if $u = r$ (Figure 15.5).

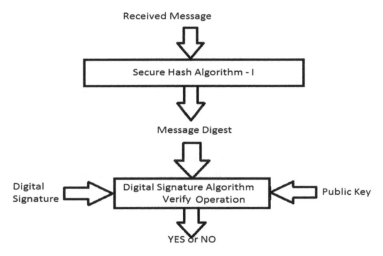

FIGURE 15.5
ECDSA signature verification process.

15.3.2 Security of ECSDA

To remain unforgeable against a message, attack is the main security moto for ECDSA. The intention of an attacker who attacks against a genuine entry S is to get a valid signature on single message m, after obtaining S's signature on a series of messages (except m) of attacker's choice.

This algorithm has not been proven as unforgeable against a message attack.

However, some other variations of DSA and ECDSA are proven secure by Pointcheval and Stern assuming that such as discrete logarithm problem is a tough and random function is used as a hash function.

ECDSA possible attacks are categorized as follows:

- Elliptic-curve discrete logarithm problem is targeted
- Used hash function is targeted
- Other attacks.

15.3.3 Comparison with RSA and DSA

Two types of operations are difficult to perform in most of the cryptographic systems. One is traceable (Gupta et al., 2004) which is known as forward operation, and the other one is non-traceable which is inverse operation. The two operations differ on the basis of the size of key pairs. The inverse operation is exponentially proportional to the key size, whereas forward operation is linearly proportional. The complexity of both the operations increases if we increase the key length. Thus, ECC is preferred because the security in this is at the same level at 160-bit key length as of 1,024-bit key length in RSA.

15.3.3.1 Comparison with RSA

1. ECC takes full exponential time, whereas RSA takes subexponential time. For example, ECC takes 9.6×10^{11} MIP years for the best-known attack, whereas RSA takes 3×10^{11} MIS years for 1,024-bit key size.
2. For smaller key sizes, ECC behaves with the same level of security.
3. In ECC, data size is larger than RSA.
4. Key size and data size are functions for the encrypted message. In RSA, key size is larger as compared with ECC; this results in a smaller encrypted message in ECC.
5. In comparison with RSA, computational power is less in ECC.

15.3.3.2 Comparison with DSA

1. Signing equation is mentioned below and is the same for both DSA and ECC. Also, both the algorithms are based on ElGamal signature scheme.

$$s = k^{-1}\{h(m) + dr\}\bmod n.$$

2. System parameters ($p, q,$ and g for the DSA; $E, P,$ and n for the ECDSA) are difficult to generate in both the algorithms.
3. SHA-1 is used as a hash function in both ECDSA and DSA (current version).

4. Some values like per-signature value k and private key p are unpredictable and statistically unique in ECDSA. These values are random in DSA (Johnson et al., 2001).

15.4 Identity-Based Signature

15.4.1 Introduction and Definition of ID-Based Signature

Identity-based signature (IBS) scheme is an open problem which is based on the difficulty of factoring integers. In 2001, two schemes were launched: one by Cocks (Cocks, 2001) whose base was quadratic residuosity problem and the other by Boneh and Franklin (Boneh & Franklin, 2004) which was based on the bilinear Diffie–Hellman problem with respect to a pairing, e.g., the Weil pairing (Yi, 2003).

Initially, the Weil pairing (Miller, 2004) was not having a good image in cryptography. It reveals that by using the Weil pairing, discrete logarithm problem in super-singular elliptic curves was reducible to that in a finite field. That resulted in dropping up of super-singular elliptic curves from cryptographic use. The situation was changed entirely by Joux's work, who published a simple tripartite Diffie–Hellman (Bellare et al., 2003) protocol based on the Weil pairing on super-singular curves. Soon after Joux's paper, a number of other applications were released, including a general signature algorithm and an identity-based encryption scheme.

The Weil pairing and other IBS schemes are also used by an identity-based public-key signature algorithm.

IBS scheme is a collection of the following four algorithms:

Setup—In this algorithm, a security parameter is taken as an input by a master entity, and the output is generated in the form of public parameters related to a master secret and scheme. The parameters are published by the master entity, and the master secret is taken care by itself.

Extract—The algorithm generates the private key d_u of u if an identity u, master secret, and parameters are given as input. This private key is then used by the master entity in order to generate private keys for all participating entities, and then private keys are distributed to their owners, respectively, through a secure and safe channel.

Sign—This algorithm generates the signature σ of identity u on m if a message (say m), an identity u, a private key d_u and parameters are given as inputs. The identity u with entity uses this algorithm for signing purpose.

Verify—This algorithm gives output and will accept it if signature σ is a valid signature on m for identity u, and reject the output otherwise where a signature σ, a message m, an identity u, and parameters are given as inputs.

Let's have a look at some notation. Bit strings are denoted as $a_1, \ldots, a_n, a_1 \| \ldots \| a_n$ is denoted as a string encoding of $a_1, \ldots, a_n$ which will be used to uniquely recover the constituents objects. $|a|$ denotes the length of a string 'a', and $|S|$ is the cardinality of set S. For A as a randomized algorithm, $y \leftarrow A(a_1, a_2, \ldots)$ implies that A has inputs $a_1, a_2, \ldots$ and the output of A is assigned to y when it runs on a new random tape.

Attacker has various resources like its number of queries and its running time asymptotically in relation to security parameter k. A function $\nu(k)$ is said to be negligible (in k) if for all $c \in N$ there exists $k_c \in N$ such that $\nu(k) < k^{-c}$ for all $k > k_c$.

The parameters for IBS scheme algorithm are *Setup, Check, Sign, and Verify*. This algorithm gives polynomial time complexity with security parameter k. The last parameter is constant, whereas the first three are randomized. Setup algorithm is executed on input 1^k by a trusted key distribution method in order to obtain a master public and private secret key pair *(mpk, msk)* (Bellare et al., 2003). (Note: For security parameter k, 1^k denotes the unary notation.) It runs the key derivation algorithm *Check* on input *msk* and *id* in order to generate secret signing key *usk* for the user with identity $id \in \{0, 1\}^*$.

Secure communication of the secret signing key is assumed. The signing algorithm Sign returns a signature σ of M by providing *usk* and a message M as input. The trusted key distribution runs the Setup algorithm on input 1^k in order to obtain a master public and private secret key pair *(pk, sk)*. (Note: $1k$ denotes the unary notation of the security parameter k.) The verification algorithm *Verify* returns 1 by providing *mpk, id, M*, and a signature σ as input. *Verify* returns 1 if σ is valid for *id* and M and returns 0 otherwise. Correctness requires that *Verify(mpk, id, M, Sign(usk, M)) = 1* with probability one for all $k \in N$ and id, $M \in \{0, 1\}^*$ whenever the keys *mpk, M, usk* are generated as indicated above.

For security reason, we consider the notion of existential unforgeability under chosen-message and chosen-identity attack (uf-cma). Security is taken care through an experiment with a forger F, and it is parameterized with security parameter k. At first, the randomized key generation algorithm *KG* on input 1^k generates a key pair *(pk, sk)*, i.e., *(pk, sk) ← KG(1^k)*. The forger F has access to the following oracles, and it runs on master key *pk*:

- Check(·): First, this oracle checks on input identity $id \in \{0, 1\}^*$ the user secret key USK[*id*] needed to be checked for id, if already generated or not. If the check returns true, then it will return the same user key; otherwise, it generates a fresh user secret key USK[*id*] ← Check(*sk, id*).
- SIGN(·, ·): Second, this oracle returns the SIG ← SIGN(USK[*id*],*m*) on input identity $id \in \{0, 1\}^*$ and message $m \in \{0, 1\}^*$, where USK[*id*] is given as an input by above oracle Check(·).

Algorithm Setup(1^k)
(SecKey,PriKey) ← setup(1^k)
Return (SecKey,PriKey)

Algorithm Check(SecKey, id)
(pk, sk) ← setup' (1^k)
new ← sign(SecKey, id || pk)
Return USK[id] ← (new, pk, sk)

Algorithm SIGN(id, USK [id], m)
Parse (cert, pk, sk) ← USK[id]
sig ← sign' (sk, m)
Return SIG = (cert, pk, sig)

Algorithm Verify(PK, id, m, SIG)
Parse (new, pk, sig) ← SIG

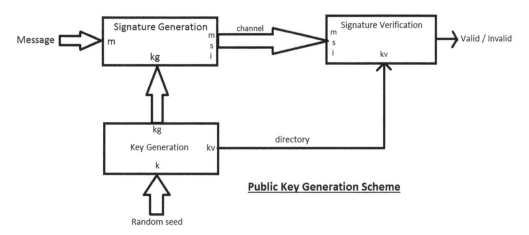

FIGURE 15.6
Public-key signature scheme.

If verify(PriKey, id ‖ pk, new) = 0 then return 0
If verify' (pk, m, sig) = 0 then return 0
Else return 1.

After completion of execution, the forger *F* results in output as a message *m**, an identity *id**, and a forged signature *sig**. The forger *F* will win if Verify (PriKey, *id**, *m**, *sig**) equals to 1 and also forger never executed about *Check(id*)* or *Sign(id*, m*)*.

Figure 15.6 shows the diagrammatic view of public-key signature scheme with the following notation:

m is the message, *s,t* is the signature, *i* is the user's identity, *n* is the product of two large primes, *e* is a large prime which is relatively prime to $\phi(n)$, *f* is a one-way function.

In public-key signature scheme, a message (*m*) is first given to signature generation module, and it takes input from key generation module (*kg*) and gives output as *msi*, i.e., a message with signature and user's identity. Now, *msi* is verified by signature verification module and gives output as valid/invalid. Signature verification module also takes input (*kv*) from key generation module.

In IBS scheme, a message (*m*) is first given to signature generation module, and it consults key generation module and verifies the identity and takes (*kg*) as input; it generates *msi*, i.e., a message with signature and user's identity as an output. Now, *msi* is verified by signature verification module by confirming it from sender's identity and gives output as valid/invalid (Figure 15.7).

15.4.2 Efficiency and Security Comparison

Table 15.1 compares and contrasts the efficiency and security properties of various IBS-based schemes. For every scheme, it displays the transformation from which that particular IBS scheme is obtained, size of signature, computational overhead time for signing and verification and the type of model, whether it is a standard model (SM) or the random-oracle model (ROM).

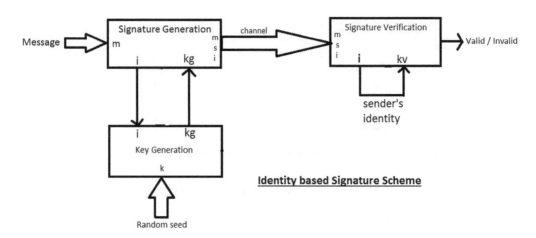

FIGURE 15.7
IBS scheme.

TABLE 15.1

Efficiency and Security Comparison of Various IBS Schemes

Scheme	Transform	Signature Size	Signing Time	Verification Time	ROM/SM
Cert-IBS	SS-2-IBS	2 *sig.* of SS 1 *pk* of SS	1 *sig.* of SS	2 Vf of SS	SM
ChCh-IBS	cSI-2-IBS	2 el. of G	2 exp. in G	2 pairings	ROM
GS-IBS	HIBE-2-IBS	2 el. of G	2 exp. in G	3 pairings	ROM
BNN-IBS	Direct	1 el. of G 1 el. of Zp	2 exp. in G	1 mexp, exp. in G	ROM
Sh-IBS	cSI-2-IBS	2 el. Z_N	2 exp. in Z_N	1 mex p in Z_N	ROM

The various IBS schemes are compared on the basis of transformation, signature size, signing time, verification time, and type of model. Depending upon these various parameters, need and the application, a particular type of IBS-based scheme is selected.

15.5 Bimodal Lattice Signature Schemes (BLISS) and Lamport Signature Scheme

15.5.1 Introduction to BLISS and Lamport

15.5.1.1 BLISS Signature Scheme

BLISS, stands for Bimodal Lattice Signature Scheme, was proposed by L. Ducas, A. Durmus, T. Lepoint, and V. Lyubashevsky in 2013. This scheme does not use the integer factorization, discrete logarithm or elliptic curve which can be broken using quantum computers. However, the BLISS is unbreakable even using quantum computers. The previous schemes are considered as pre-quantum schemes, and the BLISS is hence a post-quantum

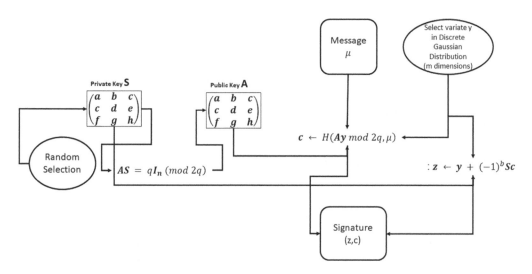

FIGURE 15.8
BLISS key generation and signature generation.

scheme (Bernstein, 2011). BLISS is considered to be efficient in terms of computation, signature size, and security (Ducas et al., 2013).

The BLISS scheme is based on lattice problems and rejection sampling from bimodal Gaussian distribution. The BLISS proposes to hide the secret key *s* with a small random *y* by choosing *y* from a narrow distribution and then performs acceptance–rejection (Casella et al., 2004) so that *s* is not leaked when *y* is added to it. Conceptually, the BLISS signs a document by drawing a sample *x* from *g* and is accepted with probability $f(x)/(M \cdot g(x))$, where *M* is some positive real. The process is repeated until the sample is accepted. In the numerical analysis, rejection sampling is a basic technique used to generate observations from a distribution. It is also commonly called the acceptance–rejection method or "accept–reject algorithm" and is a type of exact simulation method. The key generation, signature, and verification process are extracted from Ducas et al. (2013), as given in the following section (Figure 15.8).

15.5.1.2 Key Generation, Signature, and Verification Algorithms

A ring $\mathbb{Z}_q$ is identified with the interval $[-q/2, q/2] \cap \mathbb{Z}$, with *q* being an integer and for a general ring $\mathcal{R}$ and $\mathcal{R}_q$ a quotient ring $\mathcal{R}/(q\mathcal{R})$ such that $\mathcal{R}_q = \mathbb{Z}_q[x]/(x^n + 1)$, where *n* is a power of 2 and *q* is a prime number such that $q = 1 \pmod{2n}$. Hoffstein et al. proposed a ring-based public-key cryptosystem called NTRU. The BLISS was proposed to be an adaptation of NTRU (Hoffstein et al., 1998) over SIS (Ajtai, 1996) which thrives over the hardness of $R - \boldsymbol{SIS}^k \, q, n, m, \beta$ problem defined in the same paper. For simplicity, an algorithm for $\mathcal{R} = \mathbb{Z}$ is presented as in Ducas et al. (2013); however, it can be easily adapted for $\mathcal{R} = \mathbb{Z}[x]/(x^n + 1)$.

The secret key is a (short) matrix $\boldsymbol{S} \in \mathbb{Z}^{m \times n}_{2q}$, and the public key is a matrix $\boldsymbol{A} \in \mathbb{Z}^{m \times n}_{2q}$ such that $\boldsymbol{AS} = q\boldsymbol{I}_n \pmod{2q}$. A hash function *H* is modeled as a random oracle that has uniform output in $B(n, k)$, the set of binary vectors of length *n*, and weight *k* as in Fischer and Stern (1996).

15.5.1.3 Signature Algorithm

Parameters: Message μ, public key is a matrix $A \in \mathbb{Z}^{m \times n}{}_{2q}$, secret key matrix $S \in \mathbb{Z}^{m \times n}{}_{2q}$, standard deviation $\sigma \in \mathcal{R}$.

Step 1. Get a sample of discrete Gaussian D of m dimensions into y, i.e., $y = D_\sigma^m$

Step 2. Compute $c \leftarrow H(Ay \bmod 2q, \mu)$, A hash function H as a random oracle

Step 3. Choose a random bit b in $\{0,1\}$

Step 4. Compute

Step 5. Return *Signature* (z, c) if having probability, $1 \Big/ \left(Me^{-\frac{\|Sc\|^2}{2\sigma^2}} \cdot \cosh\left(\frac{\langle z, Sc \rangle}{\sigma^2}\right) \right)$ else repeat.

15.5.1.4 Verification Algorithm

The signer outputs signatures of the form $(\mathbf{z}, \mathbf{c})$, where $\mathbf{z}$ is distributed according to $Dm\sigma$. Thus, the acceptance bound $B2$ should be set a little bit higher than $m\sigma$, which is the expected value around which the output of $Dm\sigma$ is tightly concentrated.

Parameters: Message μ, public key is a matrix $A \in \mathbb{Z}^{m \times n}{}_{2q}$, Signature (z,c)

Step 1. if $\|z\| > B_2$ or $\|z\| \geq \frac{q}{4}$ then reject

Step 2. Accept the signature if and only if $c = H(Az + qc \bmod 2q, \mu)$.

15.5.1.5 Lamport Signature Scheme

The Lamport signature scheme is one more upcoming signature methodology which is deemed to be post-quantum computing. It is believed that the Lamport signatures, like lattice-based signatures, using large hash functions would be secure against quantum computing attacks. The Lamport signatures can be built from any one-way hash function, which is cryptographically secure. That is, these hash functions are usually cryptographic hash functions. One-way functions are easy to forward computation yet difficult to get its inverse (Nyberg and Rueppel, 1996). Lamport (1979) devised a schema of making signatures using one-way functions.

15.5.1.6 Key Generation Signing and Verification

The key generation procedure involves generating n bits key pairs using a random number generator, for example, if $n = 256$, then a total of $2 \times 256 \times 256$ bits (16 kb) would be needed to keep the keys. This set would be the private key. Similarly, the public key is created by hashing (f) each key pair of the private key to get a 16 kb keys. The one-way hashing functions f and g are used to create keys and signing the message, respectively.

The signing is simple; using g the message is hashed and attached with the private keys bit by bit choosing the appropriate private key of each pair. That is, if the hashed message bit is 0, then the first key of the pair is chosen, and if the hashed message bit is 1, the second key of the pair is chosen. Thus, for a 256-bit hashed message, a 256-bit signature is formed.

At the verification side, the verifier hashes the message and does the same procedure of creating α using the public key and then hashing the signature and comparing it with the generated α. Figures 15.9–15.11 illustrate the process of key generation, signing, and verification in pictorial form. After signing, the signer has to destroy the private keys completely.

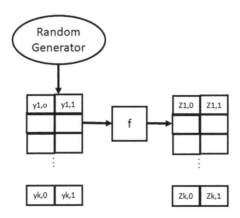

FIGURE 15.9
Lamport key generation.

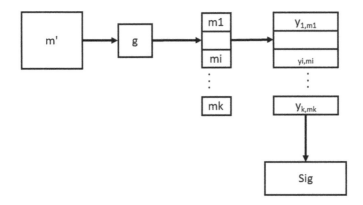

FIGURE 15.10
Lamport signing.

The process can be mathematically represented as follows:
Keys
Let k be a positive integer and let $P = \{0,1\}^k$ be the set of messages in binary form. Let $f : Y \rightarrow Z$ be a one-way hash function. For $1 \leq i \leq k$ and $\in \{0,1\}$, where k is the number of key pairs.

1. Random number $y_{i,j}$ is generated in Y
2. $z_{i,j} = f(y_{i,j})$
3. Private key pair $i = y_i$ and public key pair i is z_i where $i = 1$ to k.

Signing a message
1. For a plain message m', k bit hashed message $m = (m_1, \ldots, m_k \in \{0,1\}^k)$ where $m = g(m')$, g is an one-way function $g : Y \rightarrow Z$
2. Signature $s_{i,j} = sig(m_1, \ldots, m_k) = (m_1, y_{1,m_k}, \ldots, m_k, y_{k,m_k})$
3. Destroy the private keys $y_{i,j}$.

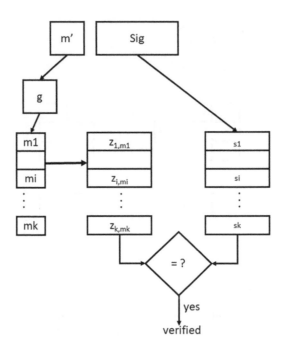

FIGURE 15.11
Lamport verification.

Verification

1. Accept if, $f(s_i) = z_i, m_i$, for all $1 \le i \le k$

Various optimizations and variants of the Lamport scheme came to light like those of short private key (seed of the random number generator), short keys, and signature and public keys for multiple messages. The last one is devised by Merkle and is called as Merkle signature scheme. This scheme is discussed in the next section.

15.5.2 Security Comparison

The correctness of both Lamport and BLISS has been presented in the respective originating and other papers (Lamport, 1979; Rompel, 1990; Ducas et al., 2013). The security aspects of Lamport depend upon the security of the one-way hash function, the length of the output, and the quality of the input. In order to forge a message in case of signed by the Lamport method, the forger has to obtain the inverse of the one-way function which has been termed very difficult in case the procedure uses large-sized inputs and outputs. John also gives a high-security rating to one-way hash signatures calling it necessary and sufficiently secure. Constructing one hash functions has also been discussed by Lamport and Rompel (see Lamport, 1979; Rompel, 1990). Lattice-based digital signatures were discussed and compared by James Howe et al. (2015). Lattice signature variants like GGH (Goldreich et al., 1997), NTRU (Hoffstein et al., 1998), GPV (Gentry et al., 2008), LYU (Lyubashevsky, 2012), and BLISS were compared and summarized by James Howe et al. (2015). Ducas presented a security analysis of BLISS in various ways of attacks namely:

a. Brute force key recovery

b. Meet in middle attack

c. Hardness of the short integer solution problem

d. Primal lattice reduction key recovery

e. Dual lattice reduction key recovery

f. Hybrid MiM Lattice key recovery.

These attack schemes are open for research and speculation. However, the BLISS models are still unbreakable in a larger sense.

15.6 Merkle Signature Scheme

Merkle signature scheme is yet another post-quantum computing signature scheme proposed by Ralph Merkle in 1979 (Merkle, 1979). In his thesis, he described a digital signature that is pre-certified in the sense of the certified underlying encryption methodology. The paper proposed a method converting any one-time signature (like Lamport's) into a convenient multiple-time signature system. The basis of the need of multiple time signature as comprehended in various papers (Merkle, 1987; Merkle, 1990; Bindel et al., 2016; Merkle, 1979) is considering the following scenario. If A transmits his public key Z to B every time before signing his document, B is not having Y previously then anyone can create a Y and claim to be A. B can be tricked in this way making him believe he had received a properly authorized signature when he had received nothing of the kind. B must somehow be able to confirm that he was sent the correct Y and not a forged Y. This has to be done with every new message m_i. That is Y_i for every message m_i needs to be authenticated each time. A solution for this was proposed as "Tree Authentication" in various papers (Merkle, 1987; Merkle, 1990; Bindel et al., 2016) and later Merkle also elaborated the Tree Authentication-based Signature Scheme in 1990 (Merkle, 1990).

The Merkle tree signature scheme key generation, signature, and verification are described below.

15.6.1 Key Generation (Merkle Tree)

Merkle scheme is used to sign a limited number of messages per public key Z. The number of messages must be in powers of 2. That is,

$$N = 2^n, N \text{ is the number of messages per } Z$$

To generate key, the following steps are needed:

1. Generate $Z = (x_i, y_i), \quad 1 \le i \le N$ as Lamport's scheme
2. A hash value of the public key $h_i = H(y_i)$ is computed
3. Build a hash tree using these hash values h_i. The tree leaves will be the $N = 2^n$ hash values and recursively hashing toward the root to form a binary tree. The value for each inner node of the tree is the hash of the concatenation of its two children, i.e.,

h_i will be leaf node $a(0, i)$

For other nodes: node $a(l, j) = H(a(l-1, i) \| a(l-1, i+1)), \quad 1 \le l \le n, 1 \le i, j \le 2^{n-l}$
Figure 15.12 is a Merkle tree where $N = 8 = 2^3$.

4. The root $a(n, 1)$ will be the public key. The $Z = (x_i, y_i)$ will be the private key.

15.6.2 Signature Scheme

1. The user signs using any one-time signing method like Lamport's using private key $Z = (x_i, y_i)$.
2. Sign the message to generate *Sig'*. Sends (Sig', y_i) along with the **authentication path**.
3. The **authentication path** consists of the nodes which are needed to create the root $a(n, 1)$, as can be seen from the example tree in Figure 15.12. If the signature *sig'* and the public key *y* are to be sent, the authentication information is also sent along with it. These will be an unused leaf and the path of the node which will be used to generate the root node. If $H(y_3)$ is used, then nodes $a(0,3)$, $a(1,1)$ and $a(2,2)$ are also sent, as depicted in Figure 15.13.

15.6.3 Verification

1. The verifier receives (Sig', y_i) along with the **authentication path**.
2. Uses the path to compute root. Compares it with the public key.
3. If the key is equal to the computed root, then the sig and public-key pair (Sig', y_i) is authentic.
4. Using the corresponding methodology (like Lamport's) to verify the signature further.

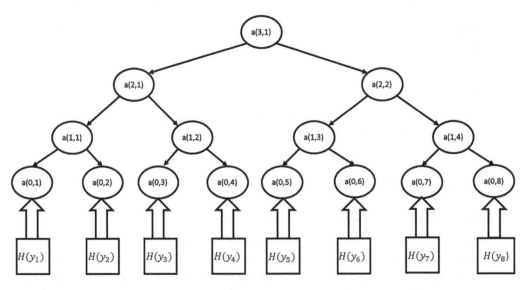

FIGURE 15.12
Merkle tree. The number of messages that can be Signed(N) is 8.

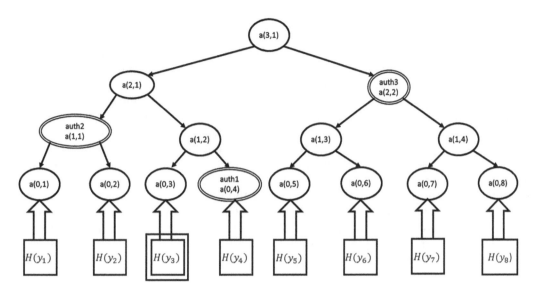

FIGURE 15.13
Authentication paths using $a(0,3)$.

Merkle trees are a fundamental part of blockchain technology (Sharma, 2017). A Merkle tree is a structure that allows for efficient and secure verification of content in a large body of data. This structure helps verify the consistency and content of the data. The fact that Lamport, BLISS, and Merkle systems are more current and outstanding is that they are based on hash-based signatures. Unlike most other signature systems, hash-based signatures are more robust against quantum computer attacks.

15.7 Other Existing Signature Schemes

Apart from the digital schemes discussed in this chapter, there are few others which are coming up and some of them are making their mark in the computing community. Taking the path from DSA to the Merkle schemes, we have many variants coming up like:

- Pointcheval–Stern
- EdDSA
- XMSS
- Aggregate signature
- BLS.

Pointcheval–Stern signature algorithm (Pointcheval and Stern, 1996) is a digital signature scheme adapted on the principles of ElGamal signature scheme. It incorporates a slight change to the ElGamal scheme and implies into an algorithm which proves highly secure against adaptive chosen-message attacks. Similar to ElGamal, it also assumes the strength of the discrete logarithm problem. Edwards-curve digital signature algorithm

(EdDSA) (Josefsson and Liusvaara, 2017) is a variant of Schnorr digital signature scheme. It is based on twisted Edwards curves designed to be faster than existing digital signature schemes without sacrificing security. An extension of the Merkle scheme was introduced in RFC (Huelsing et al., RFC 8390, 2018) as eXtended Merkle signature scheme (XMSS). XMSS is considered to be highly secure and remains unfazed even when the underlying hash function faces collisions. It is compact and simple to implement as compared to the base scheme, and naturally resists side-channel attacks. Aggregate signature (Bellare et al., 2007) is a signature scheme that supports "aggregation of users" which is the basis of the currently popular block chains and bitcoins. Multiple signatures on multiple messages from multiple users can be clubbed into a single signature having a constant size as per the number of users. This single signature will be enough to convince the verifier that the n number of users have indeed signed the n number of original messages. An Aggregate Signature scheme by Mihir Bellare and Gregory Neven may also be used with Bitcoin. The Boneh–Lynn–Shacham (BLS) (Boneh et al., 2004) signature scheme facilitates user-level verification of a signer whether the signer is not pretending to be authentic. To achieve user-level verification, the scheme uses a bilinear pairing for verification. The BLS signatures fall under (as per various elements) an elliptic-curve category.

15.8 Conclusion

This chapter covers most of the widely used schemes used for digital signature with a special emphasis on the base algorithms. Many variants of these schemes are coming up, and their viabilities are under scrutiny among the researchers. The performance of these schemes may vary depending upon the theory behind them. The base methodologies and the added variations on them are on a constant review, and many communities have tailored their way into the digitally signed world as per their requirements and adaptability. In the current quantum computing scenario, the traditional approaches have to be replaced or have to be improved upon using modern schemes or add-ons. The chapter has tried to cover those schemes and methodologies as well.

References

M. Ajtai. (1996). Generating hard instances of lattice problems (extended abstract). In *Proceedings of the Twenty-Eighth Annual ACM Symposium on Theory of Computing (STOC '96)*, ACM, New York, pp. 99–108. doi:10.1145/237814.237838.

M. Bellare, C. Namprempre, G. Neven (2007). Unrestricted aggregate signatures. L. Arge, C. Cachin, T. Jurdziński, A. Tarlecki, editors, Automata, Languages and Programming. ICALP 2007, volume 4596 Lecture Notes in Computer Science, pp. 411–422. Springer, Berlin, Heidelberg.

M. Bellare, C. Namprempre, D. Pointcheval, M. Semanko. (June 2003). The one-more-RSA-inversion problems and the security of Chaum's blind signature scheme. *Journal of Cryptology*, 16(3):185–215.

D.J. Bernstein. (2011). Post-quantum cryptography, H.C.A. van Tilborg, S. Jajodia, editors, *Encyclopedia of Cryptography and Security*, pp. 949–950. Springer, Boston, MA.

N. Bindel, J. Buchmann, J. Krämer. (2016). Lattice-based signature schemes and their sensitivity to fault attacks. In *Workshop on Fault Diagnosis and Tolerance in Cryptography (FDTC)*, Santa Barbara, CA, pp. 63–77. doi:10.1109/FDTC.2016.11.

D. Boneh. (2011). Schnorr digital signature scheme, H.C.A. van Tilborg, editors, *Encyclopedia of Cryptography and Security*. Springer, Boston, MA, pp. 1082–1083.

D. Boneh, M. Franklin. (2003). Identity-based encryption from the Weil pairing. *SIAM Journal on Computing*, 32(3):586-615. doi:10.1137/S0097539701398521.

D. Boneh, B. Lynn, H. Shacham. (2004). Short Signatures from the Weil Pairing. *Journal of Cryptology*, 17: 297–319. doi:10.1007/s00145-004-0314-9.

G. Casella, C. Robert, M. Wells. (2004). Generalized Accept-Reject Sampling Schemes. *Lecture Notes-Monograph Series*, 45: 342–347. Retrieved from http://www.jstor.org/stable/4356322.

C. Cocks. (2001). An identity based encryption scheme based on quadratic residues, B. Honary, editors, *Cryptography and Coding 2001*, volume 2260. Lecture Notes in Computer Science, pp. 360–363. Springer, Berlin, Heidelberg.

L. Ducas, A. Durmus, T. Lepoint, V. Lyubashevsky (2013). Lattice signatures and bimodal Gaussians, R. Canetti, J.A. Garay, editors, *Advances in Cryptology – CRYPTO 2013. CRYPTO 2013*, volume 8042 Lecture Notes in Computer Science, pp. 40–56. Springer, Berlin, Heidelberg.

T. Elgamal. (July 1985). A public-key cryptosystem and a signature scheme based on discrete logarithms. *IEEE Transactions on Information Theory*, 31(no. 4):469–472. doi:10.1109/TIT.1985.1057074.

FIPS PUB 186: Digital Signature Standard (DSS), 1994. qcsrc.nist.gov. Archived from the original on 2013-12-13.

FIPS PUB 186-1: Digital Signature Standard (DSS), 1998 (PDF). csrc.nist.gov. Archived from the original (PDF) on 2013-12-26.

FIPS PUB 186-2: Digital Signature Standard (DSS), 2000 (PDF). csrc.nist.gov.

FIPS PUB 186-3: Digital Signature Standard (DSS), 2009 (PDF). csrc.nist.gov.

FIPS PUB 186-4: Digital Signature Standard (DSS), 2013 (PDF). csrc.nist.gov.

J.-B. Fischer, J. Stern. (1996). An efficient pseudo-random generator provably as secure as syndrome decoding, U.M. Maurer, editors, EUROCRYPT 1996, volume 1070 Lecture Notes in Computer Science, pp. 245–255. Springer, Heidelberg.

C. Gentry, C. Peikert, V. Vaikuntanathan. (2008). Trapdoors for hard lattices and new cryptographic constructions. In *Proceedings of the Fortieth Annual ACM Symposium on Theory of Computing (STOC '08)*. ACM, New York, pp. 197–206. doi:10.1145/1374376.1374407.

O. Goldreich, S. Goldwasser, S. Halevi. (1997). Public-key cryptosystems from lattice reduction problems. *Advances in Cryptology—CRYPTO '97*, pp. 112–131. Springer, Berlin, Heidelberg. doi:10.1007/BFb0052231.

V. Gupta, D. Stebila, S. Fung, S.C. Shantz, N. Gura, H. Eberle. (2004). Speeding up secure web transactions using elliptic curve cryptography. In *Proceedings of the 11th Annual Network and Distributed System Security Symposium (NDSS 2004)*, The Internet Society, pp. 231–239. California, 5–6 February.

D. Hankerson, A. Menezes, S. Vanstone. (2004). Guide to Elliptic Curve Cryptography. Springer Professional Computing. Springer, New York. doi:10.1007/B97644.

J. Hoffstein, J. Pipher, J.H. Silverman. (1998). NTRU: A ring-based public key cryptosystem, J.P. Buhler, editors, *Algorithmic Number Theory. ANTS 1998*, volume 1423 Lecture Notes in Computer Science, pp. 267–288. Springer, Berlin, Heidelberg.

J. Howe, T. Pöppelmann, M. O'neill, E. O'sullivan, T. Güneysu. (April 2015). Practical lattice-based digital signature schemes. *ACM Transactions on Embedded Computing Systems*, 14(3), Article 41, 24 pages. doi:10.1145/2724713.

A. Huelsing, D. Butin, S. Gazdag, J. Rijneveld, A. Mohaisen. (May 2018). XMSS: Extended merkle signature scheme, Request for Comments: 8391, Internet Research Task Force (IRTF), May 2018, ISSN: 2070-1721.

D. Johnson, A. Menezes, S. Vanstone. (August 2001). The Elliptic Curve Digital Signature Algorithm (ECDSA). *International Journal of Information Security*, 1(1): 36–63. doi: 10.1007/s102070100002.

S. Josefsson, I. Liusvaara. (January 2017). Edwards-curve digital signature algorithm (EdDSA). Internet Engineering Task Force. doi:10.17487/RFC8032. ISSN 2070-1721. RFC 8032. Retrieved July 31, 2018.

L. Lamport. (October 1979). Constructing digital signatures from a one-way function. Technical Report SRI-CSL-98, SRI International Computer Science Laboratory.

V. Lyubashevsky. (2012). Lattice signatures without trapdoors. *Advances in Cryptology—EUROCRYPT*, pp. 738–755. doi:10.1007/978-3-642-29011-4_43.

A.J. Menezes, S.A. Vanstone. (1993). Elliptic curve cryptosystems and their implementation. *Journal of Cryptology*, 6:209. doi:10.1007/BF00203817.

R.C. Merkle. (1979). *Secrecy, Authentication, and Public Key Systems*. Ph.D. Dissertation. Stanford University, Stanford, CA. AAI8001972.

R.C. Merkle. (1987). A digital signature based on a conventional encryption function, C. Pomerance, editor, *Advances in Cryptology—CRYPTO '87*, volume 293 Lecture Notes in Computer Science. Springer, Berlin, Heidelberg. doi:10.1007/3-540-48184-2_32.

R.C. Merkle. (1990). A certified digital signature, G. Brassard, editor, *Advances in Cryptology—CRYPTO' 89 Proceedings. CRYPTO 1989*, volume 435 Lecture Notes in Computer Science. Springer, New York.

V. Miller. (2004). The Weil pairing, and its efficient calculation. *Journal of Cryptology*, 17:235. doi:10.1007/s00145-004-0315-8.

K. Nyberg, R.A. Rueppel. (1996). Message recovery for signature schemes based on the discrete logarithm problem. *Designs, Codes and Cryptography*, 7(1–2):61–81. doi:10.1007/BF00125076.

T. Oder, T. Pöppelmann, T. Güneysu. (2014). Beyond ECDSA and RSA: Lattice-based digital signatures on constrained devices, *51st ACM/EDAC/IEEE Design Automation Conference (DAC)*, San Francisco, CA, pp. 1–6.

D. Pointcheval, J. Stern. (1996). Security proofs for signature schemes, U. Maurer, editor, *Advances in Cryptology—EUROCRYPT'96*, volume 1070 Lecture Notes in Computer Science. Springer, Berlin, Heidelberg. doi:10.1007/3-540-68339-9_33.

D. Pointcheval, J. Stern. (2000). Security arguments for digital signatures and blind signatures. *Journal of Cryptology*, 13:361. doi:10.1007/s001450010003.

R.L. Rivest, A. Shamir, L. Adleman. (February 1978). A method for obtaining digital signatures and public-key cryptosystems. *Communications of the ACM*, 21(2):120–126. doi:10.1145/359340.359342.

J. Rompel. (1990). One-way functions are necessary and sufficient for secure signatures. In *Proceedings of the Twenty-Second Annual ACM Symposium on Theory of Computing (STOC '90)*, Harriet Ortiz (Ed.). ACM, New York, pp. 387–394. doi:10.1145/100216.100269.

C.P. Schnorr. (1991). Efficient signature generation by smart cards. *Journal of Cryptology*, 4:161. doi:10.1007/BF00196725.

T.K. Sharma. (August 21, 2017). *What Is Merkle Tree & Merkle Root In Blockchain?* Blockchain-council article, retrieved from:https://www.blockchain-council.org/blockchain/what-is-merkel-tree-merkel-root-in-blockchain/.

M.M.J. Stevens. (June 19, 2012). *Attacks on Hash Functions and Applications*, Doctoral Thesis, Leiden University, http://hdl.handle.net/1887/19093.

S.A. Vanstone. (2003). Next generation security for wireless: elliptic curve cryptography. *Computers and Security*, 22(No. 5):412–415. doi:10.1016/S0167-4048(03)00507-8.

X. Yi. (February 19, 2003). An identity-based signature scheme from the Weil pairing. *IEEE Communications Letters*, 7(2):76–78, doi:10.1109/LCOMM.2002.808397.

16

Applications

M. A. Rizvi and Ifra Iqbal Khan

National Institute of Technical Teachers' Training and Research

CONTENTS

16.1 Introduction..274
16.2 Web Security...274
    16.2.1 Need of Security 275
    16.2.2 Threats to Web..275
    16.2.3 Measures for Achieving Web Security 276
16.3 Database Security 277
    16.3.1 Classification of Database Security 277
    16.3.2 Threats to Database 278
    16.3.3 Securing the Database..278
    16.3.4 Design to Security..279
16.4 Cloud Security..279
    16.4.1 Cloud and Its Deployment Models 280
    16.4.2 Threats to Data on Cloud 280
    16.4.3 Securing a Cloud...280
16.5 CAPTCHA 281
    16.5.1 Origin of CAPTCHA 281
    16.5.2 Working of CAPTCHA 282
    16.5.3 Types of CAPTCHA 282
    16.5.4 Using Hard AI Algorithms for CAPTCHA 284
16.6 E-mail Security...285
    16.6.1 Simple Mail Transfer Protocol 285
    16.6.2 Threats...285
    16.6.3 Attacks...286
    16.6.4 Measures to Attacks and Threats 286
16.7 SMS Security...287
    16.7.1 Encrypting Short Messages...287
    16.7.2 SMS Threats 288
    16.7.3 Security Measures 288
16.8 Lightweight Cryptography Applications in IoT, Sensor Networks,
    and Cyber-Physical Systems..289
    16.8.1 Lightweight Security Key Management System 289
    16.8.2 Threats...290
    16.8.3 Attacks...291
    16.8.4 Requirement for Security 291
16.9 Conclusion 291
References...292

16.1 Introduction

As the growing use of technology is encountered by this generation, more and more threats are being observed. To overcome the threats, different types of architectures, models, and schemes are being practiced by security professionals. Hindrances to security involve a vast domain of web services, e-mails, SMS, Cloud Databases, etc. In this chapter, we will look at some applications concerning the above-mentioned domains. Applications are the most commonly used software nowadays; they need some security measures to be taken by manufacturers and administrators. Organizations have encountered security threats and the measures to prevent them. In this chapter, we will talk about applications that will be concerning data or information on the web that are databases, cloud, e-mail, SMS, etc. Security measures that stop AI interference of bots on the Internet are known as CAPTCHA; it secures the sites from being encountered by automated URLs. CAPTCHA conducts the test which decides whether the user is human or not. Another application providing security is lightweight cryptography; it extends its use in technology-constraint environments such as RFID tags, sensors, contact lenses, smart cards, and healthcare devices.

Web security issues with the security of the user's system and his or her accounts on the net it should embrace online banking systems, net browsers, social networking, and so on. It primarily involves encoding that may be a secured key for authenticating users; this is often conjointly referred to as cybersecurity. Info security involves securing large knowledge as data processing and large knowledge is trending lately the mining of information by unauthorized users, its purpose of concern for the organizations. During this, the drawback is a lot of knowledge which is being provided than it's required. Security for it involves imposing access controls and regular Audits for the info manipulations and implementation of many models which is able to be effective in providing a lot of security. This is often conjointly referred to as laptop security, info security, and risk management. Cloud security involves securing the cloud because it contains info which might be accessed from anyplace notwithstanding platform, system, and geographical locations and they're terribly bulk storage servers. Most organizations and firms have moved to the cloud. Threats involve knowledge leakages and account hijacking. Security measures deal with encoding technology and management of keys; this could be classified underneath identity and risk management issue.

E-mail Security is often thought to be being relied only on Simple Mail Transfer Protocol (SMTP); it provides the sure extent of security however not several options that issues stopping hackers and malicious users to send e-mails containing viruses. Mail system was the primitive methodology to speak on the Internet; it developed another reasonably stress of receiving spam and phishing e-mails because it is employed as a weapon for hackers and attackers. Thus, it's most significant to eliminate threats and improvise the measures to secure e-mail systems.

16.2 Web Security

The web is witnessing threats to security since the 1980s, and the Internet is most vulnerable by the malware. Most of the information is on the Internet from personal information on social networking to data stored in our national identities and bank credentials which is most sensitive of all. Government is making it more difficult by uploading data on unsecured web portals and making it online which is giving privileges to hackers to attain personal information of its citizens; more and more skilled workers are required in the field of security.

Web security concerns with the security of the user's system and his or her accounts on the Internet, which may include online banking systems, web browsers, and social networking. It mainly involves encryption which is a secured key for authenticating users, which is also known as cybersecurity. The need for securing data prevails at every level, including network infrastructure level, user level, and middleware (Webroot survey finds concern over Web 2.0 collaboration, 2009).

16.2.1 Need of Security

Need for web security felt because of sniffing attacks, cross-site scripting, SQL injection, request encoding, DOS attacks these are names which poses a security threat to the Internet. Web security is mostly needed in business, e-commerce, e-banking, and several web-based applications. Past year almost all systems of the world were hacked by a virus known as ransom-ware; it took over most of the computers and locked them all so that no one was able to access their own files; it was observed that only systems that were on the Internet got affected by this; it asked people to pay, if they wanted their systems to unlock. Our systems on the Internet are in definite need of security as integrity and confidentiality are basic rights which should be provided to the data on electronic devices (Webroot survey finds concern over Web 2.0 collaboration, 2009) (Figure 16.1).

16.2.2 Threats to Web

1. DOS attack: It is used for making a machine or network corrupt, by making it unavailable to the user, simply denying services to the user. That is debarring users from being able to login to the system and, on the other hand, taking all authority over the system.

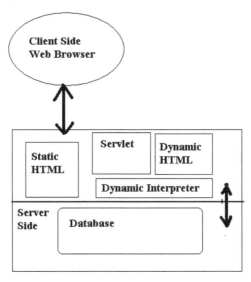

FIGURE 16.1
Web application.

2. Request encoding attack (XSS): This is known for inducing malicious viruses into a trustworthy link. The most important way to embed a virus into a system is through a trustable source. The malicious code gets into the system and validates file by allowing security breach.

3. SQL injection: This attack inserts an SQL code through an input bar on the web so as to get information about input sources.

4. Sniffing (request encoding): When data are transmitted over a network, these can be hacked by the third party; this is called sniffing attack. There are well-developed sniper networks, even college graduates experiment and develop new methods to intrigue safety, making Internet unsafe for genuine users. Most sniffing attacks can detect or extract our bank details and cases of online money theft being observed, without the owner being aware of it.

5. Directory traversal: It occurs when attackers hack the secured system and access the data from restricted file systems; they send HTTP request to bypass web and applications and servers.

6. Web-based phishing attacks: These attacks are mostly done through the URLs and links which attackers send through e-mails or social networks. Sometimes they even target a particular person or group so as to extract information. These types of attacks are so advanced these days that it may even copy the photographs and other details of users. Most vulnerable areas are banking sites, social networking, train, and airplane ticketing system.

7. Click-jacking attack: It is attack encountered on the web pages when we click a certain area on a web page which routes it to another page; it is a malicious button on a trusted website which gets the system locked and exposes it to malware.

16.2.3 Measures for Achieving Web Security

1. Vulnerability scanner: Most popularly known as web vulnerability scanner, it computes the vulnerability that what is the amount of threat present from the website. It scans and determines the weakness, loopholes (points which are easily intrigued by attackers/hackers) present which can get attacked easily. This tool is mainly devised for systems and internal network which are connected to the Internet so as to rule out all possibility of systems being threatened by rogue software (Lomte, 2012).

2. Web firewalls: The firewalls mostly provide protection against the known vulnerabilities and threats. These are available in both software and hardware forms. It has better filtering efficiency; it can inspect elements of data (like header and web-content) (Kallapur & Geetha, 2011).

3. Application-level gateways: This is another type of firewall which is more concerned with TCP modules. It is supported to authentication at the user level, and it is also capable of analyzing data packets with respect to the payload. One of its drawbacks is its slower process. When it comes to filtering, it becomes more slowly than packet filtering, also every time support to the connection cannot be achieved (Lomte, 2012).

4. Antivirus: These antiviruses, which we use like MacAfee Windows Essential and many more, provide a certain amount of immunity from threats. Nowadays, in all the systems, most of the users are installing them to safeguard their systems. In fact, all systems are on the Internet these days. So, it can be recommendable to

update it periodically, and it is the easiest method to protect our PCs from damage (Kallapur & Geetha, 2011).

5. Vulnerability patching: The patching technique is applied when we know the vulnerability of a particular attack. The attackers with prior knowledge of vulnerability can exploit the system and user. If patching is not done, then the system can get affected by viruses and malwares (Lomte, 2012).

6. Network firewalls: They protect the web from web application attacks. It only blocks attacks dealing with network components and also performs activities like blocking of Internet packets with respect to address and port number, and it monitors the network traffic.

7. Intrusion detection system: It helps in detecting the infected traffic. It also blocks the path and suggests ways to tackle the malware. One of its drawbacks is that false alarm can be generated, so there is non-existence of a fully accurate Intrusion Detection System (IDS) (Kallapur & Geetha, 2011).

8. Intrusion prevention system: These are network security components which monitor network and system activity, and blocks for malicious data packets. For further investigation, it reports the data that are prevented from entering into the system, but it is still not efficient enough to detect undisclosed vulnerabilities (Kallapur & Geetha, 2011).

16.3 Database Security

Data mining is related to the database. When it comes to an enormous amount of data, mining is an add-on advantage of databases. Another aspect related to the database is the sharing of information; databases can be of any type and when mining accompanied by sharing can result in a breach of data security. The mining can impose the threat of data being used for a harmful purpose i.e. published data can be used undesirably. The managing and ordering of bulk data in a very large amount are known as mining. It provides facility to cluster or group the data so that sorting becomes easy. Its main use is to find information from big data clusters in that way it helps in detecting records. The mining techniques involve the use of dependencies, which is also known as pattern mining. The process of data mining involves the use of sorting algorithms. This was coined after people started taking advantage of mining for personal benefits rather than maintaining integrity in the system such as spoofing, fishing, and mining unnecessary information.

16.3.1 Classification of Database Security

Securing the database was quite difficult, but now after the advent of cloud, the databases started to have easy backups. It is classified into several parts like distributed and cloud data. The common thing is the host and location of the database (Denning, 1984). So, it is classified into physical and logical as follows (Malik & Patel, 2016):

1. Physical database security: Security is involved in recovery mechanism so as to revive the data, if its cloud servers are safe. The physical location of data plays an important role as it concerns for the location where data are physically present.

The locations at which data are preferred to be located must be less prone to natural calamities such as earthquakes and floods, which can damage the data-storing servers.

2. Logical and system issues: These issues exist in the operating systems, and security here is far more difficult to achieve. These issues consist of infection of malware or data theft. It can be a very hard task to secure data from unwanted access.

16.3.2 Threats to Database

1. Distributed systems: This may be due to the widespread use of the distributive system and distributed client/server architecture and database being exposed to the threat of unreliable networks. Internet and intranet increase the risks.

2. Excessive privileges: This may result in exploitation of data by which integrity hindered; for example, we can take of a banking system where the customers expect their data to be confidential, but if there is threat of data leak, the bank may lose its customers.

3. Not authentication of data: It poses another threat to database security. Flexibility in authentication may lead to data theft and transforming its credentials. If anyone gets access to data other than the original user, then it is fully capable of attacking other systems as well.

4. Loss of data integrity: Loss of integrity to data can result in corrupting of it; when an integrity breach is observed, it tends to disrupt the workflow causing delay. When integrity is lost, its confidentiality is also lost (Malik & Patel, 2016).

5. SQL injections: The information entered by user acts as data and is comprised of the database. SQL injections are SQL statements that are injected by attackers to input fallacious information (Malik & Patel, 2016).

6. Malware: The viruses impose a threat on all systems; if the database gets infected, then its access by authorized users is hindered, and they are not able to use the data, system, and files on the server.

16.3.3 Securing the Database

There is always a need to build a bolster system which provides a better architecture to support the security of the system. It is advisable that normalization should be done prior to use, and updating should be avoided at the time when data are fed for further use. The database can be secured by following certain security aspects while using bulk databases (Malik & Patel, 2016):

1. Limiting access: Providing access only to authorized and concerned users, i.e., restricting the parts of the database being visible to all public users. They can be categorized under private and protected. Rather, of the large sample, the person gets abridged data by which risk to whole data is reduced.

2. Fuzzing the data: Fuzz means to blur and add bit distortions. Introducing a little unreal data provides ambiguity to existing values, like changing some values or changing the set of values to keep original data secure and covert. It is the addition of misleading data to keep the probability of misusing data zero.

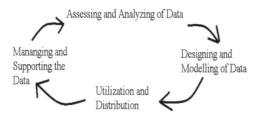

FIGURE 16.2
Design adopted for securing database.

3. Elimination of grouping: Data often provide an identification number to its users. Suppose that we are registering for a national ID and we get a notification through a unique number if that number has got groupings like first three digits representing the locality of the individual, this can lead to malicious mining of data by unwanted elements. This imposes a threat to information being secure. After considering the above example, by such information anyone can search user's locality leading to further more information being available to unauthorized users. The example given here resembles the social security number in the USA; they have first three numbers encoded as an office from which ID is issued.

4. Providing secure authentication: Normally, security involves authentication of data and identification of the authorized users. The methods which involved only passwords in the past, but now advancements in technology led to several other methods like biometric identification and analysis of signature, etc. Data documents that need authentication are signed by authentication authorities and then sent to desired recipients.

16.3.4 Design to Security

The design includes simple steps which can provide availability, authentication, integrity, and security. The four components here concern with an analysis of data like how prone to threats it is, how vulnerable it is, and adopting proper solutions for the same. After this, designing of security algorithms and incorporating them in the actual databases which provide security in order to maintain confidentiality. Then providing proper management and support of the data by proper and regular audits and checking on malware and weak points which serve as loopholes to be worked upon while data are being practically used. All these measures can be summed up in a simple flow diagram (Figure 16.2).

16.4 Cloud Security

Cloud security involves securing the cloud as it contains information which can be accessed from anywhere regardless of platform, system, and geographical locations; these are very bulk storage servers. Most organizations and companies have moved to the cloud. Threats involve data leakages, account hijacking, and others. Security measures deal with encryption technology and management of keys; this can be categorized under identity and risk management issue (Padhy & Patra, 2012).

16.4.1 Cloud and Its Deployment Models

Cloud is being used as the most reliable source for the data storage. It provides facility to retrieve or view for data files from anywhere and on any computing devices like mobiles, laptops, etc. Google Cloud, i.e., drive, provides cloud storage; it allows storage of data on the Internet. The cloud servers are placed in a safe place from any physical damage. These servers have more capacity than the usual servers (Padhy & Patra, 2012; Rao & Selvamani, 2015).

Cloud servers are in three forms, i.e., it can be used as Software as a Service, Platform as a Service, and Infrastructure as a Service. These three models are being adopted by corporate vendors for providing cloud services to their users (Rao & Selvamani, 2015).

1. Software as a Service (SaaS): These are application-based models, which can make a login and operate Internet and applications like Google, Facebook, etc.; it is provided by Application Service Providers (APS). For example, Google apps vendor manages the infrastructure so as to provide full maintenance.

2. Platform as a Service (PaaS): It provides services which include the platform that does not involve installation and other software configuration, such as Microsoft Azure and RedHat open-Shift, which provides services like J2EE and DotNet. It provides good connectivity and backup which is more favorable to developers.

3. Infrastructure as a Service (IaaS): It consists of the virtualization phenomenon, which consists of a virtual interface which helps in remotely working on any systems and controlling servers, which may include disk allocation and other related operations. It is mostly concerning the interaction with network devices like switches, host-routers, etc. All of these can be summed as sharing hardware components by making use of virtualization.

16.4.2 Threats to Data on Cloud

1. Authorization and access: Malicious users can breach security and get into a cloud environment.

2. Weak authentication: It can occur by not adopting robust methods top allowing access to data, by using weak passwords and primitive methods to allocate access to data to users.

3. Vulnerabilities and misconfiguration on network: As data is on cloud, 24 hours online and is exposed to all kinds threats which are present on a network like sniffing and Hacking, etc.

4. Exposing backup: As the cloud is mostly used as a backup, then it can simply be vulnerable to attacks and being exposed to the malicious user. This can happen while transferring data; it depends on how data are useful.

16.4.3 Securing a Cloud

Security issues include data transmission, integrity, network security, privacy, availability, location, and segregation. All these issues can be checked by security policies and compliance audits. The services on the cloud do not belong to the user itself; the data of the user are being saved on a third base. It does not provide any control over queries of what can happen to our data. Valuable information may also be threatened by any kind

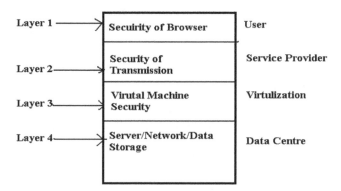

FIGURE 16.3
Different layers of security in Cloud Models.

of security breach. Therefore, a service-providing layer having components maintains security like service-level agreement monitoring, policy management, scheduler, and dispatcher (Figure 16.3).

Also, security issues like data integrity are handled by virtual machine layer, which defines regularity and legal conducts. Infrastructure maintenance involves identity and access management by data center layer which has memory storage of CPU and server. Sometimes it is referred to as IaaS as it provides security at the network layer.

All security issues have got a solution through data encryption like Kerberos authentication; data on the cloud can be safe by an encrypted key and similar mechanism followed for encryption can be RSA algorithm.

16.5 CAPTCHA

Due to increasing increased intervention of automation in industries, it is implicit that attacks related to robots have increased. A new threat troubling users is robot-calls which are spam and sometimes fraudulent. As a result, some humans have experienced loss related to life and business. CAPTCHA is a test which we come through when entering into a site as a security we have to input the correct string so that site can recognize as humans and not robots. CAPTCHA stands for "Completely Automated Public Turing test to tell Computers and Humans Apart". It was designed to stop the interference of vague URLs and bots into sites; in other words, CAPTCHA can be referenced as an application which provides security.

16.5.1 Origin of CAPTCHA

The origin of CAPTCHA started in 1997 when producers of AltaVista, a search engine, were looking for solutions to block automated URLs. Andrie Broder found an algorithm which uses an image that only humans can understand to pass the code. When humans recognize the correct letters and input it, they got a reply from the website in the form of text; if it was some robot, then it was not able to guess the CAPTCHA. Later in 2001,

Andrie got its technology patented. After 2 years, Luis Von Ahn, Manuel Blum, Nicholas Happer of CMU, and John Langford came up with the flawless algorithm and termed it CAPTCHA.

16.5.2 Working of CAPTCHA

CAPTCHA is nothing but a Turing test, which was developed by Alan Turing (1950); it tested machine's ability to show intelligent behavior like that of human beings. Turing's philosophy was to test that a machine can think like humans or not. This gave birth to a new judgmental algorithm which tested humans and computer apart. It is a simple task which involves users. A JPEG or GIF image is displayed before directing us to a particular web page asking a human to identify it; this image is not able to be read by robots because they can only read the source code, not the image. For making it easier to recognize the letters, the option to request a new CAPTCHA is always there. It is the test which humans come across when every time they log in to an Internet website. It is a kind of challenge-response test in which we had to guess the blurred and twisted combination of alphabets and numbers. After providing input humans can proceed further to view the content on the website. Nowadays, new CAPTCHA is to distinguish between human and bots (robots). It is an acronym for "Completely Automated Public Turing Test to tell Computers and Humans Apart". The administer of this test is a computer. Though it provided security, it made everyday working slow, but it was required so that anyone cannot interfere into the website system (Soni & Bonde, 2017).

CAPTCHAs are saved on servers in the form of images and audio files and in HTML form. The web pages generate CAPTCHA dynamically in a submit form type; it can be of any programming language like PHP, JAVA, Ruby, JSP, etc. Now the integration of CAPTCHA takes place with the use of a secret key, i.e., common to web servers and CAPTCHA server. Also, a string is generated within the link, which eventually generates a password, which is computed by CAPTCHA to display the image and website server here validates the user's input. The CAPTCHAs that are available to millions of users are due to the backup servers (Soni & Bonde, 2017).

16.5.3 Types of CAPTCHA

For ease of humans and making it quiet time efficient, programmers came up with several types of CAPTCHAs that can be elaborated as follows (Gafni & Nagar, 2016; Mankhair, Raut, Mohimkar, Sukal, & Khedekar, 2016).

1. TEXT CAPTCHA: It is most widely and commonly used, i.e. re-CAPTCHA, where the letter and number series is shown in a distorted manner. Mostly, the shape of letters and numbers are whimsical. The user needs to decode it by identifying correct alphabets and numbers to get access to that particular web page. The pattern observed is a combination of number and alphabets, i.e., an alphanumeric string (Figure 16.4).

2. Audio-based CAPTCHA: In this form, the text message is saved in the form of an audio signal, specially developed for vision-impaired people. User is required to listen to alphanumeric sequence and fill it in the dialog box given.

3. No CAPTCHA re-CAPTCHA: It was started by Google in the year 2013 which conducted a pre-analysis of the behavior of the user's browser in regard to CAPTCHA

(a)

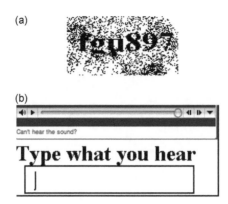

(b)

FIGURE 16.4
(a) Example of simple CAPTCHA. (b) Audio CAPTCHA.

whether to tell that user is human or machine. This service was used when there was no reason for a user to be a robot; later in 2014, it was not used for Google services which were for public use (Figure 16.5).

4. Arithmetic operations CAPTCHA: It demands the user to solve a simple mathematical query or expression, for example, the CAPTCHA Displayed will be $1 + 4 =$ and the user has to solve and as input must write numeral 5 in top the text box given in order to get CAPTCHA right (Figure 16.6).

5. Picture-based CAPTCHA: In this test, several pictures are displayed which ask users to select a particular type of image or categorize them into one field of or example; a series of roads and restaurants are displayed in a single frame and CAPTCHA demands to select the picture that displays only roads. Some other tests require us to adjust a given bar so as to correct the distorted image, which solves as clear image to pass CAPTCHA. As shown in Figure 16.7, it may ask to click on images with hills.

6. Game-based CAPTCHA: It is a tedious but a rational task which involves user to perform activities, for example, puzzles and interactive game. There are several types of research going on user experiences on such games; it is not much used these days but has scope in near future (Figure 16.8).

FIGURE 16.5
Re-CAPTCHA by Google.

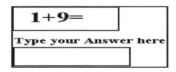

FIGURE 16.6
Arithmetic CAPTCHA.

FIGURE 16.7
Image-based CAPTCHA.

FIGURE 16.8
Game-based CAPTCHA.

16.5.4 Using Hard AI Algorithms for CAPTCHA

It is possible to make a program which can work like the human brain, known as AI; engineers deploy it to behave as human, and so there is a possibility that an intelligent computer can solve problems which humans can solve. Inventors of CAPTCHA, Von, Blum, Hopper, and Langford, have taken an approach which incorporates techniques similar to that of cryptography. It helped in reducing the time taken to solve a problem. It is being concerned that AI being widespread, if used by malicious users, can affect the protocol, i.e., CAPTCHA. The solution is being stirred up with an analysis of modern cryptographic practice.The assumptions made for security are that all problems that are hard will eventually be solved as a P not equal to NP; if a user solves it once, then it can solve it twice and as expectedly for the third time as well. This method is used because it helps in identifying the gap between human and computer (Ahn, Blum, Hopper, & Langford, 2003).

The CAPTCHA involving hard AI must be automated so that every produced test must not be solved by robots. For example, a Turing test owing a secret textbook picks a random paragraph as a test and asks the user whether it makes sense or not. Thus, we can conclude that robots cannot recognize whether line makes sense or not; it is quite a tough test for humans too. These tests will surely be able to distinguish humans from computers (Ahn, Blum, Hopper, & Langford, 2003).

16.6 E-mail Security

E-mail security can be regarded as being relied solely on SMTP. It provides the certain extent of security but not many features which concerns stopping hackers and malicious users to send e-mails containing viruses. Mail system was the primitive method to communicate on the net; it developed another kind of stress of receiving spam and phishing e-mails as it is used as a weapon for hackers and attackers, so it is most important to eliminate threats and ameliorate the measures to secure e-mail systems (Banday, 2011).

16.6.1 Simple Mail Transfer Protocol

It is the protocol which was originally designed for communication happening between small groups. Initially, no security was provided, but after experiencing some breaches, the changes were incorporated in protocols, accompanied by technological advancements (Banday, 2011) (Figure 16.9).

The changes included reducing the size of the e-mail, filtering, verification of header, machine learning for filtering SPAM, and scrupulous mails. SMTP is regarded as the standard protocol used for sending and receiving e-mails. Some security issues faced by SMTP are security and integrity. Since data of mail are not decrypted, it is present in the plaintext form (Jin, Takabi, & Joshi, 2010).

16.6.2 Threats

1. SPAM: It is also called as junk. Such e-mails mostly contain viruses and malicious links. Sometimes just by opening the mail, it can hack the entire system. Users are unaware of what a mail contains; it is recommended to not open such e-mails which are from unknown recipients. Many steps have been taken for filtering such e-mails which filter suspicious senders' mail getting into the inbox (Kamthe & Nalawade, 2015).

2. Viruses: Virus is the high-risk destroyer that can enter into the system through e-mails. Some virus example can be of the pop-up e-mails on our screen; now, it is rare to see pop-up e-mails. The threat hovering over our e-mails can be injected into our known recipient e-mails and also through our SPAM e-mails.

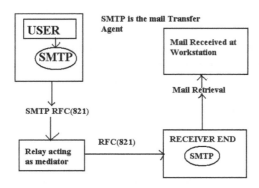

FIGURE 16.9
E-mail system incorporating SMTP.

This has been a long-time issue, and humans are more aware now; they are cautious while opening any e-mails and debar themselves from clicking any image in the mails. Viruses are long-term destroyer; they affect data immediately, which makes it tedious and hard to recover the lost data. Also, more labor and money are required to recover the loss by viruses as they are not easily removed by antiviruses (Kamthe & Nalawade, 2015).

16.6.3 Attacks

1. Passive attacks: The attackers make use of data that they observe, as for simple systems that they have plain text. The attackers intervene in the system making the mail available to the attacker. They use such techniques to send mail to unintended receiver—an example of such attack is "Man-in-the-Middle Attack".

2. Active attack: This kind of attacks causes interruption. Attackers debar the sender from accessing the site, and a modification of the content can be done. Most changes observed were during transmission of the mails. Without any authentication, data can be fabricated including the creation of false accounts.

3. DOS attacks: This attack can easily be launched on the network; it is so powerful that it can shut down the whole network system. By such an attack, the computer gets infected without the user being aware. This attack mainly checks availability to send a request on the network so as to attack bandwidth and TCP connection, CPU cycle being the main part of the system gets affected by this attack (Kamthe & Nalawade, 2015).

16.6.4 Measures to Attacks and Threats

Talking of current times, all sorts of remedies are provided but privileged due to less understanding of backend processes by a layman. Many companies just release software for security without much integrity toward users. Many free software are available, but in reality, they do not provide 100% solution to virus problem. Some frameworks and methods that are currently being used are discussed below.

1. It was developed for authentication of sender. Meng Wing was the developer behind the framework. In this the sender's domain is recorded, that get checked by SPF (address) that from which domain machine it came, then IP address is matched for further authentication. If it doesn't match, then the mail is declared as SPAM.

2. Encryption: It is a method of preventing the intended user from gaining the information in the middle. The key is provided to the receiver from a sender which authenticates the data. An example can be taken of Kerberos authentication. In e-mails, while transmitting the data, various techniques are used such as HTTPs and SHTTP. To prevent sniffing attacks, encryption is used with the virtual private network (VPN); it detects the malware and viruses, and debars them from getting into the network through the use of a firewall.

3. Measures of DOS attack: Systems such as intrusion detection, enhanced routers, and firewalls are used as a defense against such attacks. Generally, the analysis of incoming and outgoing traffic is used for protection at TCP level which detects the traffic at baseline; for example, attack such as TCP SYN flooding can be prevented

only if the system is protected by its own private individual firewall. It is like working on real-time evaluation. So, as a measure we can use back-servers and involve security in routers; it is observed that by resolving heavy traffic of attacks, we can prevent system from these attacks.

4. Abolishing phishing and SPAM: To avoid phishing and SPAM, the most common method is to blocking. Nowadays, we can label mail as SPAM which acts as an exemplar to block the same kind of other e-mails which may pose threat to the user, by use of this mail directly go to spam mail list of our inbox.

16.7 SMS Security

Short Message Services (SMS) are most widely used nowadays. Chat applications came into existence, but before that, the SMS was the fastest and cheapest way to security to the devices and no valid identification of the user. Many SMS contains malicious and suspicious links which may lead to corrupting of devices. The protocols of SMS are laid ETS ITS 03.48 which provide a key identifier, SPI (Security Parameter Index), and integrity values. Digital signatures and redundancy checks are used for authentication and validation. SMS security has been a concern for Telecom companies and most probably for the mobile users as attacks related to snooping; flooding, faking, SMS interception, and modification are most common. The SMS gave rise to messengers in which Trojans and several viruses can sneak into our systems easily. Its security can be achieved by encryption and user authentication to make it safe for users (Reaves et al., 2016) (Figure 16.10).

16.7.1 Encrypting Short Messages

Encryption is the process of allocating security keys to the sender and the receiver. It uses an algorithm to provide accessibility of the message to the person who is having the key to decrypt the message. The advantage of encryption is to provide security to the message in the system and also when it is getting transmitted in the network, as physically securing the whole network is difficult. The term encryption emerges from the Greek word "Kryptos" meaning hidden or secret. Thus, it is clear from the above meaning that encryption hides the data; it changes data into ciphertext from that is only understood by

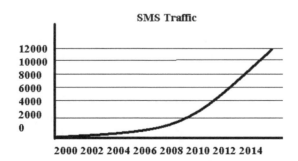

FIGURE 16.10
Graph to depict increasing SMS traffic.

authorized users. Encryptions primary use involves protection of the message's confidentiality. It saves data being interfered as pseudorandom encryption key generated by an algorithm which only can decrypt the message. The first use of encryption was observed in the form of RSA and Diffie–Hellman Key algorithm. Protocols such as SSH, S/MIME, and SSL/TLS are used to protect data, and also Digital Rights Management protects data using the concept of encryption. There are algorithms which are being used for SMS encryption; they are devised according to the operating system of the device. We have named some encryption algorithms present like XTR, IDEA, Blowfish, RSA, DES, MD5, and SHA. It provides a different amount of security levels. RSA and DES are popular algorithms; XTR is derived from XTR cryptosystem, and using this technique for encryption is the fastest way to secure message transmission. XTR is another form of ECSTR system, i.e., efficient and compact subgroup trace representation. It uses discrete logarithmic computation so as to provide fast processing of the system (Nanda & Awasthi, 2012; Reaves et al., 2016).

16.7.2 SMS Threats

1. Message availability: As encryption is not present by default, they can be read or hacked by malicious users during the flow in the network. They are usually saved as plain text that can be easily viewed in the exchange servers by persons who have access to messaging systems. Programs that are used for snipping and spying can automatically record all messages. Also, they can be saved on external/remote servers for further studies.

2. Spamming: Mostly, advertisement messages are sent in bulk. These messages have data that can harm the user's device system. The facility of bulk messages allows spams to circulate in the network, waiting to attack.

3. Flooding: This is sending bulk messages on a particular phone; for example, we can assume that 200 messages being sent per second will disrupt or clog the mobiles or system. It can even make the device inaccessible; these attacks exploit the SMS protocol.

4. SMS phone crashers: Sending such messages results in total data loss or phone being corrupted. These attacks can hang and infect the cell not allowing the user to control. It can also result in the hacking of the phone so that the phone becomes inactive.

5. SMS viruses: As the technological changes are toward the advancing path, more threats are being detected and reported as robust as the previous versions of them. They have got the tendency to install themselves in just one click on the message which leads to malfunction of mobile phones.

6. SMS phishing: Users are being led astray by false messages, tricking users into downloading, and activating services which are malicious in nature. The text inside the message directs the screen to the unknown website; sometimes the process is so fast that the user is not able to recognize what happens to the system.

16.7.3 Security Measures

1. Transmission: The message can be protected while sending through the Internet by activating secure Socket Layer (SSL) for mobile platform encryption is recommended in order to stop the snooping messages, so as to provide end to end security for messaging. Crypto SMS uses a strong encryption algorithm to help protect against threats.

2. Storing the message: While the phone is exposed to unknown users, it is advisable to keep the contact list and personal data being safe and protected according to privacy protocols.

3. Authenticating the user: SSL provides secure transmission through login ID or User ID accompanied with the password and will increase the security aspect in sending messages. It will be easier for service providers to provide end-to-end encryption after users get authenticated.

4. Protection of systems: For sending messages, it is advisable not to use public Internet terminals, as it can cause DOS attacks to spread through unattended systems.

16.8 Lightweight Cryptography Applications in IoT, Sensor Networks, and Cyber-Physical Systems

Internet of Things (IoT) and Cyber-Physical Systems (CPS) are two newly introduced scenarios which play an important role for communication among wireless devices. IoT refers to the interactions happening between RFID systems sensors and smart devices. These all are categorized under IoT. These are also most vulnerable to attacks; threat is always observed for data integrity. The new emerging small-sized devices such as RFIDs and small sensor networks working under environments confined under different kinds of constraints include memory spaces, energy requirements, and networks. The systems connected to small networks known as Wireless Sensor Networks (WSN); they need shorter code size algorithms which take less memory space. This is the reason they are named as lightweight. A lightweight security key is generated by using algorithms made to work efficiently in the constrained environments. Examples of such protocols are TSL and IPsec. They provide security and efficiency trade-offs. It also uses symmetric keys for an end-to-end communication as they are applicable low-resource devices of smaller footprints. Lightweight cryptography applications in IoT, Sensor Networks, and CPS refer to the requirements in security which leads to the system in which lightweight security key is used; the systems in which such kind of algorithms used are RFIDs WSNs and CPS. We will look at lightweight cryptography, its threats, how it can be made secure, what is the mechanism that lightweight security key generation follows.

Cyber-Physical Systems—It is a control and monitoring mechanism with the computer-based algorithm which is closely integrated with the Internet and its usage. These are built by integration of physical components that are deeply inter-wind; they work on different spatial and temporal scales. They show a modular behavior and interact using n number of ways that can change according to usage. Examples of such systems are a smart grid, autonomous, automobile systems, medical monitoring, process control systems, robotic systems, and automatic pilot avionics. They process properties such as adaptability, scalability, resilience, security, and usability that provide a foundation to the infrastructural systems which will emerge smart services for future.

16.8.1 Lightweight Security Key Management System

In this system, several nodes are considered first. All nodes are chosen by a self-running algorithm by which they divide themselves among service node, alternative service node,

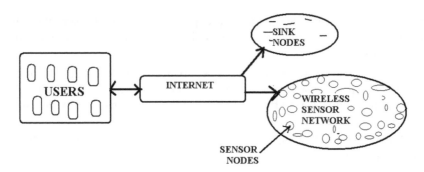

FIGURE 16.11
IoT nodes interaction.

and worker node. After this selection process, service nodes generate space for keys for worker nodes. Second, worker nodes and the associated service node establish a secure network between them. Now, these will have a shared key for equivalent service nodes. As IoT consists of heterogeneous networks, sensor is a categorization tool for sensing the devices and networks to which a particular device is exposed to. When we take a bulky network with smart devices connected to it, that is distributed randomly; each node plays the role of the sensor. It was Rabins Cryptosystem to ensure information security of the key. Key is generated pairwise using cryptography that is polynomial and matrix-based cryptography. Each node is assumed to have Ipv6 address UID which is unique in nature. In this system, each node is expected to select itself as a service node before being fed into the wireless network. These are self-configuration algorithms—here the connection is established between the service node and worker node which uses the primary numbering scheme which uses the cryptographical system to generate the key. After this, with the help of UID allocation, a secure connection is established in the lightweight key management system. IoT comprises a small wireless device that is connected to the Internet, they are popular and widely used technical products. The main reasons for being commonly utilized are low power, high dependability, easy to mobilize, need of low storage space, and large number of masses connected through the IoT. This makes devices more stressful and vulnerable with security threats present. The most concerned topic is the device's security, such as weak passwords, hacked accounts, and no or less security updates. Application of IoT involves automation, security devices, wired and wireless sensors, self-driving cars, smart homes, and similar products which always stay on network or online always prone to the threat of being corrupted. These systems are connected through thousands of sensors to cover a wider area range which works with the use of the battery. Sensors are exposed to sporadic attacks; they have such a mechanism where encryption is confined, and the complex cryptographic algorithm will not be enough to maintain pace among the IoT devices. Therefore, it is required to adopt lightweight cryptographic system (Chen, Chang, Sun, Jia, & Wang, 2012) (Figure 16.11).

16.8.2 Threats

1. Nodes intersection: The interaction between nodes from inside the WSN and outsider nodes can lead to anomalous behavior.

2. Interface insecurity: The web portals to which our devices are exposed to pose a great threat of cross-site scripting, credential exposure, SQL Injection, and weak

account lockout settings. These can be saved by strong password policy and by introducing three to four account lockouts.

3. Authentication and authorization: The authority to the user must be checked; it should not provide unnecessary access to everything.

4. Unsecured network: It may result in unauthorized giving of privileges to the user to extract information or data while on the network. these attacks include DOS, DOS via network fuzzing, and Buffer Overflow.

16.8.3 Attacks

1. Passive attacks: This means monitoring of packets of packets as they get exchanged over WSNs; it may give birth to eavesdropping attack, so a close monitoring is recommended.

2. Active attacks: Sometimes false data stream can interrupt the flow of data resulting in interference and also existing data stream gets modified sometimes.

3. The attack on node: The threat can be posed by nodes which are similar to network nodes and on attack they can consist of a more robust and powerful mechanism attack the WSN.

16.8.4 Requirement for Security

1. Network availability: Even in adverse condition of DOS attacks, the network overflow must be available so as to ensure uninterrupted transmission.

2. Authorization: The sensors which are authorized by known sources will only be allowed in the network System to interact in CPS.

3. Authenticity: Network nodes must be sure of its truthfulness so as to prevent any unidentified and scrupulous node to interfere in the network system.

4. Encryption: Mechanism of the well-designed algorithm must be followed which cannot be deciphered by anyone other than the intended user; in other words, the data must be confidential.

5. Integrity: It ensures that the data are untouched or unmodified by the third unknown malicious user via the intermediate nodes.

6. Non-repudiation: Node should be able to send data which it had earlier transmitted that is non-denial of sending a message which it had previously sent (Tausif, Ferzund, Jabbar, & Shahzadi, 2017).

16.9 Conclusion

Security is the feature which needs keen attention in every field. The computer applications that are used widely in the world are under unknown attacks from the hackers and malicious users. There are different kinds of threats in different kinds of applications such as CPS, cloud, e-mail, SMS, Internet, and databases. Anything which is online or confidential, i.e., on servers, is under the hovering concerns of security. Hence, the security measures discussed in the chapter deal with the type of security issues including the threats and measures to deal with the threats. The security measures suggested in the chapter use artificial intelligence and lightweight cryptography systems as the recent advancements being incorporated for the system and network security.

References

Ahn, L. V., Blum, M., Hopper, N. J., & Langford, J. (2003). *CAPTCHA: Using Hard AI Problems for Security.* Lecture Notes in Computer Science. Advances in Cryptology—EUROCRYPT 2003, Springer, Berlin, Heidelberg.

Banday, M. T. (2011). Effectiveness and limitations of e-mail security protocols. *International Journal of Distributed and Parallel Systems*, 2(3), 38–49. doi:10.5121/ijdps.2011.2304.

Chen, D., Chang, G., Sun, D., Jia, J., & Wang, X. (2012). Lightweight key management scheme to enhance the security of internet of things. *International Journal of Wireless and Mobile Computing*, 5(2), 191. doi:10.1504/ijwmc.2012.046773.

Denning, D. E. (1984). Cryptographic checksums for multilevel database security. *1984 IEEE Symposium on Security and Privacy.* doi:10.1109/sp.1984.10011.

Gafni, R., & Nagar, I. (2016). CAPTCHA—Security affecting user experience. *Proceedings of the 2016 InSITE Conference*, 13, 63–77. doi:10.28945/3469.

Jin, L., Takabi, H., & Joshi, J. B. (2010). Security and privacy risks of using e-mail address as an identity. *2010 IEEE Second International Conference on Social Computing.* doi:10.1109/socialcom.2010.134.

Kallapur, P. V., & Geetha, V. (2011). Web security: A survey of latest trends in security attacks. Lecture Notes in Electrical Engineering. *Advances in Computer, Communication, Control and Automation*, 405–415. doi:10.1007/978-3-642-25541-0_52.

Kamthe, P. S., & Nalawade, S. P. (2015). Email security: The challenges of network security. *International Journal on Recent and Innovation Trends in Computing and Communication*, 3(6), 3505–3509. Retrieved May 9, 2018, from https://ijritcc.com/index.php/ijritcc/article/download/4481/4481.

Lomte, V. M. (2012). A secure web application: E-tracking system. *International Journal of UbiComp*, 3(4), 1–18. doi:10.5121/iju.2012.3401.

Malik, M., & Patel, T. (2016). Database security—Attacks and control methods. *International Journal of Information Sciences and Techniques*, 6(1/2), 175–183. doi:10.5121/ijist.2016.6218.

Mankhair, S., Raut, A., Mohimkar, M., Sukal, K., & Khedekar, A. (2016). Secured CAPTCHA password verification using visual cryptography. *International Journal of Engineering Science and Computing*, 6(5), 5247–5251. doi:10.4010/2016.1286.

Nanda, A. K., & Awasthi, L. K. (2012). XTR cryptosystem for SMS security. *International Journal of Engineering and Technology*, 4(6), 836–839. doi:10.7763/ijet.2012.v4.495.

Padhy, R. P., & Patra, M. R. (2012). Evolution of cloud computing and enabling technologies. *International Journal of Cloud Computing and Services Science (IJ-CLOSER)*, 1(4), 182. doi:10.11591/closer.v1i4.1216.

Rao, R. V., & Selvamani, K. (2015). Data security challenges and its solutions in cloud computing. *Procedia Computer Science*, 48, 204–209. doi:10.1016/j.procs.2015.04.171.

Reaves, B., Scaife, N., Tian, D., Blue, L., Traynor, P., & Butler, K. R. (2016). Sending out an SMS: Characterizing the security of the SMS ecosystem with public gateways. *IEEE Symposium on Security and Privacy (SP)*, 339–356. doi:10.1109/sp.2016.28.

Soni, S., & Bonde, P. (2017). CAPTCHA: A security review. *International Journal of Computer Applications*, 168(4), 44–47. doi:10.5120/ijca2017914377.

Tausif, M., Ferzund, J., Jabbar, S., & Shahzadi, R. (2017). Towards designing efficient lightweight ciphers for internet of things. *KSII Transactions on Internet and Information Systems*, 11(8). doi:10.3837/tiis.2017.08.014.

Webroot survey finds concern over Web 2.0 collaboration. (2009). Infosecurity, 6(4), 10. doi:10.1016/s1754-4548(09)70074-9.

17

Hands-On "SageMath"

Uma N. Dulhare

Muffakham Jah College of Engineering and Technology

Khaleel Ahmad

Maulana Azad National Urdu University

CONTENTS

17.1   Introduction 293
17.2   Installation 294
17.3   Programming Fundamentals 296
     17.3.1   Objects, Values, and Expressions 296
     17.3.2   Operators 297
     17.3.3   Operator Precedence 297
     17.3.4   Variables 297
     17.3.5   Print Statement 297
     17.3.6   Comments 298
     17.3.7   if Statement 298
     17.3.8   Functions of the Statement That Is Used to Define a Function Is
           Called Def and Its Syntax 301
17.4   Interfaces 301
17.5   Plotting 303
     17.5.1   Two-Dimensional Plots 303
     17.5.2   Three-Dimensional Plots 307
17.6   Conclusion 307
References 308

17.1 Introduction

The main aim of the sage software is to experiment and explore the knowledge with mathematical concepts and related applications in a fast way. It also makes an interactive experiment with mathematical objects. Nowadays Sage can run on various other platforms which have a connection with the internet, such as MP3 players, cellular phones (Sang-Gu Lee, 2012). SageMath is built on many existing packages such as matpoltib, Numpy, and highly optimized mature software, viz. GMP, NTL, etc (Shravan, 2017). SageMath software is more popular due to the following points (http://doc.sagemath.org/):

- Easy to compile: It provides more flexible and easy to compile windows, Linux, and OS X users.

- Cooperation: It provides robust interfaces to other computer algebra systems, including GAP, PARI, Singular, KASH, Maxima, Maple, Magma, and Mathematica.
- **Extensible**: It can be used to define derive from built-in types or new data types and also use code written in various languages.
- **User-friendly**: It is easy to understand the object functionality, source code, view documentation, etc.

Sage uses the well-known and powerful language Python for new code and makes high-quality free software packages:

GAP: Groups, Algorithms, Programming

Maxima: general purpose Computer Algebra System

Pari/GP: Number Theory Calculator

R: Statistical computing

Singular: fast commutative and non-commutative algebra (http://doc.sagemath.org/).

As well as: ATLAS, BLAS, Bzip2, Cddlib, Common Lisp, CVXOPT, Cython, mwrank, F2c, Flint, FpLLL, FreeType, G95, GD, Genus2reduction, Gfan, Givaro, GMP, GMP-ECM, GNU TLS, GSL, JsMath, IML, IPython, LAPACK, Lcalc, Libgcrypt, Libgpg-error, Linbox, M4RI, Matplotlib, Mercurial, MoinMoin Wiki, MPFI, MPFR, ECLib, NetworkX, NTL, Numpy, OpenCDK, PALP, Pexpect, PNG, PolyBoRi, PyCrypto, Python, Qd, Readline, Rpy, Scipy, Scons, SQLite, Sympow, Symmetrica, Sympy, mpmath, Tachyon, Termcap, Twisted, Weave, Zlib, ZODB

17.2 Installation

Sage can run on Linux, Mac, and Windows. The following steps can be used to install the SageMath:

1. Based on your OS version, download the pre-built binary from **http://www.sagemath.org/ download.html**
2. untar the compressed binary once you downloaded
3. By running ./sage you could.

From the extracted folder, you could launch sage by running ./sage. The sage interactive shall get launched. Sage shell is a python like a shell and almost every python command can be run here.

The VMware Virtual Machine Distribution of SAGE Step by step instruction for easy installation

1. Download sage-vmware-x.y.z.zip from the website "https://download.cnet.com/Sagemath/3000-2053_4-10967002.html"
2. Extract it on a system. It prompts a window, where we need to specify the destination, where the file has to be extracted. You can search the destination folders using the "Browse" tab. Once you specify the older, click on "Extract" button at the bottom of the page.

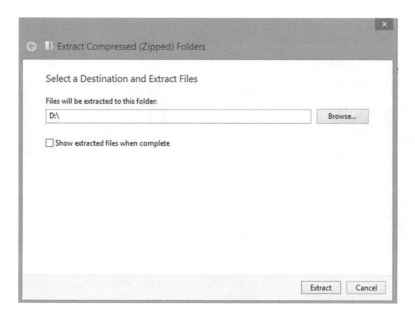

3. Make sure you have installed the free VMware program: www.vmware.com/products/player/

4. Double click on sage_vmx in the file directory, where you have extracted the file to run SAGE.

5. The window of the VMware station appears.

6. Click on SAGE, which opens into a sage interactive shell, where we are supposed to give our credentials.

7. After you give the login id and password, you can use the powerful sage.

8. You can also use the web version of SAGE, from your browser. Type notebook() in the command prompt.

Successfully installation process will be completed by using the above steps (Table 17.1).

17.3 Programming Fundamentals

17.3.1 Objects, Values, and Expressions

SAGE itself is built with objects and the data that SAGE program processes are also represented as objects. Type () command tell us what the type of the objects (Kosan, 2008).

TABLE 17.1

Packages Included with Sage-4.4.2 (Eröcal & Stein, 2010)

atlas	gap	libgcrypt	palp	Scipy_sandbox
Has	gd	libgpg_error	pari	scons
boehm_gc	gdmodule	libm4ri	pexpect	setuptools
boost	genus2reduction	libpng	pil	singular
cddlib	gfan	linbox	polybori	sphinx
cliquer	ghmm	matplotlib	pycrypto	sqlalchemy
cvxopt	givaro	maxima	pygments	sqlite
cython	gnutls	mercurial	pynac	symmetrica
docutils	gsl	moin	python	sympow
ecl	iconv	mpfi	python_gnutls	sympy
eclib	im1	mpfr	r	tachyon
ecm	ipython	mpir	ratpoints	termcap
f2c	jinja	mpmath	readline	twisted
flint	jinja2	networkx	rubiks	weave
flintqs	lapack	ntl	sagenb	zlib
fortran	lcalc	numpy	sagetex	zn_poly
freetype	libfplll	opencdk	scipy	zodb3

17.3.2 Operators

The characters +, −, *, /, ^ are called operators, and it gives the information about SAGE what operations to perform on the objects in the given expression.

17.3.3 Operator Precedence

SAGE uses a set of rules called operator precedence to determine the order in which the operators are applied to the objects in the expression

^ Exponents use for evaluating from right to left.

*, %, / multiplication, remainder, and division operations are evaluated left to right.

+, − addition and subtraction are evaluated from left to right.

SAGE evaluates the operations in this expression according to the precedence rules:

6 + 9*24/18 − 2^3	^ operator first so we get 8
6 + 9*24/18 − 8	* operator is executed next we get 216
6 + 216/18 − 8	/ operator is executed and get 12
6 + 12 − 8	+ & then—result we get 10
18 − 8	
10	

17.3.4 Variables

When the expression returns an object, the object is assigned to the variable. Create variables in SAGE through the assignment. SAGE variables are also case sensitive.

Example: A = 5 a variable called A is created and assigned the number 7.

If you want to see the contents of a variable, type its name into a blank cell and then evaluate the cell:

A

|

7

Variables that are created in a given cell in a worksheet are also available to the other cells in a worksheet. Variables can be saved before a worksheet is closed. This means that SAGE takes into account the case of each letter in a variable name when it is deciding if two or more variable names are the same variable or not. For example, the variable name B and the variable name b are not the same variables because the first variable name starts with an upper case "B" and the second variable name starts with a lower case "b" (Kosan, 2008).

17.3.5 Print Statement

SAGE allows the results of expressions to be displayed regardless of where they are located in the cell by print statement.

Strings string object consists of text which is enclosed within either double quotes or single quotes.

17.3.6 Comments

SAGE allows comments to be added to source code by placing a pound sign '#' to the left of any text or for multiple-line comments use in a set of triple quotes for enclosing the text

> Conditional Operators y == z,y <>z,y != z,y < z,y <= z,y > z,y >= z
>
> y = 5
>
> z = 5
>
> print y, "==", z, ":", y == z
>
> |
>
>   5 == 5: True

17.3.7 if Statement

In SAGE, an object is "true" if it is nonzero or nonempty and it is "false" if it is zero or empty. A simplified syntax specification for the if statement is as follows:

> if <expression>:
>
> <statement>
>
> <statement>
>
> <statement>
>
> .
>
> .
>
>
> Example
>
> x = 4
>
> if x < 5:
>
>   print x
>
>   print "Smaller"
>
> |
>
>   4
>
>   Smaller
>
> Looping with while Statement
>
> The syntax specification for the while statement is as follows:
>
> while <expression>:
>
> <statement>
>
> <statement>
>
> <statement>
>
> .

The following example program uses a while loop to print the Fibonacci series:

```
sage:
sage: a,b= 0,1
sage: while b< 10:
....:       print b
....:       a,b = b, a+b
....:
1
1
2
3
5
8
sage:
```

Looping with for statement

for <target> in <object>:
<statement>
<statement>
<statement>
.
.

Example 1. Display the array elements

for x in [20,21,22,23]:
        print x
|
        20
        21
        22
        23

Example 2. Matrix multiplication if [A] = ({[2,2,5],[1,4,1],[6,3,8]})
    And [B] = ({[6,2,1],[9,4,4],[1,2,4]})

```
sage:
sage:
sage: A= matrix ([[2,2,5], [1,4,1], [6,3,8]])
sage: A

[2 2 5]
[1 4 1]
[6 3 8]
sage: B= matrix ([[6,2,1], [9,4,4], [1,2,4]])
sage: B

[6 2 1]
[9 4 4]
[1 2 4]
sage: A * B

[35 22 30]
[43 20 21]
[71 40 50]
sage:
```

Example 3. Program to find the determinants of a matrix:

```
sage:
sage:
sage: A= matrix ([[2,2,5], [1,4,1], [6,3,0]])
sage: A

[2  2  5]
[1  4  1]
[6  3  0]
sage: B= matrix ([[6,2,1], [9,4,4], [1,2,4]])
sage: B

[6  2  1]
[9  4  4]
[1  2  4]
sage: A * B

[35  22  30]
[43  20  21]
[71  40  50]
sage: A.determinant()
-51
sage: B.determinant()
-2
sage:
```

Example 4. Program to print the first 20 whole numbers:

```
sage:
sage: i=0
sage: while i< 20:
....:     print(i)
....:     i +=1
....:
0
1
2
3
4
5
6
7
8
9
10
11
12
13
14
15
16
17
18
19
sage:
```

E xample 5. Program to print even numbers below 20:

```
19
sage: i=0
sage: while i < 20:
....:     if i%2 ==1:
....:         i +=1
....:         continue
....:     print(i)
....:     i+=1
....:
0
2
4
6
8
10
12
14
16
18
sage:
```

Example 6. Program to print Fibonacci series below 10:

```
sage:
sage: a,b= 0,1
sage: while b< 10:
....:     print b
....:     a,b = b, a+b
....:
1
1
2
3
5
8
sage:
```

17.3.8 Functions of the Statement That Is Used to Define a Function Is Called Def and Its Syntax

The specification is as follows:

```
def <function name>(arg1, arg2,... argN):
<statement>
.
.
.
def addnums(num1, num2, num3):
answer = num1 + num2 + num3
return answer
#Call the function and have it added.
a = addnums(2, 3, 4)
print a
|
9
```

17.4 Interfaces

Sage makes to use a wide range of mathematical software packages together by providing a unified interface which handles data conversion automatically. The interfaces can also run code from libraries written in the interpreted language of another program. Table 17.2 represents the list of interfaces provided by Sage (http://doc.sagemath.org/).

The corresponding software requires for each interface which is installed on the computer. The most common type of interface is called a **pexpect** interface which communicates with another command line program by reading and writing strings to a text console,

TABLE 17.2

Sage Interfaces to Mathematical Software

Pexpect	axiom, ecm, fricas, frobby, g2red, gp, Gnuplot, gap, gfan, kash, lisp, lie, magma, maple, macaulay2, mathematica, mwrank, maxima, mupad, Matlab, phc, polymake, povray, qepcad, qsieve, r, rubik, scilab, singular, tachyon, octave
C Library	eclib, fplll, gap (in progress), iml, linbox, maxima, ratpoints, r (via rpy2), singular, symmetrica
C Library arithmetic	int, mpir, pari, polybori, ntl, singular, pynac

as if another user was in front of the terminal. It is the default method of communication with some mathematics software, including commercial and open-source programs, viz. Maple (non-free), Mathematica (non-free), Magma (non-free), KASH, or GAP. Sage provides a framework to represent elements over these interfaces perform arithmetic with them or apply functions to the given object as well as using a file to pass the data if the string representation is too big, viz., % magma, % gap, etc.

If Z is an interface object, Z.interact() allows interacting with it. It is different from Z.console() and starts a completely new copy of whatever program Z interacts with it. By using sage(expression) Z, pull the data into an interactive session. Both console and interact methods of an interface do different things.

An example using gap:

1. gap.interact(): It is an easy way to interact with a running gap instance that may be "full of" Sage objects. Sage objects can import into this gap. Due to the console function, we get the actual program.
2. gap.console(): It is another program, e.g., gap/magma/gp Here Similar to bash, Sage is serving more than one convenient program.

Common interface functionality through pexpect

clear_prompts()—Returns the command used in this interface.

detach ()—subprocess forget to run but pretend that it's no longer running.

eval (code, strip = True, synchronize = False, locals = None, allow_use_file = True, split_lines = 'nofile',**kwds)

    code – text to evaluate

    strip – bool; strip output prompts check, etc.

    locals – None (ignored); It is used for compatibility with Sage generic
      system interface.

    allow_use_file – bool (default value: True);

      If code exceeds an interface-specific threshold and True, then the code will be communicated via a temporary file rather than the character-based interface.

    Otherwise code will be communicated via the character interface.

    split_lines – Tri-state (default value : "nofile");

      If "nofile" then code is sent line by line until it gets communicated through temporary file.

If True then code is sent line by line, but also individually some lines might be sent through temporary file otherwise at once the whole block of code is evaluated.

**kwds – All other arguments are passed onto the_eval_line method.

An example is reformat = False

is_running() - Return True, currently is running itself

pid() - Return the PID of the underlying sub-process

quit(verbose = True) - running the sub-process and if False- Quit running the sub-process

server () – indicates the server used in this interface

17.5 Plotting

Sage can be used to plot two-dimensional and three-dimensional.

17.5.1 Two-Dimensional Plots

Circles, lines, and polygons can be drawn by Sage: contour plots, polar plots, vector field plots, and also plots of functions in rectangular coordinates in two dimensions (Kosan, 2008). The symbols [sin(x), cos(x)] are an example of a list in Sage (Bard, 2018).

Program for 2D plotting for below expressions

```
a = x^2
b = plot(a,0,10)
type(b)
show(b)
```

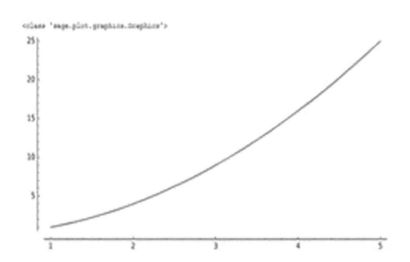

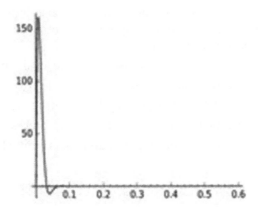

FIGURE 17.1
2D plotting for expression 500*e^(−100*x)*sin(100*x) using figsize[4,3].

2D plotting of symbolic expression using parameters: xmin, xmax, ymin, ymax, and fig size
(Figure 17.1)

V = 500*e ^(−100*x)*sin(100*x)
Show(plot(v,0,.1),xmin = 0.6,figsize[4,3])
V = 300*e ^(−100*x)*sin(100*x)

Show(plot(v,0,.1), xmin = 0, xmax = 0.05, ymin=0, ymax=100,0.6, figsize[4,4]) (Figure 17.2)
   Program to combine six plots into a single plot. Each plot can be shown with a different color

var('x')
p1 = x/4E5

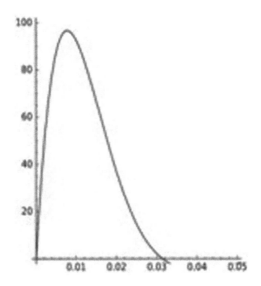

FIGURE 17.2
2D plotting using figsize[4,4].

```
p2 = (6*(x − 9)/2 − 10)/1000000
p3 = (x − 10)/400000
p4 = 0.0000004*(x − 20)
p5 = 0.0000004*(x − 20)
p6 = −0.0000006*(6 − 3*(x − 40)/2)
g1 = plot(p1,0,5,rgbcolor = (0, .2, 1))
g2 = plot(p2,5,10,rgbcolor = (1,0,0))
g3 = plot(p3,10,15,rgbcolor = (0,.7,1))
g4 = plot(p4,15,25,rgbcolor = (.3,1,0))
g5 = plot(p5,25,30,rgbcolor = (1,0,1))
g6 = plot(p6,30,60,rgbcolor = (.2,.5,.7))
show(g1 + g2 + g3 + g4 + g5 + g6, xmin = 0, xmax = 50, ymin = −.00001, ymax = .00001)
|
```

Output: The color of each plot can be changed using the **rgbcolor** parameter. RGB stands for Red, Green, and Blue and the tuple that is assigned to the **rgbcolor** parameter contains three values between 0 and 1 (Figure 17.3).

Program to create Graphics Object

It is useful to combine various kinds of graphics into one image. In the following example, six points are plotted along with a text label for each plot.

Plot the following points on a graph:

A (−10,13)

B (8,15)

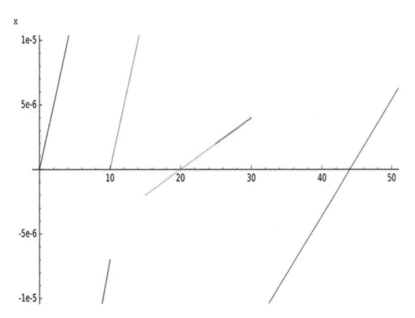

FIGURE 17.3
2D six plots into a single plot.

C (0,0)

D (20,–13)

E (–8,–10)

F (–2,–4)

```
g = Graphics()
points = [(–10,13), (8,15), (0,0), (20,–13), (–8,–10), (–2,–4)]
g += point(points)
for (pnt,letter) in zip(points,['A','B','C','D','E','F']):
g += text(letter,(pnt[0] – 1.5, pnt[1] – 1.5))
show(g,figsize = [5,4])
|
```

An empty Graphics object is instantiated and a list of plotted points is created using the point() function. These plotted points are added to the Graphics object using the += operator. Then a label for each point is added to the Graphics object using a **for** loop. Lastly, the Graphics object is displayed in the worksheet using the show() function (Figure 17.4).

```
g = Graphics()
points = [(–10,13), (8,15), (0,0), (20,–13), (–8,–10), (–2,–4)]
g += point(points)
for (pnt,letter) in zip(points,['A','B','C','D','E','F']):
g += text(letter,(pnt[0] – 1.5, pnt[1] – 1.5))
show(g,figsize = [5,4])
g += line([(–13,22),(25,–10)])
show(g)
|
```

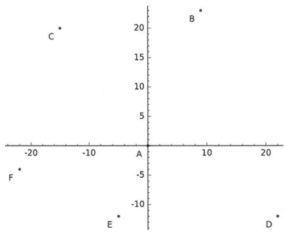

FIGURE 17.4
Create Graphics object.

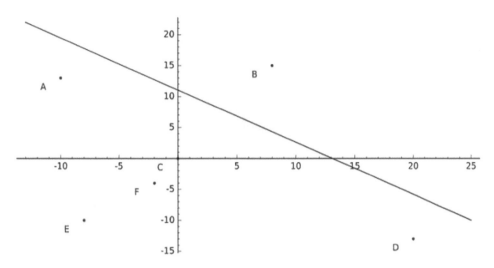

FIGURE 17.5
Line drawn between two points.

If a line needed to be drawn between points A and D, the above code can be executed (Figure 17.5).

17.5.2 Three-Dimensional Plots

There are numerous ways to plot in 3D in Sage. Specifically, import the code in order to use the command new plot3d. In order to do this, it depends either working in SageMathCloud or SageMathCell (Bard, 2018; Documentation, n.d.).

1. SageMathCell: copy the code for new plot3d from one of the sources listed above, and paste it into your SageMathCell. It is relatively inconvenient.
2. SageMathCloud: Sage and place it in the same folder on SageMathCloud as a worksheet. Second, put the following commands toward the start of your SageMathCloud worksheet.%auto

Plot a 3D graph for the expression x^2 + y^2 (Figure 17.6)

Program[7]:

        var ('x y')
        plot3d(x^2 + y^2, (x, –2,2), (y, –3,3))

17.6 Conclusion

In this chapter, we discussed the installation process of "SageMath," which is an open-source software licensed under the GPL. We discussed the programming fundamentals of "SageMath" such as Objects, Expressions, Operators, Variables, and Print Statements. Apart

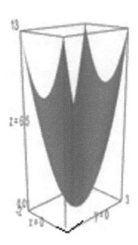

FIGURE 17.6
Graph for the expression x^2+y^2.

from this, we also gave the practical demonstration of "SageMath" tool like how to create the Graphics objects in two-dimensional and three-dimensional through programming.

References

Bard, G. V. (2018). Sage for undergraduates online electronic-only appendices about color and 3D plotting. Retrieved from http://www.gregorybard.com/papers/appendices.pdf.

Documentation. (n.d.). Retrieved from http://doc.sagemath.org/.

Eröcal, B., Stein, W. (2010). The sage project: Unifying free mathematical software to create a viable alternative to Magma, Maple, Mathematica and MATLAB. *ICMS 2010*. Lecture Notes in Computer Science, vol 6327. Springer, Berlin, Heidelberg.

Kosan, T. (2008). SAGE for newbies. Retrieved from http://verso.mat.uam.es/~pablo.angulo/laboratorio/sage_for_newbies_v1.23.pdf.

Lee, Sang-Gu (2012) Linear algebra with Sage-Math and the smartphone, 12th International Congress on Mathematical Education, ICME-12, 2012.

Sage Tutorial—SageMath Documentation. (n.d.). Retrieved from http://doc.sagemath.org/pdf/en/tutorial/SageTutorial.pdf.

Shravan, V. S. (2017). Mathematics simplified with SageMath—researchgate.net. Retrieved from https://www.researchgate.net/publication/317098317_Mathematics_simplified_with_SageMath.

Index

A

ABE, *see* Attribute-based encryption (ABE)
Abelian group, 14, 122
Abolishing phishing, 287
Abstract algebra, 12, 25
  field, 16–20
  group, 13–15
  integers, 68
  polynomial, 21–22
  preliminaries, asymmetric cryptosystems, 122–123
  ring, 15–16
Abstract syntax notation 1 (ASN.1), 208
Access Control Server (ACS), 223
Acquirer domain, 223
Active attack
  e-mail security, 286
  lightweight security, 291
Additive inverses, 4, 16
Advanced Encryption Standard (AES), 90–91, 94, 108–110, 235, 240–242
Aggregate digital signature scheme, 270
Agrawal–Kayal–Saxena (AKS), 27, 37–38
AKA, *see* Authentication key agreement (AKA)
AKS, *see* Agrawal–Kayal–Saxena (AKS)
Algebraically Homomorphic Scheme (AHS), 112–113
Algebraic expression, 21, 22
Algebraic geometry, 8
Algebraic polynomial, 21–22, 25
Alice and Bob, 116, 129, 130, 162, 163, 175, 176, 253
American National Standards Institute (ANSI), 255
Antivirus
  e-mail security, 286
  web security, 276–277
Application-level gateways, 276
Application service providers (APS), 280
Arithmetic operations
  CAPTCHA, 283
  number theory, 2–3
ASN 1, *see* Abstract syntax notation 1 (ASN.1)

Associative property, of addition and multiplication, 16
Asymmetric cryptography, 119–121
  chaos-based, 132–136
  general structure of, 120
  key management, 199
  preliminaries
    algebra, 122–123
    basic number theory, 123–124
    Euclidean algorithm, 121
    modular arithmetic, 123
  RSA algorithm, 124–126
Attribute-based encryption (ABE)
  algorithm of, 187–188
  applications of, 193
  ciphertext-policy, 190–192, 194
  data sharing system architecture, 185–187
  Fuzzy identity-based encryption, 179
  key-policy, 188–190, 192–194
  theory of, 183–184
Audit log, 193
Authentication key agreement (AKA), 173–175
Authentication of message, *see* Message authentication

B

Baby-step giant-step algorithm, 49–51, 54
Basic number theory, 123–124
BBG scheme, *see* Boneh, Boyen, and Goh (BBG) scheme
BB-HIBE scheme, *see* Boneh–Boyen HIBE (BB-HIBE) scheme
BDHP, *see* Bilinear Diffie–Hellman problem (BDHP)
Berlekamp–Massey algorithm, 85–86
Bezout's theorem, 8
Bilinear Diffie–Hellman problem (BDHP), 167, 168
Bimodal lattice signature scheme (BLISS), 262–267
Binary composition, *see* Binary operation
Binary operation, 12–17, 25, 122
Biometric authentication method, 216–218
BLAKE (hash function), 244–245

BLISS, *see* Bimodal lattice signature scheme
          (BLISS)
Block ciphers
    Data Encryption Standard, 87
    lightweight, 115
    message authentication code, 235–236, 239
    serpent algorithm, 108
    symmetric
        algorithms, 82–84
        key primitives, 81
    Twofish, 109, 110
    Whirlpool, 241
Blowfish algorithm, 101–104
BLS signature scheme, *see* Boneh–Lynn–
          Shacham (BLS) signature scheme
Bob and Alice, 116, 129, 130, 162, 163, 175, 176, 253
Boneh, Boyen, and Goh (BBG) scheme, 168
Boneh–Boyen HIBE (BB-HIBE) scheme, 168–169
Boneh–Boyen IBE scheme, 164–167
Boneh–Franklin IBE scheme, 161–162
Boneh–Lynn–Shacham (BLS) signature scheme,
          170–171, 270
Boolean functions, in RIPEMD, 243
Brute force attack, 101, 112
Buchmann–Williams key exchange protocol, 202

C

CA, *see* Certificate authority (CA)
Caesar cipher, 198
CAPTCHA, *see* Completely Automated Public
          Turing test to tell Computers and
          Humans Apart (CAPTCHA)
CAST-128 (CAST5) algorithm, 104–106
CBC mode, *see* Cipher Block chain (CBC) mode
Central controlling unit, biometric systems, 217
Certificate authority (CA), 205, 207–208
    see also Certification authority (CA)
Certificate directories, 205
Certificate management system, 205
Certificate revocation list (CRL), 205, 209
Certification authority (CA), 170
    see also Certificate Authority (CA)
CFM, *see* Cipher feedback mode (CFM)
ChaCha stream cipher, 244
Challenge-handshake authentication protocol
          (CHAP), 218–219
Challenge response mode, 215, 216
Chaos-based asymmetric cryptography,
          132–136
Chebyshev polynomials, 133–136
Chinese remainder theorem, 40–42, 54
Chor–Rivest knapsack cryptosystem, 130, 132

Cipher-based message authentication (CMAC),
          239–240
Cipher block chain (CBC) mode, 82, 235–236
Cipher feedback mode (CFM), 83, 84
Ciphertext-policy attribute-based encryption
          (CP-ABE) scheme, 190–192, 194
Click-jacking attack, 276
Clock-controlled generators, 87
Close vector problem (CVP), 149
Cloud security, 274, 279–281
CMAC, *see* Cipher-based message
          authentication (CMAC)
Cocks IBE scheme, 160–161
Code-based cryptography, 142–146
Collision-resistant hash functions, 115, 153, 185,
          229–230, 251, 252
Communication theory of secrecy systems, 81
Commutative cipher, 220
Commutative property, 16, 122, 220, 221
Completely Automated Public Turing test to
          tell Computers and Humans Apart
          (CAPTCHA), 274, 281–284
Composite numbers, 30, 31, 59, 60, 74, 75
Computing security, 98
Confusion and diffusion, 81, 133
Congruent modulo, 123
Coppersmith method, 22–23
Co-prime numbers, 29–30
Coset of group, 14
Counter mode, block ciphers operation, 83
CP-ABE scheme, *see* Ciphertext-policy attribute-
          based encryption (CP-ABE) scheme
CPS, *see* Cyber-physical systems (CPS)
CRL, *see* Certificate revocation list (CRL)
Cryptanalysis, 80
    of Blowfish, 101
    of Twofish, 112
CTR mode, 83
CVP, *see* Close vector problem (CVP)
Cyber-physical systems (CPS), 289–291
Cyclic group, 13–14, 48, 49, 55
Cyclomatic test, *see* Agrawal–Kayal–Saxena
          (AKS)

D

Data authentication algorithm (DAA), 239
Database security, 277–279
Database storage, 218
Data confidentiality, 185
Data encryption standard (DES), 87–89, 94,
          99, 239
Data integrity, 278, 281

Data mining, 277
Data origin authentication, 227–228
Data owner, 186
Data sharing system architecture, 185–187
Data Storing Centre, 186
Data transmission, 80
Decoding attacks, 145
Decryption algorithm
    Blowfish, 101
    Boneh–Boyen HIBE, 169
    Chaos-based asymmetric cryptography, 136
    ciphers, 198–199
    definition of, 80
    Diffie-Hellman, 163
    ElGamal public-key, 130
    identity-based encryption, 164
    International Data Encryption Algorithm, 92
    KeyGen(), 124, 125, 189
    Massy–Omura cryptosystem, 221, 222
    Rabin cryptosystem, 126–128
    scheme of, 100
    three-pass protocol, 220
    triple DES, 89
Def function, 301
DES, *see* Data encryption standard (DES)
Deterministic algorithm, 28
DHE protocol, *see* Diffie–Hellman key
        exchange (DHE) protocol
Diffie–Hellman key agreement scheme, 173
Diffie–Hellman key algorithm, 288
Diffie–Hellman key exchange (DHE) protocol,
        200–201
Digital signature algorithm (DSA), 121, 228,
        250–270
Digital tampering, hash-based-algorithm, 246
Dilemma, 116
Directory traversal, 276
Discrete dynamical system, 132, 133
Discrete logarithm problem (DLP)
    baby-step giant-step algorithm, 49–51, 54
    Boneh–Boyen scheme, 167
    description of, 47–48
    ElGamal signature scheme, 251
    Elliptic-curve digital signature algorithm,
        258, 259
    function field sieve, 51
    index calculus algorithm, 51–53
    number field sieve, 53
    number theory, 124
    Pohlig–Hellman algorithm, 53–54
    Pollard's
        Kangaroo algorithm, 55
        rho algorithm, 54–55

    in subgroup of Z_p, 48–49
Distributed systems, 278
Divisibility rules, 33–35
DLP, *see* Discrete logarithm problem (DLP)
DOS attack, 275, 286–287
DSA, *see* Digital signature algorithm (DSA)
Dynamical systems, 132–133

E

EAP, *see* Extensible authentication protocol
        (EAP)
ECB, *see* Electronic code book (ECB)
ECC, *see* Elliptic curve cryptosystems (ECC)
ECDH key exchange, *see* Elliptic Curve Diffie–
        Hellman (ECDH) key exchange
ECDSA, *see* Elliptic-curve digital signature
        algorithm (ECDSA)
Edwards-curve digital signature algorithm
        (EdDSA), 269–270
Electronic code book (ECB), 82
Electronic signature, 228, 250
ElGamal digital signature scheme, 251–255, 269
ElGamal public-key algorithms, 128–130
Elliptic curve cryptosystems (ECC), 139–140,
        148, 201, 255, 258
Elliptic curve Diffie–Hellman (ECDH) key
        exchange, 201–203
Elliptic-curve digital signature algorithm
        (ECDSA), 255–259
Elliptic curve method, 60, 64–65
E-mail security, 274, 285–287
Encryption algorithm
    algebraically homomorphic scheme, 112, 113
    attribute-based, 183–194
    Blowfish, 103
    Chaos-based asymmetric cryptography, 135
    cipher block chaining, 82
    definition of, 80
    ElGamal public-key, 129
    Goldreich–Goldwasser–Halevi, 148, 149
    identity-based, 160–179
    key generation, 124
    knapsack, 131, 132
    lightweight security, 291
    output feedback, 83
    Rabin cryptosystem, 127–128
    short message services, 287, 288
    symmetric key, 98, 100
    symmetric searchable, 116–117
    three-pass protocol, 220
    triple data, 89
    word auto key, 93

Energy consumption, 114
Entity authentication, 213–223
Euclidean algorithm, 6–9, 121
Euler's theorem, 32, 43, 123
Euler's totient function, 31–32
Excessive privileges, 278
Expansion
    Blowfish, 102
    Feistel function, 88, 89
    fraction, 25
    polynomial, 37, 38
Extended Euclidean algorithm, 8–9, 121
eXtended Merkle signature scheme (XMSS), 270
Extensible authentication protocol (EAP), 219
Extract algorithm, 259, 260

F

Face recognition, 217
Feature extraction, 218
Federal information processing standard
        (FIPS), 251, 253
Feistel function (*f*), 88–89
Fermat's factorization method, 60, 61
Fermat's little theorem, 30–31, 123
Fields, abstract algebra, 16–20
Fine-grained access control, 185
Fingerprint authentication, 216–217
Flooding, 288
Fraction expansion, 25
Function field sieve, 51
Fuzzy identity-based encryption (FIBE)
        scheme, 178–179, 184

G

Galois fields, 17–20
Gaussian elimination method, 66, 151
Gauss–Jordan elimination, 144
General number field sieve (GNFS) method, 68
General-purpose algorithm methods
    number field sieve, 60, 68–71
    quadratic sieve, 60, 67–68
    random squares, 60, 65–67
Generator matrix, 141
Goldreich–Goldwasser–Halevi (GGH)
        cryptosystem, 148–150
Graphics object, 305, 306
Graph Laplacian, 245, 246
Greatest common divisor (GCD), 7–10, 121
Group, abstract algebra, 13–15
Grover's algorithm, 140

H

Hamming distance, 141, 246
Hamming weight, 141
Hash-based cryptography, 153–155
Hash functions
    advantage of, 228
    application of, 113
    BLAKE, 244–245
    ECDSA, 258
    lightweight, 115
    Message Digest algorithm, 230–232, 241
    pre-image resistance, 229
    properties of, 229
    roles in, 226
    Whirlpool, 241–242
Hash message authentication code (HMAC),
        237–239
Heterogeneity, 199
Hierarchical identity-based encryption (HIBE)
        scheme, 167, 168, 177
Hierarchical identity-based signature (HIBS)
        scheme, 171–173
Homomorphic encryption system, 112–113

I

IaaS, *see* Infrastructure as a Service (IaaS)
IBE, *see* Identity-based encryption (IBE)
IBS scheme, *see* Identity-based signature (IBS)
        scheme
IDEA, *see* International Data Encryption
        Algorithm (IDEA)
Identity-based encryption (IBE)
    advantage and disadvantage, 162
    history of, 160–162
    key terminology and definitions, 163–169
    schemes, 170–179
    working block diagram, 163
Identity-based encryption with wildcards
        (WIBE), 176–177
Identity-based signature (IBS) scheme, 170, 171,
        259–262
IDS, *see* Intrusion detection system (IDS)
IEEE, *see* Institute of Electrical and Electronics
        Engineers (IEEE)
IEEE 802 standard, 219
IFP, *see* Integer factorization problem (IFP)
Index calculus algorithm, 51–53, 171
Infinite group, 122
Infrastructure as a Service (IaaS), 280
Initialization vector (IV), 81–84

Initial permutation, 88
Institute of Electrical and Electronics Engineers
    (IEEE), 255
Integer arithmetic, 2
Integer factorization problem (IFP), 59–60
    algorithms
        general-purpose, 60, 65–71
        prime factorization, 73
        special-purpose, 60–65
    records, 73–75
    square root approximation, 71–73, 75
Integers, 2, 10, 27, 68, 121, 123, 124
Integral domain, 122
Integrity, 226–229, 250, 278, 291
Interface
    insecurity, lightweight, 290
    SageMath software, 301–303
International Data Encryption Algorithm
    (IDEA), 91–93, 101
International Standards Organization (ISO), 255
Internet of Things (IoT), 289–290
Interoperability domain, 223
Intrusion detection system (IDS), 277
Intrusion prevention system, 277
IoT, *see* Internet of Things (IoT)
Iris recognition, 217
ISO, *see* International Standards Organization
    (ISO)
Issuer domain, 223
Iterative function, 54, 55
IV, *see* Initialization vector (IV)

J

Junk (e-mails), 285

K

Key agreement, identity-based, 173–175
Key distribution center (KDC), 204, 205
Key distribution, in cryptosystem, 198, 204–209
Key exchange
    learning with errors, 202–203
    protocols, 200–204
    schemes, 121
    symmetric key, 98
Key generation centre (KGC), 170, 187
Key generation/KeyGen(), 189
    of ABE and KP-ABE scheme, 189
    Bimodal lattice signature scheme, 263
    BLISS, 263
    in Blowfish algorithm, 102

chaos-based asymmetric cryptography, 135
data sharing system architecture, 187
digital signature scheme, 228, 250
ECDSA, 256
ElGamal public-key algorithm, 129
Knapsack algorithm, 132
lightweight cryptography, 115
Merkle tree signature scheme, 267–268
Niederreiter cryptosystem, 146
Rabin cryptosystem, 127
RC algorithms, 107–108
RSA algorithms, 124–126, 134
and signature generation, 251–252, 263
signing and verification, 264–266
Key management system
    challenges of, 199
    description of, 198–199
    importance of, 199
    key distribution, 198, 204–209
    key exchange, 200–204
    lightweight security, 289–291
Key mixing, Feistel function, 89
Key-Policy attribute-based encryption
    (KP-ABE) scheme, 188–190, 192–194
Keystream, 81, 83, 84, 87, 93
KGC, *see* Key generation centre (KGC)
Knapsack algorithm, 130–132, 145

L

Lagrange's theorem, 15
Lamport–Diffie one-time signature scheme
    (LD-OTS), 153–154
Lamport signature scheme, 153, 262–267
Last-In First-Out (LIFO) method, 220
Latency, lightweight cryptography, 114
Lattice-based cryptosystems, 141, 146–150
Learning With Errors (LWE) key exchange,
    202–203
Lemma (Bezout), 66–67
Lightweight cryptography, 113–116, 289–291
Linear code, 141
Linear congruence, 6
Linear feedback shift register (LFSR), 85–87, 94
Logarithm, *see* Discrete logarithm problem
    (DLP)
Logical and system issues, database security, 278

M

MAC, *see* Message authentication code (MAC)
McEliece public-key cryptosystem, 142–146

Malware, 274, 278
Massey–Omura cryptosystem, 221–222
Matching algorithm, 217
Mathematical software, Sage interfaces, 302
Matrix, 5
MD algorithm, *see* Message Digest (MD) algorithm
Merchant plug-in (MPI), 223
Merkel tree-like structure, 232
Merkle–Hellman cryptosystem, 130
Merkle signature scheme, 267–270
Mersenne prime/numbers, 39–40
Message authentication, 227–229
Message authentication code (MAC), 226–227, 235–237
Message availability, 288
Message Digest (MD) algorithm, 154, 228–232, 241, 242
Miller–Rabin for primality testing, 36–37
MK-HIBS, *see* Multi-keys HIBS (MK-HIBS)
Modular arithmetic, 3–4, 17, 18, 123
Modulo operation, 30–31, 47
Monomial, 22
MPI, *see* Merchant plug-in (MPI)
Multi-keys HIBS (MK-HIBS), 171–172
Multiple polynomial quadratic sieve (MPQS), 74
Multiple signatures, 270
Multiplicative inverses, 4–5, 16
Multiplicative property, Fermat's theorem, 30, 31
Multivariate cryptography schemes, 150–153

N

National Institute of Standards and Technology (NIST), 90, 210, 233, 242, 244, 247, 253
Network availability, 291
Network firewalls, 277
"New Directions in Cryptography," 119
Niederreiter cryptosystem, 145–146
Nonlinear dynamical systems, 133
Nonlinear filter, 86–87
Non-repudiation, 228, 291
Non-residue modulo, 123
NTRU algorithm, 146–148
Number field sieve (NFS) method, 51, 53, 60, 68–71
Number theory, 2–3, 9–10, 27, 123–124

O

Oil and vinegar (OV) signature scheme, 150
One-dimensional discrete dynamical systems, 132–133

One-Time Password (OTP) authentication, 215–216
One-way hash functions, 48, 229, 233, 264, 266
Output feedback mode (OFM), 83

P

PaaS, *see* Platform as a Service (PaaS)
Palm authentication, 217
PAP, *see* Password authentication protocol (PAP)
Parity-check matrix, 141, 145
Passive attacks, 286, 291
Password authentication protocol (PAP), 214
Pattern mining, 277
PCT, *see* Polar cosine transformation (PCT)
PEKS scheme, *see* Public-key encryption with keyword search (PEKS) scheme
Permutation box (P-Box), 89
Permutation network, 81
Personal health record (PHR), 193
Pexpect interface, 301
Phone crashers, 288
PHR, *see* Personal health record (PHR)
PHT, *see* Pseudo Hadamard transform (PHT)
Physical database security, 277–278
PKG, *see* Private-key generator (PKG)
PKI, *see* Public-key infrastructure (PKI)
Platform as a Service (PaaS), 280
Plotting, SageMath software, 303–307
Pohlig–Hellman algorithm, 53–54
Pointcheval–Stern signature algorithm, 269
Point-to-point protocol (PPP), 214, 218, 219
Polar cosine transformation (PCT), 245, 246
Pollard's Kangaroo algorithm, 55
Pollard's p method, 60, 62–64
Pollard's $p - 1$ method, 60–62
Pollard's rho algorithm, 54–55
Polynomial(s), 21–22
    Agrawal–Kayal–Saxena primality test, 37–38
    algebraic, 21–22
    Chebyshev, 133–135
    expansion, 37, 38
    quadratic, 150, 151
    ring, 142
Post-quantum cryptography, 139–155
PPP, *see* Point-to-point protocol (PPP)
Pre-image resistance, of hash functions, 229
Preliminaries
    asymmetric cryptosystems, 121–124
    post-quantum cryptography, 141–142
Primality testing
    Agrawal–Kayal–Saxena, 37–38
    Miller–Rabin, 36–37

square root, 35–36
Prime factorization algorithms, 59, 73
Prime numbers, 6, 7, 27–29, 59
Private-key generator (PKG), 161, 162, 164, 167, 172, 173, 184
PRNG, *see* Pseudorandom number generators (PRNG)
Proxy re-encryption scheme, 176
Pseudo Hadamard transform (PHT), 111
Pseudorandom number generators (PRNG), 83–84
Public announcement, of public keys, 204, 206, 207
Public available directory, 204, 206–207
Public-key authority, 207
Public-key certificate scheme, 207–208
Public-key cryptography
    attribute-based encryption, 184
    Buchmann-William key exchange protocol, 202
    chaos-based, 132–136
    description of, 119–121
    general structure of, 120
    key management system, 206–208
    multivariate cryptography, 150
    preliminaries
        algebra, 122–123
        basic number theory, 123–124
        Euclidean algorithm, 121
        modular arithmetic, 123
    RSA algorithm, 124–126
    X.509 standard, 208
Public-key encryption with keyword search (PEKS) scheme, 175
Public-key infrastructure (PKI), 170, 205, 206, 250
Public-key signature scheme, 261

Q

Quadratic character, 71
Quadratic polynomials, 150, 151
Quadratic sieve (QS) method, 60, 67–68
Quantum computational complexity, 140
Quantum computing, on cryptographic algorithms, 139–140
Quantum parallelism, 140
Quartz signature scheme, 150, 152–153
Queen of mathematics, 2–3

R

R (statistical computing), 294
Rabin cryptosystem, 126–128

RACE Integrity Primitives Evaluation Message Digest (RIPEMD), 242–243
Rainbow signature scheme, 150–152
Random squares method, 60, 65–67
RC algorithms, 106–108
Reed–Solomon codes, 145
Registered authority, 205
Relatively prime numbers, 6–7, 29, 121
Request encoding attack (XSS), 276
Retina biometrics, 217
RFC 5280, 208
RFID systems sensors, 289
Ring, abstract algebra, 15–16
Ring Learning With Errors (RLWE) key exchange, 203–204
RIPEMD, *see* RACE Integrity Primitives Evaluation Message Digest (RIPEMD)
RSA algorithm
    asymmetric cryptography, 124–126
    in Chebyshev polynomials, 134–136
    Elliptic-curve digital signature algorithm, 258–259
    identity-based encryption, 160
    integer factorization problem, 60, 73–74

S

SaaS, *see* Software as a Service (SaaS)
SageMath (software)
    installation process, 294–296
    interfaces, 301–303
    packages of, 294
    plotting, 303–307
    program process, 296–301
    purpose of, 293
SageMathCell, 307
SageMathCloud, 307
S-Box, *see* Substitution box (S-Box)
Scalability, 185, 199
Schnorr digital signature scheme, 270
Schoof–Elkies–Atkin (SEA) algorithm, 20–21
Searchable encryption scheme, 116–117
Secure hash algorithm (SHA), 153, 229, 233–235
Secure socket layer (SSL), 288, 289
Security, key management system, 199
Self-synchronous stream ciphers, 84, 88
Serpent algorithm, 108–109
Setup algorithm, 259, 260
SH, *see* Spectral hash (SH)
SHA, *see* Secure hash algorithm (SHA)
Shamir's three-pass protocol, 220–221
Shor's algorithm, 140
Short message services (SMS) security, 287–289

Sign algorithm, 259, 260
Signature authentication, 217
Signature generation
    Bimodal lattice signature scheme, 263
    Elliptic-curve digital signature algorithm
        verification, 256–257
    and key generation, 251–252
    Lamport–Diffie one-time signature scheme,
        153
    and verification algorithm, 256–257
    Winternitz one-time signature scheme, 154
Simple mail transfer protocol (SMTP), 274, 285
SMS security, *see* Short message services (SMS)
    security
SMTP, *see* Simple Mail Transfer Protocol (SMTP)
SNF method, *see* Special number field sieve
    (SNF) method
Sniffing (request encoding), 276
Software as a Service (SaaS), 280
Solovay–Strassen test, 43
SPAM e-mails, 285, 287
Spamming, 288
Special number field sieve (SNF) method, 68
Special-purpose algorithms
    elliptic curve method, 64–65
    Fermat's method, 60, 61
    Pollard's
        p method, 60, 62–64
        $p – 1$ method, 60–62
        trial division, 61
Spectral hash (SH), 245–247
Splitting fields, 18
SQL injections, 276, 278
Square root
    approximation, factorization method based,
        71–73, 75
    primality test, 35–36
SSE, *see* Symmetric Searchable Encryption
    (SSE)
S/Sky mode, 216
SSL, *see* Secure socket layer (SSL)
Stability coefficient, 133
Stream ciphers, 81, 84–87, 94
Structural attacks, McEliece cryptosystem, 145
Sub-key generation, RC5algorithm, 107–108
Subset sum problem, 130
Substitution box (S-Box), 81, 89, 91, 93, 108
Substitution network, 81
Surreptitious writing, 100
Symmetric key cryptography
    block ciphers, 82–84
    block diagram, 99

classification of, 81
computing security, 98
DHE protocol, 200
key distribution, 204–205
key management, 199
objective of, 80
stream ciphers, 84–87
substitution box, 81
Symmetric searchable encryption (SSE),
    116–117
Synchronous stream ciphers, 84, 88

T

Targeted broadcast, 193
3DES, *see* Triple Data Encryption Standard
    (3DES)
Three-dimensional plots, 307
Three-Domain (3D) secure protocol, 222–223
Three-pass protocol (TPP), 220–222
Threshold gate, 188
Time synchronization mode, 215
Trapdoor multivariate quadratic, 150
Trapdoor one-way function, 120, 150
Tree authentication scheme (Merkle), 153, 267
Trial division method, 60–62
Triple data encryption standard (3DES), 89, 94
Trusted authority, 205–207
Trust factor, 245, 246
Two-dimensional plots, 303–307
2D plotting, 303–305
Two-factor authentication (TFA), 215–216
Twofish algorithm, 100, 109–112

U

Unsecured network, 198, 291
User authentication, 223, 287
User gateway, 217
User revocation, 185

V

Verification algorithm, 256–257, 259, 260, 264
Virtual private network (VPN), 286
Viruses
    e-mail, 285, 286
    SMS, 288
VMware software, 294
Voice authentication, 217
Vulnerability patching, 277
Vulnerability scanner, 276

W

Web-based phishing attacks, 276
Web firewalls, 276
Web security, 274–277
Whirlpool (hash function), 241–242
Wiener's theorem, 23–25
Wildcards key derivation (WKD-IBE) scheme, 177–178
Winternitz one-time signature scheme (W-OTS), 154–155

Wireless sensor networks (WSN), 289
Word Auto Key Encryption (WAKE), 93, 94

X

X.509 Standard Certificate, 208–209
XOR operation, 83, 89
XMSS, *see* eXtended Merkle signature scheme (XMSS)
XTR cryptosystem, 288